PURE MATHEMATICS

2

S. L. PARSONSON

Senior Mathematics Master
Harrow School

CAMBRIDGE UNIVERSITY PRESS

CAMBRIDGE

LONDON · NEW YORK · MELBOURNE

Published by the Syndics of the Cambridge University Press
The Pitt Building, Trumpington Street, Cambridge CB2 1RP
Bentley House, 200 Euston Road, London NW1 2DB
32 East 57th Street, New York, NY 10022, USA
296 Beaconsfield Parade, Middle Park, Melbourne 3206, Australia

ISBN: 0 521 08032 0

First published 1971
Reprinted 1973 1975

Printed in Great Britain
at the
University Printing House, Cambridge
(Euan Phillips, University Printer)

Contents

Preface

This book completes the course designed to cover the work required for modern 'A' level pure mathematics syllabuses (including probability), particularly the M.E.I. syllabus. As explained in the Preface to Volume 1, the calculus has been excluded, since it is already adequately covered by existing texts, but a knowledge of the subject is demanded throughout the book, and particularly in Chapter 25, on continuous probability.

In places the subject is carried beyond the bare limits of 'A' level requirements: most chapters contain a small proportion of work which would probably best be left to a second reading, while Chapters 20, 24, 28, 29 contain a substantial proportion of 'S' level work.

As in Volume 1, the book is liberally supplied with exercises for the student. Most questions are straightforward applications of the bookwork, though a few harder questions may be found in the Miscellaneous Exercises. The questions marked Ex. occurring in the text illustrate the associated bookwork: those marked with an asterisk should be regarded as obligatory.

The same nomenclature as in Volume 1 has been used to indicate the source of examination questions and I am grateful to the Examination Board of the University of London and the Oxford and Cambridge Schools Examination Board for permission to reproduce their questions. I should also like to record my thanks to Dr N. A. Routledge and Mr A. J. Moakes who read the book and made many valuable suggestions and to my wife who once again lent invaluable assistance in checking the answers.

S.L.P.

17. *Complex numbers (1)*

1. AN EXTENSION OF THE REAL NUMBER SYSTEM

We have seen (Chapter 1, *et seq.*) that, by broadening the meaning of the term 'number', we are able to ascribe solutions to an increasing range of problems. To take a simple example, consider the solution of certain equations. If we restrict 'number' to mean 'positive integer', the equation

$$x-2 = 0$$

has a root, but the equation

$$x+2 = 0$$

has not. The latter equation does have a root, however, if we postulate the existence of negative integers. Thus, by augmenting the set Z^+ of positive integers into the set Z of integers we are able to solve a wider variety of equations. Z^+ is a subset of Z, and the familiar laws of algebra governing the combination of elements of Z^+, that is

(i) closure: if $a, b, \in Z^+$, then $a+b, \quad ab \in Z^+$;
(ii) commutativity: $a+b = b+a, \quad ab = ba$;
(iii) associativity: $(a+b)+c = a+(b+c), \quad (ab)\,c = a(bc)$;
(iv) distributivity: $a(b+c) = ab+ac, \quad (b+c)\,a = ba+ca$;
(v) the existence of a multiplicative identity: $1a = a$

hold for the elements of Z. But we also have two new laws which hold for Z but not for the subset Z^+:

(vi) the existence of an additive identity: $0+a = a$;
(vii) the existence of additive inverses: $(-a)+a = 0$.

In a similar way, an equation such as

$$2x+1 = 0,$$

which has no solution in Z, has a solution if we augment Z into the set Q of rational numbers. Again, Z is a subset of Q, and laws (i)–(vii) above hold in Q, together with an additional law

(viii) the existence of multiplicative inverses: $a^{-1}a = 1 \ (a \neq 0)$.

Laws (i)–(viii) define a *field*; that is, any set of numbers combined by the operations of addition and multiplication and satisfying laws (i)–(viii) above constitute a field.

Ex. 1. Show that the set

$$S = \{x = a+b\sqrt{2}: a, b \text{ rational}\}$$

constitute a field under the usual operations of addition and multiplication. (For example, to verify (iv) you have to show that

$$(a_1+b_1\sqrt{2})\,(a_2+b_2\sqrt{2})$$

is of the form

$$a_3+b_3\sqrt{2},$$

where $a_1, \ldots, b_3$ are all rational numbers. Check, in a similar way, the other seven laws.)

The extension of the number system to the set Q vastly increases the number of equations with a solution, but it is not difficult to formulate equations in terms of the elements of Q which have no rational solution: consider, for example, the equations

$$x^2-2 = 0 \quad \text{or} \quad 2\sin x = 1 \quad \text{or} \quad 10^x = 5.$$

Many such equations acquire solutions if we augment the set Q into the set R of all real numbers. Although it is possible to define R in terms of the elements of Q, we have contented ourselves with the intuitive geometrical concept of R as 'completing the number line' but, since real numbers may be approximated arbitrarily closely by rational numbers, it seems reasonable to assume that laws (i)–(viii) above hold in R as well.

Since we have 'filled in the gaps' of the number line it might seem reasonable to suppose that our work is done and that any equation involving elements of R as coefficients should have elements of R as solutions. Such is not the case however; indeed, if we modify the three equations of the previous paragraph only slightly to

$$x^2+2 = 0 \quad \text{or} \quad \sin x = 2 \quad \text{or} \quad 10^x = -5,$$

none of these equations has a solution in R. However, these equations have solutions provided we extend the number systems still further. It will be the purpose of the remainder of this chapter to introduce such an extension in a reasonably informal manner. The resulting augmented set we shall call the *set of complex numbers* and denote it by C. Since C will contain R as a subset, it will be necessary to verify that the laws of combination (i)–(viii), suitably redefined, hold in C also.

Up to this point, each extension of the number system has led inexorably to a further extension. However, it has been shown (by C. F. Gauss in the *Fundamental Theorem Algebra*) that no further such extensions are needed, in the sense that any polynomial equation with its coefficients in C will have all its roots lying in C as well. But the real significance of the set C goes deeper than simply enabling solutions to be found for any polynomial equation: it reveals an underlying unity among mathematical concepts that would otherwise remain unsuspected, for example between trigonometric and exponential functions, suitably redefined.

2. COMPLEX NUMBERS AND THEIR MANIPULATION

The extension of the real number system that was foreshadowed in Section 1 is effected as follows: we introduce a new symbol j and define a *complex number* to be an expression of the form $a+b\mathrm{j}$ where a, b are *any* real numbers. (The sign + is used at the moment simply to unite the two components a and $b\mathrm{j}$ of the complex number: the operation + has only been defined so far for combining real numbers.) For the moment, j is to be regarded purely as a new symbol: an interpretation of j will appear as soon as we have defined operations between complex numbers. Before we do so we must define *equality*: two complex numbers $a+b\mathrm{j}$ and $c+d\mathrm{j}$ are *equal* if and only if $a = c$ and $b = d$.

Addition, subtraction and *multiplication* of complex numbers is defined to proceed precisely as if each complex number were a polynomial in j, *where* j *is subject to the condition* $\mathrm{j}^2 = -1$.

Thus for example,

$$(3+\mathrm{j})+(5-2\mathrm{j}) = 8-\mathrm{j}$$

$$(3+\mathrm{j})(5-2\mathrm{j}) = 15-6\mathrm{j}+5\mathrm{j}-2\mathrm{j}^2$$

$$= 17-\mathrm{j} \text{ (since } \mathrm{j}^2 = -1).$$

Ex. 2. The result of dividing the polynomial $P(x)$ by x^2+1 may be written in the form

$$P(x) \equiv (x^2+1)\,Q(x)+ax+b.$$

Discuss the relation of this polynomial identity to the concept of a complex number. (Put $x = \mathrm{j}$.)

Ex. 3. Simplify: (i) $(2+\mathrm{j})-(3-\mathrm{j})$, (ii) $(7-7\mathrm{j})+(4+3\mathrm{j})$, (iii) $(2-\mathrm{j})(2+3\mathrm{j})$, (iv) $(1+3\mathrm{j})^2$, (v) $(1+3\mathrm{j})^3$.

In the same way that constant polynomials were identified with real numbers, so a complex number of the form $a+0\mathrm{j}$ can be identified with the real number a. Thus we write

$$2+0\mathrm{j} = 2 \quad \text{and} \quad 2(3-4\mathrm{j}) = (2+0\mathrm{j})(3-4\mathrm{j}) = 6-8\mathrm{j},$$

as might have been expected. $0+0\mathrm{j}$ is written as 0 (the *complex number zero*). Similarly, we abbreviate $0+b\mathrm{j}$ to $b\mathrm{j}$.

The existence of a multiplicative inverse for any non-zero complex number has already been postulated in Section 1. To find an explicit form for such an inverse we use the fact that

$$(a+b\mathrm{j})(a-b\mathrm{j}) = a^2-(b\mathrm{j})^2 = a^2+b^2.$$

Thus
$$(2-\mathrm{j})^{-1} = \frac{1}{2-\mathrm{j}}\frac{2+\mathrm{j}}{2+\mathrm{j}} = \frac{2+\mathrm{j}}{4+1} = \tfrac{2}{5}+\tfrac{1}{5}\mathrm{j};$$
more generally,
$$(a+b\mathrm{j})^{-1} = \frac{1}{a+b\mathrm{j}}\frac{a-b\mathrm{j}}{a-b\mathrm{j}} = \frac{a}{a^2+b^2}-\frac{b}{a^2+b^2}\mathrm{j} \quad (a, b, \text{ not both zero}).$$

Ex. 4. Show that the multiplicative inverse of $(2-\mathrm{j})$ obtained above is unique; that is, show that
$$(2-\mathrm{j})^{-1} = x+y\mathrm{j} \Rightarrow x = \tfrac{2}{5}, \quad y = \tfrac{1}{5}.$$

Division of complex numbers by complex numbers now follows: $z \div w$ means zw^{-1}. For example
$$\frac{4-\mathrm{j}}{1+3\mathrm{j}} = \frac{(4-\mathrm{j})(1-3\mathrm{j})}{(1+3\mathrm{j})(1-3\mathrm{j})} = \frac{1-13\mathrm{j}}{10} = \tfrac{1}{10}-\tfrac{13}{10}\mathrm{j}.$$

Ex. 5. If $z_1 = 2+3\mathrm{j}$, $z_2 = 3-4\mathrm{j}$, express in the form $a+b\mathrm{j}$:
(i) $2z_1-3z_2$, (ii) $(2z_1-z_2)^2$, (iii) $(z_1-2\mathrm{j}z_2)^2$.

Ex. 6. Express as complex numbers in the standard form $a+b\mathrm{j}$:
(i) $1/(2+\mathrm{j})$, (ii) $(2+3\mathrm{j})/(1+\mathrm{j})$, (iii) $\{(1+\mathrm{j})(1+2\mathrm{j})\}/(1+3\mathrm{j})$.

**Ex.* 7. Taking $0 = 0+0\mathrm{j}$ as the additive identity and $1 = 1+0\mathrm{j}$ as the multiplicative identity, verify that laws (i)–(viii) of Section 1 hold for complex numbers.

The interpretation of $a+0\mathrm{j}$ as a real number enables us to speak meaningfully of the square roots of a negative real number. Thus, if $a \in R$, then the real number $-a^2$, regarded as an element of C, is the square of the complex number $a\mathrm{j}$ and also of the complex number $-a\mathrm{j}$.

Given a complex number $z = a+b\mathrm{j}$ we call a the *real part* of z and b the *imaginary part* of z:
$$a = \mathrm{Re}\,(z), \quad b = \mathrm{Im}\,(z).$$
(Note that the imaginary part of a complex number is *real*!)

If z is non-zero, we have seen that the reciprocal of z, $1/(a+b\mathrm{j})$, is obtained as follows:
$$z^{-1} = \frac{1}{a+b\mathrm{j}} = \frac{1}{a+b\mathrm{j}}\frac{a-b\mathrm{j}}{a-b\mathrm{j}} = \frac{a-b\mathrm{j}}{a^2+b^2}.$$

The number $z^* = a-b\mathrm{j}$, obtained by writing $-\mathrm{j}$ for j in z, is called the *complex conjugate* of z. The real number $|z| = \sqrt{(a^2+b^2)}$ is called the *modulus* of z. Thus we have shown that, for any non-zero complex number z,
$$z^{-1} = \frac{z^*}{|z|^2}.$$

Ex. 8. Write down the real and imaginary parts of the following complex numbers:
(i) $3-2j$, (ii) $(3-2j)^2$, (iii) $(3-2j)^{-1}$.

Ex. 9. Show that Re, Im may be interpreted as functions $C \to R$. Find

$$\text{Re}\,[\text{Im}\,(a+bj)] \quad \text{and} \quad \text{Im}\,[\text{Re}\,(a+bj)],$$

R being regarded as a subset of C.

Ex. 10. If $z_1 = 2-j$, $z_2 = 3+2j$ find:
(i) z_1^*, (ii) z_2^*, (iii) $(z_1+z_2)^*$, (iv) $(z_1z_2)^*$, (v) $(1/z_1)^*$, (vi) $(z_1+3z_2)^*$, (vii) $(z_1^2+z_2^2)^*$.

Ex. 11. The conjugacy function $f: C \to C$ is defined by $f(z) = z^*$. For what subset of C is f the identity function?

Ex. 12. Find:
(i) $|3+4j|$, (ii) $|2j|$, (iii) $|1/(1-j)|$, (iv) $|\cos\theta + j\sin\theta|$, (v) $|1-\cos\theta + j\sin\theta|$.

Ex. 13. The word 'modulus' has, prior to its use in this chapter, been used in the context of real numbers only, to mean $\sqrt{a^2}$. Show that the two uses coincide if R is regarded as a subset of C.

**Ex.* 14. If z_1, z_2 are two complex numbers, and $z_2 \neq 0$, prove that:
(i) $(z_1+z_2)^* = z_1^*+z_2^*$, (ii) $(z_1z_2)^* = z_1^*z_2^*$, (iii) $z_1z_1^* = |z_1|^2$,
(iv) $\left(\dfrac{1}{z_2}\right)^* = \dfrac{1}{z_2^*} = \dfrac{z_2}{|z_2|^2}$, (v) $z_1+z_1^* = 2\,\text{Re}\,(z_1)$,
(vi) $z_1 - z_1^* = 2j\,\text{Im}\,(z_1)$.

The results proved in Ex. 14 are of great importance and should be committed to memory.

The concept of modulus enables us, in a sense, to order complex numbers, but this is not entirely analogous to the ordering of real numbers, since different complex numbers can have the same modulus. It is not possible to order the complex numbers in the same way as we order the reals.

Exercise 17(*a*)

1. If $z_1 = 1+2j$, $z_2 = 2-j$, $z_3 = 4+5j$, express as complex numbers in the standard form $a+bj$:
(i) $z_1+z_2+z_3$; (ii) $3z_1-z_2+2z_3$; (iii) z_1z_2; (iv) $z_2^2+2z_3$; (v) $z_2z_3+z_3z_1+z_1z_2$;
(vi) $z_1z_2z_3$; (vii) $(z_1-jz_2)/(z_2+2jz_3)$; (viii) $\dfrac{1}{z_1}+\dfrac{1}{z_2}$; (ix) $(z_2+z_3)^3$;
(x) $(z_1-z_2)(z_2+jz_3)$; (xi) $j/(z_1^2+jz_2^2)$; (xii) $(z_1^2-z_2z_3)/(z_2^2-z_3z_1)$.

2. If $z_1 = \cos\theta_1 + j\sin\theta_1$, $z_2 = \cos\theta_2 + j\sin\theta_2$, show that

$$z_1z_2 = \cos(\theta_1+\theta_2) + j\sin(\theta_1+\theta_2),$$

and find z_1/z_2. Evaluate z_1^2 and $(z_1z_2)^2$.

3. Solve the following equations:
(i) $x^2-4x+5 = 0$; (ii) $2x^2-2x+1 = 0$; (iii) $x^2-5x+7 = 0$;
(iv) $jx^2-2x-2j = 0$.

4. If $(5-12j) = (a+bj)^2$ where $a > 0$, find a, b. Find similarly complex numbers which when squared give (i) j; (ii) $3+4j$.

5. Solve the equation $x^2-(4+j)x+5-j = 0$, giving each root in the form $a+bj$.

6. Given that $A(3+2j)-B(1-j)-(5+2j) = 0$, find A and B: (i) if A, B are both real; (ii) if A, B are conjugate complex numbers.

7. If $z = 1-\cos\theta-j\sin\theta$, write down the value of z^*. Express z^{-1} as a complex number in standard form.

8. If a is a real number and z is a complex number, prove that $(1+az)^* = 1+az^*$. Deduce that, in standard form, $(1+az)^{-1} = (1+az^*)/(1+2a\,\text{Re}\,(z)+a^2|z|^2)$.

9. Given $z \in C$, show that z is real $\Leftrightarrow z = z^*$.

10. What is the conjugate of z^*? By considering the product $z_1 z_2^*$ for any pair of complex numbers z_1 and z_2, prove that $z_1 z_2^* + z_1^* z_2$ is real. What can you say about the complex number $z_1 z_2^* - z_1^* z_2$?

11. Show, by constructing an example, that *non-real numbers a* and *b* may be found so that the quadratic equation

$$x^2-ax+b = 0$$

has a real root.

Is it possible for the equation to have two real roots if a, b are non-real?

12. Form the quadratic equation whose roots are:

(i) $2+j, 2-j$; (ii) $2-3j, 2+3j$; (iii) $2-j\sqrt{3}, 2+j\sqrt{3}$; (iv) $1+2j, 2+j$; (v) $4, 1+j$.

Can you conjecture any general result about the coefficients of a quadratic equation and its roots?

13. By writing x^2+y^2 in the form $x^2-(jy)^2$, split x^2+y^2 into linear factors with coefficients in C.

Factorize into linear factors with coefficients in C:

(i) x^2-2x+2; (ii) x^2+4y^2; (iii) x^2+3y^2; (iv) $j(x^2+x+1)-1$; (v) $jx^2-4xy-4jy^2$; (vi) x^3+1; (vii) x^4+1.

(Hint for (vii): the expression may be written $(x^2+1)^2-2x^2$.)

14. By putting x equal to j in the identity

$$x^4-3x^3-x+2 \equiv (x^2+1)\,Q(x)+ax+b$$

determine the numerical values of a and b.

Find the remainder, on division by x^2+1, of (i) x^8+1; (ii) x^9+1.

15. Evaluate j^n in the four cases $n = 4m, 4m+1, 4m+2, 4m+3$ (m integral).

Find the sum of the series

$$\sum_{r=0}^{n} j^r.$$

16. If $(1+j)^n = a+bj$, prove that $a^2+b^2 = 2^n$:

(i) by taking the complex conjugate of each side of the original expression;

(ii) by mathematical induction.

17. Evaluate the determinant

$$\begin{vmatrix} 2 & 1+\mathrm{j} & 2 \\ 1-\mathrm{j} & -1 & 4-\mathrm{j} \\ 2 & 4+\mathrm{j} & 3 \end{vmatrix},$$

where $\mathrm{j}^2 = -1$.

Explain how without evaluation, it could have been concluded that the value of the determinant was real. (M.E.I.)

3. THE ARGAND DIAGRAM

As forecast in Section 1, we may set up a 1–1 correspondence between the set C of complex numbers and the points in a plane by associating the complex number $x+y\mathrm{j}$ with the point P whose coordinates are (x, y) referred to a given pair of rectangular coordinate axes. The x axis corresponds to the set R of real numbers, the y axis to the set *of pure imaginary numbers*, while a general complex number $a+b\mathrm{j}$, $a \neq 0$, $b \neq 0$ lies off the axes in one of the four quadrants.

Ex. 15. Plot the points corresponding to the complex numbers $1+2\mathrm{j}$, $1-2\mathrm{j}$, $-\mathrm{j}, +\mathrm{j}$. What is the geometrical relationship that exists between the points representing z and z^*?

Demonstrate geometrically the result

$$z = z^* \Leftrightarrow z \in R.$$

Such a geometrical representation of complex numbers is generally referred to as the *Argand diagram* (J. R. Argand, 1768–1822)† or as the *complex plane.*

As we saw in Chapter 3, there is a 1–1 correspondence between the set of points $P(x, y)$ and the set of position vectors $\mathbf{OP} = x\mathbf{i}+y\mathbf{j}$. Thus, the complex number $x+y\mathrm{j}$ may be alternatively represented in the Argand diagram by the *point P* or by the *position vector of P*. Both representations have their value, and we shall use both freely.

**Ex.* 16. If P represents the complex number $a+b\mathrm{j}$ (P is the *affix* of the complex number $a+b\mathrm{j}$) show that $|a+b\mathrm{j}| = |\mathbf{OP}|$.

Consider now two complex numbers $z_1 = a_1+b_1\mathrm{j}$ and $z_2 = a_2+b_2\mathrm{j}$, with affixes P_1 and P_2. The sum $z_1+z_2 = (a_1+a_2)+(b_1+b_2)\,\mathrm{j}$; but

$$\mathbf{OP}_1+\mathbf{OP}_2 = (a_1+a_2)\,\mathbf{i}+(b_1+b_2)\,\mathbf{j}$$

† The first exposition of the geometrical treatment of complex numbers was in fact published in 1797 by a Norwegian surveyor, Casper Wessel (1745–1818). For a translation of his paper see *The Treasury of Mathematics: 2* by Henrietta Medonick (Pelican).

and thus *the position vector representing the sum of two complex numbers is the sum of the position vectors which separately represent the two numbers* (see Figure 17.1).

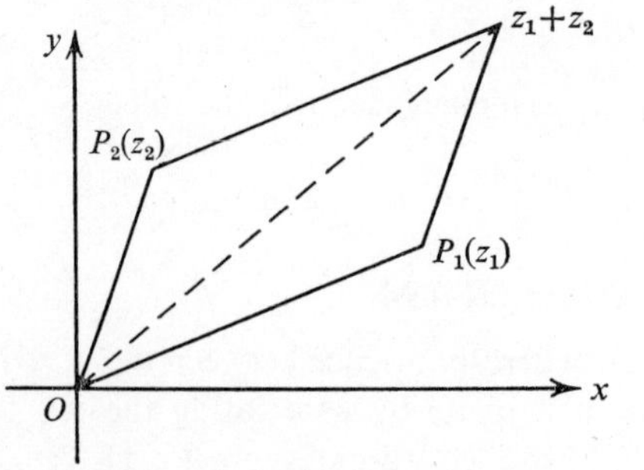

Fig. 17.1

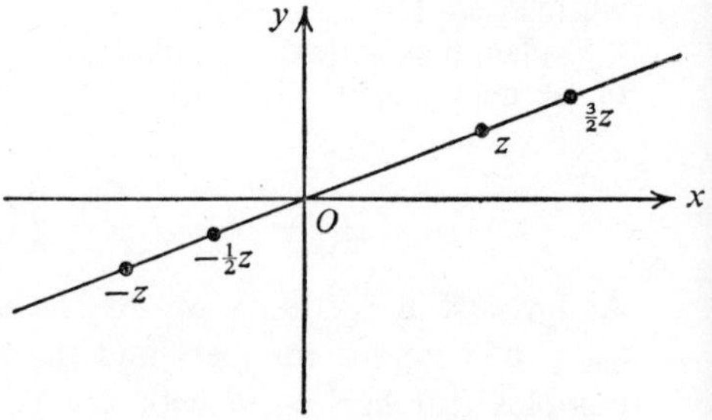

Fig. 17.2

**Ex*. 17. Describe the vector representing the complex number z_1-z_2.

Ex. 18. Mark in the Argand diagram the affixes of the complex numbers $2-\mathrm{j}$, $2(2-\mathrm{j})$ and $\frac{3}{2}(2-\mathrm{j})$. Interpret multiplication of a complex number z by the real number (i) $a > 0$, (ii) $b < 0$, in terms of an operation upon the vector **OP** representing z in the Argand diagram.

Ex. 18 shows that multiplication of a complex number z by a positive real number a is represented geometrically by an enlargement of **OP** (possibly by a factor less than 1). Multiplication by a negative real number $-a$ both enlarges **OP** and rotates it through an angle π (see Figure 17.2).

Ex*. 19. Mark in the Argand diagram the affixes of the complex numbers $3+2\mathrm{j}$, $\mathrm{j}(3+2\mathrm{j})$, $-(3+2\mathrm{j})$, $-\mathrm{j}(3+2\mathrm{j})$. Interpret multiplication of a complex number z by the pure imaginary number j in terms of an operation upon the vector **OP representing z in the Argand diagram. Show that multiplication of z by the pure imaginary number $b\mathrm{j}$ is represented in the Argand diagram by an enlargement of magnitude $|b|$ followed by (i) an anticlockwise rotation through $\frac{1}{2}\pi$ if $b > 0$, or (ii) a clockwise rotation through $\frac{1}{2}\pi$ if $b < 0$.

Ex. 20. Interpret geometrically the statement $\mathrm{j}^2 = -1$.

To interpret geometrically the multiplication of two general complex numbers, it is convenient to introduce a new concept, the *argument* of a complex number z.

Given a complex number $z = x+y\mathrm{j}$, represented by the vector **OP**, we have $x = r\cos\theta$, $y = r\sin\theta$, where $r = |z| = \sqrt{(x^2+y^2)}$ (see Figure 17.3).

The two real numbers, r and θ, determine the complex number z uniquely; conversely r is determined uniquely by z, but there is an infinite number of values of θ corresponding to z, differing from one another by integral multiples of 2π. Each such value of θ is called *an argument of* z. Just one of these values will lie in the interval $-\pi < \theta \leqslant \pi$, and this value is called the *principal argument of* z, and is written $\arg z$.

If $\arg z = \phi$, we may write

$$z = r(\cos\phi + \mathrm{j}\sin\phi).$$

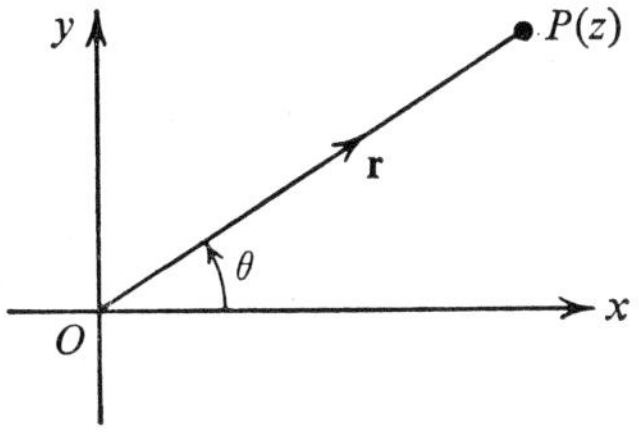

Fig. 17.3

A complex number may always be stated in this *modulus-argument form*. For example

$$1+\mathrm{j} = \sqrt{2}\left(\frac{1}{\sqrt{2}}+\frac{1}{\sqrt{2}}\mathrm{j}\right) = \sqrt{2}(\cos\tfrac{1}{4}\pi+\mathrm{j}\sin\tfrac{1}{4}\pi),$$

$$1-\mathrm{j} = \sqrt{2}(\cos[-\tfrac{1}{4}\pi]+\mathrm{j}\sin[-\tfrac{1}{4}\pi]),$$

$$-1+\mathrm{j} = \sqrt{2}(\cos\tfrac{3}{4}\pi+\mathrm{j}\sin\tfrac{3}{4}\pi),$$

$$-1-\mathrm{j} = \sqrt{2}(\cos[-\tfrac{3}{4}\pi]+\mathrm{j}\sin[-\tfrac{3}{4}\pi]).$$

(Notice that we must always be careful to select the angle to lie in the range $-\pi < \arg z \leqslant \pi$.)

Example 1. *Express in modulus-argument form the complex numbers*

(i) $-\sqrt{3}+\mathrm{j}$, (ii) $2-\mathrm{j}$, (iii) $1+\mathrm{j}\cot\alpha$ $(-\pi < \alpha < \pi$ and $\alpha \neq 0)$.

(i) $|-\sqrt{3}+\mathrm{j}| = 2$ and we have

$$-\sqrt{3}+\mathrm{j} = 2\left(-\frac{\sqrt{3}}{2}+\frac{1}{2}\mathrm{j}\right)$$

$$= 2(\cos\tfrac{5}{6}\pi+\mathrm{j}\sin\tfrac{5}{6}\pi).$$

(ii) $|2-\mathrm{j}| = \sqrt{5}$ and we have

$$2-\mathrm{j} = \sqrt{5}\left(\frac{2}{\sqrt{5}}-\frac{1}{\sqrt{5}}\mathrm{j}\right)$$

$$= 5(\cos\alpha+\mathrm{j}\sin\alpha),$$

where α is the angle in the range $-\frac{1}{2}\pi < \alpha < 0$ such that $\tan\alpha = -\frac{1}{2}$.

(iii) First observe that

$$|1+j\cot\alpha| = \sqrt{(1+\cot^2\alpha)} = |\mathrm{cosec}\,\alpha|.$$

If $0 < \alpha < \pi$, $|\mathrm{cosec}\,\alpha| = \mathrm{cosec}\,\alpha$ and we write

$$\begin{aligned} 1+j\cot\alpha &= \mathrm{cosec}\,\alpha\,(\sin\alpha+j\cos\alpha) \\ &= \mathrm{cosec}\,\alpha\,(\cos\phi+j\sin\phi), \end{aligned}$$

where ϕ is chosen so that

$$-\pi < \phi \leqslant \pi$$

and $$\sin\phi = \cos\alpha$$

and $$\cos\phi = \sin\alpha.$$

Thus, for $0 < \alpha < \pi$

$$1+j\cot\alpha = \mathrm{cosec}\,\alpha[\cos(\tfrac{1}{2}\pi-\alpha)+j\sin(\tfrac{1}{2}\pi-\alpha)].$$

If $-\pi < \alpha < 0$, $|\mathrm{cosec}\,\alpha| = -\mathrm{cosec}\,\alpha$ and we write

$$\begin{aligned} 1+j\cot\alpha &= |\mathrm{cosec}\,\alpha|\,(-\sin\alpha-j\cos\alpha) \\ &= |\mathrm{cosec}\,\alpha|(\cos\theta+j\sin\theta), \end{aligned}$$

where θ is chosen so that

$$-\pi < \theta \leqslant \pi$$

and $$\sin\theta = -\cos\alpha$$

and $$\cos\theta = \sin\alpha.$$

The reader should verify that this gives

$$1+j\cot\alpha = |\mathrm{cosec}\,\alpha|[\cos(-\tfrac{1}{2}\pi-\alpha)+j\sin(-\tfrac{1}{2}\pi-\alpha)]$$

for $-\pi < \alpha < 0$.

Ex. 21. Find the modulus and principal argument of each of the following complex numbers, and write each in modulus-argument form

(i) $1+\sqrt{3}j$; (ii) $2+2j$; (iii) -1; (iv) j; (v) $-j$; (vi) $-\sqrt{3}-j$; (vii) $\sqrt{3}-j$; (viii) $1+j\tan\alpha$; (ix) $\tan\alpha+j$; (x) $1-\cos\alpha-j\sin\alpha$ (use half angle formulae).

**Ex.* 22. Show that $\arg z = -\arg z^*$ and hence find $\arg z^{-1}$ in terms of $\arg z$.

Ex. 22. By writing z in the modulus-argument form $r(\cos\theta+j\sin\theta)$, discuss the geometrical interpretations for the multiplication of the complex number z by: (i) the real number $a > 0$; (ii) the real number $a < 0$; (iii) the pure imaginary number bj, $b > 0$; (iv) the pure imaginary number bj, $b < 0$.

Returning now to the question of the geometrical interpretation of the multiplication of two complex numbers z and w, let us write z, w in the modulus-argument form:

$$z = r(\cos\theta+j\sin\theta), \quad w = s(\cos\phi+j\sin\phi).$$

Then

$$wz = sr(\cos\phi + j\sin\phi)(\cos\theta + j\sin\theta)$$
$$= sr[(\cos\phi\cos\theta - \sin\phi\sin\theta) + j(\cos\phi\sin\theta + \sin\phi\cos\theta)]$$
$$= sr[\cos(\phi+\theta) + j\sin(\phi+\theta)],$$

and thus the product, wz, is a complex number with modulus $|w||z|$ and an argument $\arg w + \arg z$. In words, the modulus of a product is the product of the moduli and an argument of the product is the sum of the arguments. Geometrically, if the affix of z is P, multiplication by w (i) enlarges **OP** by a factor $|w|$, and (ii) rotates **OP** anticlockwise through an angle $\arg w$.

Ex. 24. If z is any complex number and $w = u + vj$ (u, v real and positive), plot in the Argand diagram, the affixes of z, jz, uz, jvz, $uz + jvz$. Deduce that

$$|wz| = |w||z| \quad \text{and} \quad \arg(wz) = \arg w + \arg z.$$

Is it necessary to modify your demonstration in any way if either u or v is negative?

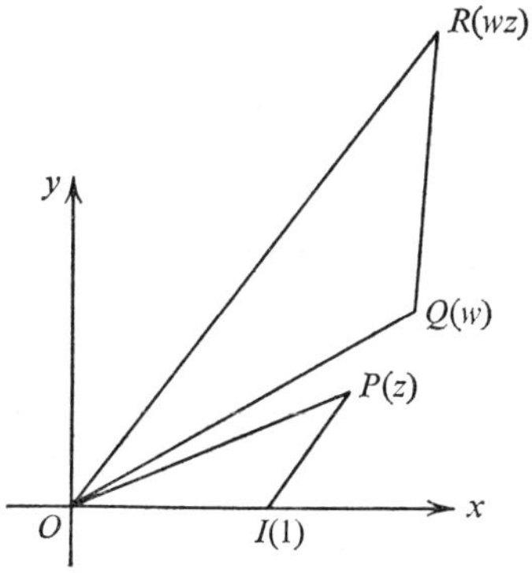

Fig. 17.4

In Figure 17.4, I, P, Q are the affixes of 1, z, w respectively. If each side of triangle IOP is enlarged by a factor $|w|$ and the triangle is then rotated anticlockwise through an angle $\arg w$ about O, I is brought into coincidence with Q. Let P be brought into coincidence with R (so that the triangles OIP, OQR are similar). Then

$$|OR| = |w|\,|OP| = |w||z|$$

and $I\hat{O}R = \arg w + \arg z$; thus R is the affix of the complex number wz. (Notice that **IP**, which represents the complex number $z-1$, is rotated into the displacement **QR**, which represents the complex number $w(z-1)$.)

Ex. 25. Form the product $(1+j)(\sqrt{3}+j)$ and interpret your result geometrically. What are the values of cos 75° and sin 75°?

Now suppose that w, z_1 and z_2 are any three complex numbers. Let P, Q, be the affixes of z_1 and z_1+z_2 respectively (Figure 17.5). Then **OP** represents z_1 and **PQ** represents z_2. Now enlarge triangle OPQ linearly by a factor $|w|$ and rotate through an angle arg w to bring it into coincidence with the triangle ORS. Then **OR** represents wz_1, **RS** represents wz_2 and **OS** represents $w(z_1+z_2)$. But **OS** = **OR**+**RS** and we have thus verified the distributive law

$$w(z_1+z_2) = wz_1+wz_2.$$

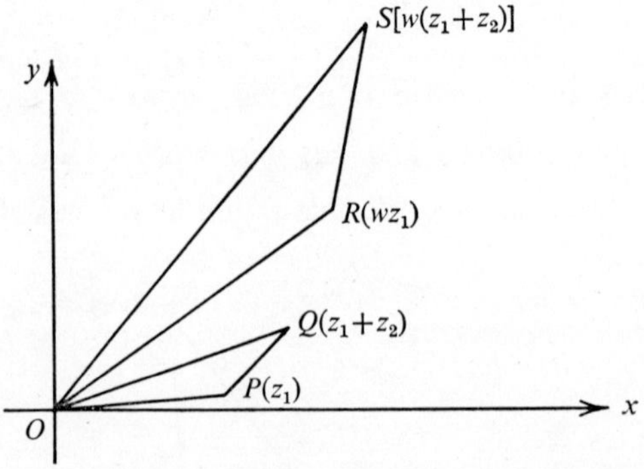

Fig. 17.5

Ex. 26. If $w \neq 0$, show that

$$\left|\frac{z}{w}\right| = \frac{|z|}{|w|} \quad \text{and that arg} \left(\frac{z}{w}\right) = \arg z - \arg w,$$

or differs from $\arg z - \arg w$ by 2π.

**Ex.* 27. Draw diagrams similar to that of Figure 17.4 to illustrate the construction from the affix of the complex number z, of the affixes of the complex numbers (i) z^2; (ii) z^3; (iii) z^{-1}.

Example 2. *z_1 and z_2 are arbitrary complex numbers. Give a geometrical verification of the triangle inequalities*

$$|z_1+z_2| \leqslant |z_1|+|z_2|,$$

$$|z_1-z_2| \geqslant ||z_1|-|z_2||.$$

Describe geometrically the set $S = \{z \in C : |z-1-j| \leqslant 1\}$ *and show that if* $z \in S$, *then*

$$\sqrt{2}-1 \leqslant |z-2| \leqslant \sqrt{2}+1.$$

In Figure 17.6, **OP** represents z_1, **PQ** represents z_2 and thus **OQ** represents z_1+z_2 and **RP** represents z_1-z_2, where $OPQR$ is a parallelogram. But, in ΔOPQ, $|\mathbf{OQ}| \leqslant |\mathbf{OP}|+|\mathbf{PQ}|$ (with equality only if OPQ is a straight line) and thus

$$|z_1+z_2| \leqslant |z_1|+|z_2|.$$

Again, in ΔOPR,

$$|\mathbf{OP}| \leqslant |\mathbf{OR}|+|\mathbf{RP}| \quad \text{and} \quad |\mathbf{OR}| \leqslant |\mathbf{OP}|+|\mathbf{RP}|;$$

thus,
$$|z_1| \leqslant |z_2|+|z_1-z_2| \quad \text{and} \quad |z_2| \leqslant |z_1|+|z_1-z_2|,$$

which combine together to give

$$|z_1-z_2| \geqslant ||z_1|-|z_2||.$$

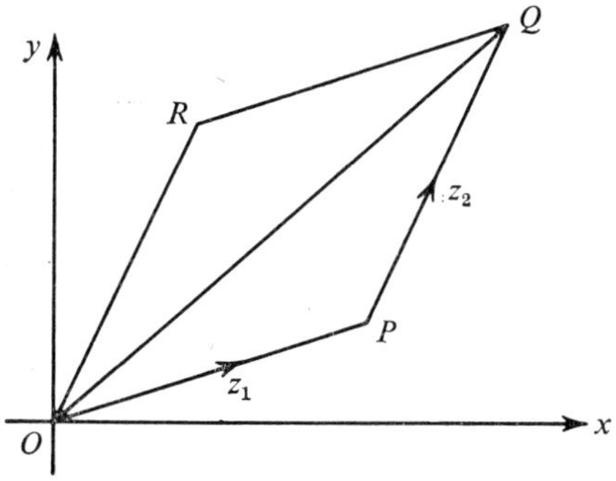

Fig. 17.6

For the second part of the question, we know that $|z-w|$ gives the distance between the affixes of z, w in the Argand diagram, and the set S is thus represented by the circumference and interior of the circle, centre $(1+\mathrm{j})$ and radius 1.

We demonstrate two methods of deducing the final part of the question.

Method (i) (*Analytical*)

From the inequalities

$$|z_1-z_2| \geqslant ||z_1|-|z_2||, \quad |z_1+z_2| \leqslant |z_1|+|z_2|,$$

we have

$$|z-2| = |(z-1-\mathrm{j})-(1-\mathrm{j})| \geqslant ||z-1-\mathrm{j}|-|1-\mathrm{j}|| \geqslant \sqrt{2}-1,$$

$$|z-2| = |(z-1-\mathrm{j})+(-1+\mathrm{j})| \leqslant |z-1-\mathrm{j}|+|-1+\mathrm{j}| \leqslant \sqrt{2}+1.$$

Method (ii) (*Geometrical*)

Let C be the centre of the circle $|z-1-\mathrm{j}| = 1$ (which *touches* the real axis at A, say) and let B be the affix of the number 2. Let BC cut the circle at P, Q (see Figure 17.7).

For given real $a > 0$, $|z-2| = a$ represents a circle, centre B and radius a. Of all such circles, the smallest containing a point of S (that is, the circumference and interior of the circle $|z-1-\mathrm{j}| = 1$) has radius BP and the largest a radius of BQ. But, by Pythagoras, $BC = \sqrt{2}$ and thus $BP = \sqrt{2}-1$, $BQ = \sqrt{2}+1$ and the result follows.

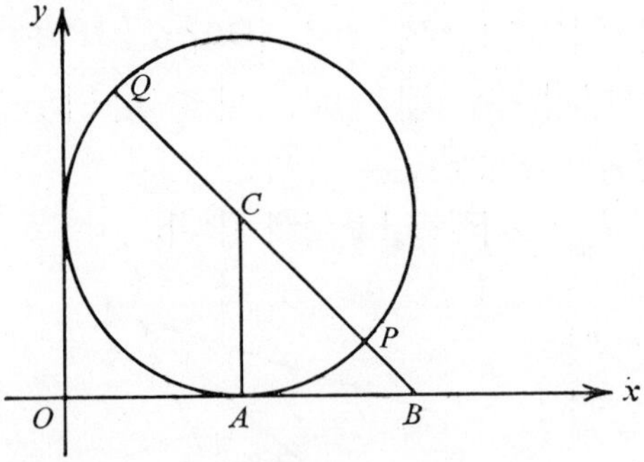

Fig. 17.7

Ex. 28. The triangle inequalities stated in the Example above may be deduced without recourse to geometry. Show that

$$z_1 z_2^* + z_1^* z_2 = 2\,\mathrm{Re}\,(z_1 z_2^*)$$

and deduce that $$|z_1+z_2|^2 = |z_1|^2 + 2\,\mathrm{Re}\,(z_1 z_2^*) + |z_2|^2.$$

Show also that $$\mathrm{Re}\,(z_1 z_2^*) \leqslant |z_1||z_2|$$

and deduce that $$|z_1+z_2| \leqslant |z_1| + |z_2|.$$

Prove the second triangle inequality in a similar fashion.

Consider now the effect of multiplying the complex numbers 0, 1, $1+\mathrm{j}$, j by $a+b\mathrm{j}$: we get

$$0, \quad a+b\mathrm{j}, \quad (a-b)-(a+b)\,\mathrm{j}, \quad -b+a\mathrm{j}$$

—the vertices of another square (Figure 17.8).

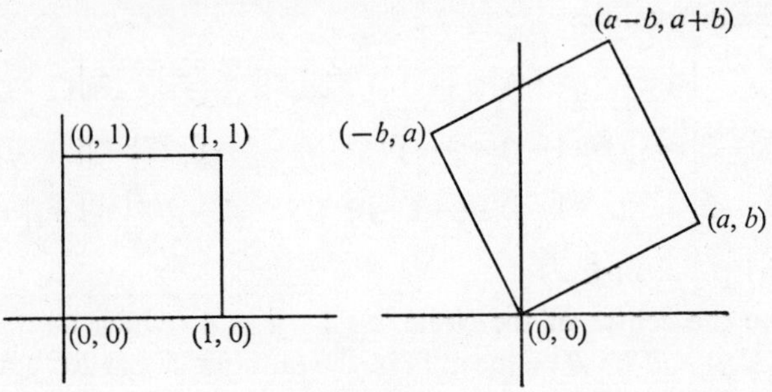

Fig. 17.8

The result is strongly reminiscent of the linear transformation with matrix

$$\begin{pmatrix} a & -b \\ b & a \end{pmatrix} = a\begin{pmatrix} 1 & 0 \\ 0 & 1 \end{pmatrix} + b\begin{pmatrix} 0 & -1 \\ 1 & 0 \end{pmatrix}$$
$$= a\mathbf{I} + b\mathbf{J}, \text{ say.}$$

The result suggests that it might be profitable to associate the number 1 with the matrix **I**, the number j with the matrix **J** and the general complex number $a+b\mathrm{j}$ with the matrix $a\mathbf{I}+b\mathbf{J}$.

Consider now two complex numbers $a+b\mathrm{j}$, $c+d\mathrm{j}$ and the associated matrices

$$\begin{pmatrix} a & -b \\ b & a \end{pmatrix}, \quad \begin{pmatrix} c & -d \\ d & c \end{pmatrix}.$$

We have

(i) $(a+b\mathrm{j})+(c+d\mathrm{j}) = (a+c)+(b+d)\,\mathrm{j}$

and $$\begin{pmatrix} a & -b \\ b & a \end{pmatrix} + \begin{pmatrix} c & -d \\ d & c \end{pmatrix} = \begin{pmatrix} a+c & -(b+d) \\ b+d & a+c \end{pmatrix};$$

(ii) $(a+b\mathrm{j})\,(c+d\mathrm{j}) = (ac-bd)+(bc+da)\,\mathrm{j}$

and $$\begin{pmatrix} a & -b \\ b & a \end{pmatrix}\begin{pmatrix} c & -d \\ d & c \end{pmatrix} = \begin{pmatrix} ac-bd & -(bc+da) \\ bc+da & ac-bd \end{pmatrix}.$$

The 1–1 correspondence between the set C of complex numbers on the one hand and the set of all 2×2 matrices of the form

$$\begin{pmatrix} a & -b \\ b & a \end{pmatrix} \quad (a, b \in R)$$

on the other is thus seen to preserve the structures of addition and multiplication.

Ex. 29. Confirm, by direct computation of $(a+b\mathrm{j})^{-1}$ and the inverse of the matrix $\begin{pmatrix} a & -b \\ b & a \end{pmatrix}$ that the structure of division is also preserved.

**Ex.* 30. Interpret the correspondence between the complex number $\cos\theta+\mathrm{j}\sin\theta$ and the matrix $\begin{pmatrix} \cos\theta & -\sin\theta \\ \sin\theta & \cos\theta \end{pmatrix}$ in the light of their geometrical properties.

Ex. 31. If C, R have their usual meanings and M is the set of all matrices of the form $\begin{pmatrix} a & -b \\ b & a \end{pmatrix}$, $a, b \in R$, determine the nature of the function $f\colon M \to R$ which corresponds to the modulus function $C \to R$.

To summarize, we have the following representations of the complex number $a+b\mathrm{j}$:

(i) the point (a, b) in the Argand diagram;

(ii) the vector $a\mathbf{i}+b\mathbf{j}$;

(iii) the matrix $a\mathbf{I}+b\mathbf{J}$ where

$$\mathbf{I} = \begin{pmatrix} 1 & 0 \\ 0 & 1 \end{pmatrix}, \quad \mathbf{J} = \begin{pmatrix} 0 & -1 \\ 1 & 0 \end{pmatrix}.$$

Representation (i) is, as it were, a static representation: it is useful in that it enables us to discuss *sets* of complex numbers in geometrical terms, but by itself, it lacks the additive and multiplicative structure of complex numbers.

Representation (ii) is, as it were, dynamic: addition of vectors and addition of complex numbers have the same structure. However, multiplication of complex numbers has to be interpreted by a new operation on vectors: *dot products of vectors on the one hand and products of complex numbers on the other do NOT have the same structure.*

Representation (iii) gives a *full structural representation* of the complex numbers in terms of matrices: addition, subtraction, multiplication and division of complex numbers have their exact analogue in the corresponding operations on the associated matrices.

Example 3. *The affixes of the complex numbers* $a = 3+\mathrm{j}$ *and* $b = 1+2\mathrm{j}$ *are A and B respectively. Find complex numbers* p, p', q, q' *with affixes* P, P', Q, Q' *such that ABQP and ABQ'P' are squares.*

Find also complex numbers r, r' *with affixes* R, R' *such that ABR, ABR' are equilateral triangles.*

$$\mathbf{AB} = b-a$$
$$= -2+\mathrm{j}.$$

To obtain **AP** and **AP′** we must rotate **AB** through $\pm\frac{1}{2}\pi$; that is, we must multiply $(b-a)$ by $\pm\mathrm{j}$ (see Ex. 19). Thus

$$\mathbf{AP} = \mathrm{j}(-2+\mathrm{j})$$
$$= -1-2\mathrm{j}$$

and

$$\mathbf{AP'} = -\mathrm{j}(-2+\mathrm{j})$$
$$= 1+2\mathrm{j}.$$

Thus

$$p = a+(-1-2\mathrm{j})$$
$$= 3+\mathrm{j}-1-2\mathrm{j}$$
$$= 2-\mathrm{j}$$

and

$$p' = 3+\mathrm{j}+1+2\mathrm{j}$$
$$= 4+3\mathrm{j}.$$

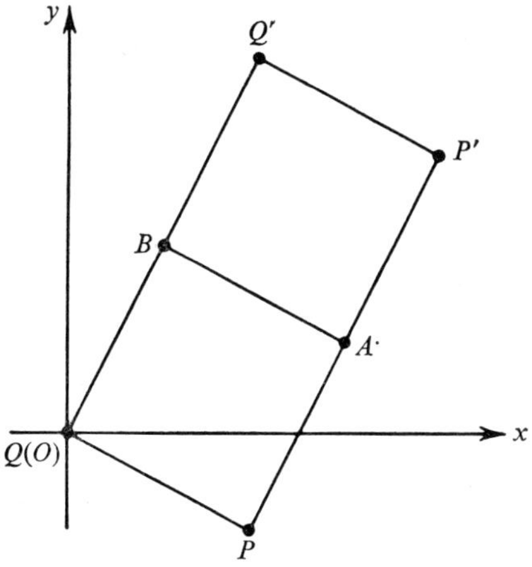

Fig. 17.9

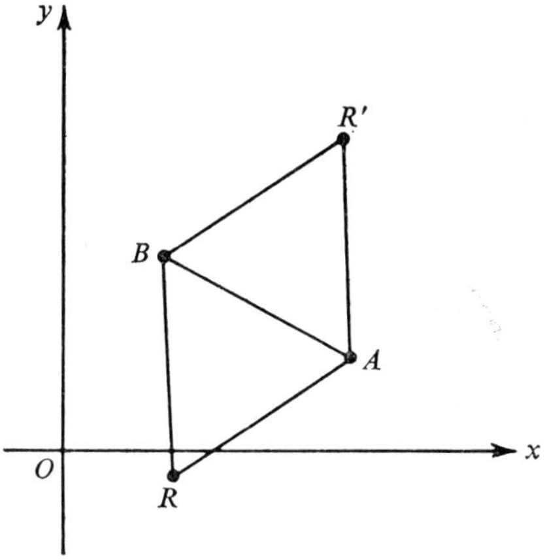

Fig. 17.10

Similarly
$$q = b+(-1-2\mathrm{j})$$
$$= 1+2\mathrm{j}-1-2\mathrm{j}$$
$$= 0$$
and
$$q' = 1+2\mathrm{j}+1+2\mathrm{j}$$
$$= 2+4\mathrm{j}.$$

To rotate AB through an angle $\pm\frac{1}{3}\pi$ we seek complex numbers z such that $|z| = 1$ and $\arg z = \pm\frac{1}{3}\pi$; thus

$$z = \cos\tfrac{1}{3}\pi \pm \mathrm{j}\sin\tfrac{1}{3}\pi$$
$$= \tfrac{1}{2}(1\pm \mathrm{j}\sqrt{3}).$$

It follows that
$$\mathbf{AR} = \tfrac{1}{2}(1+\mathrm{j}\sqrt{3})\,(b-a)$$
and thus
$$r = a+\tfrac{1}{2}(1+\mathrm{j}\sqrt{3})\,(b-a)$$
$$= 3+\mathrm{j}+\tfrac{1}{2}(1+\mathrm{j}\sqrt{3})\,(-2+\mathrm{j})$$
$$= 2-\tfrac{1}{2}\sqrt{3}+(\tfrac{3}{2}-\sqrt{3})\,\mathrm{j}.$$
Similarly
$$\mathbf{AR}' = \tfrac{1}{2}(1-\mathrm{j}\sqrt{3})\,(b-a)$$
and thus
$$r' = 3+\mathrm{j}+\tfrac{1}{2}(1-\mathrm{j}\sqrt{3})\,(-2+\mathrm{j})$$
$$= 2+\tfrac{1}{2}\sqrt{3}+(\tfrac{3}{2}+\sqrt{3})\mathrm{j}.$$

Exercise 17(b)

1. Determine the modulus and principal argument of each of the following complex numbers, where necessary leaving the modulus in surd form and giving the argument in *degrees*, correct to the nearest degree: (i) $3-4j$; (ii) $2+3j$; (iii) $-1-2j$; (iv) $\frac{1}{2}+j$; (v) $-2+5j$; (vi) $-2-j$; (vii) $12-5j$; (viii) $-9-40j$.

2. Simplify the following complex numbers:
(i) $(\cos\theta+j\sin\theta)^3$; (ii) $(\cos\theta+j\sin\theta)/(\cos\phi-j\sin\phi)$;
(iii) $(\cos\theta+j\sin\theta)(\sin\theta+j\cos\theta)$; (iv) $(\cos\frac{1}{8}\pi+j\sin\frac{1}{8}\pi)^2$;
(v) $(\cos\frac{1}{12}\pi+j\sin\frac{1}{12}\pi)(\cos\frac{5}{12}\pi+j\sin\frac{5}{12}\pi)$;
(vi) $(\cos\frac{1}{12}\pi+j\sin\frac{1}{12}\pi)(\cos\frac{5}{12}\pi-j\sin\frac{5}{12}\pi)$;
(vii) $(\cos\frac{5}{12}\pi+j\sin\frac{5}{12}\pi)/(\cos\frac{1}{12}\pi-j\sin\frac{1}{12}\pi)$;
(viii) $(1+\cos\theta+j\sin\theta)(1-\cos\phi-j\sin\phi)$.

3. Determine the principal argument of the complex number

$$(\sin\theta+j\cos\theta)(\cos\phi+j\sin\phi)$$

(i) when $-\frac{3}{2}\pi < \phi-\theta \leqslant \frac{1}{2}\pi$; (ii) when $\frac{1}{2}\pi < \phi-\theta \leqslant \frac{5}{2}\pi$.

4. If P is the affix of the complex number z, show how to construct geometrically the affixes of: (i) $1+z$; (ii) $1-2z$; (iii) $2j+3z$; (iv) $1-jz$; (v) $(1-j)z$; (vi) $j+z^2$; (vii) $z+z^2$; (viii) $(1+j)(1-z)$.

5. If P is the affix of the complex number z, show how to construct the affixes of: (i) z^*; (ii) $z-z^*$; (iii) $1+2/z$; (iv) $(1+z)^*$; (v) $1/(1+z)$; (vi) $j/(j-z)$; (vii) $z/(1-z)$; (viii) $(1-z)^{-2}$.

6. If P, Q are respectively the affixes of the complex numbers z, w show how to construct the affixes of: (i) $2z+w$; (ii) $\frac{1}{3}(z+2w)$; (iii) $\frac{1}{2}(3z-w)$; (iv) $z+jw$; (v) $(1+j)(z+w)$; (vi) $(z+1)(w+1)$; (vii) w/z; (viii) $j(z-w)$.

7. Describe geometrically the set $S = \{z \in C: |z-2-j| \leqslant 1\}$ and prove that, for all $z \in S$, $\sqrt{5}-1 \leqslant |z| \leqslant \sqrt{5}+1$.

8. Describe geometrically the set $S = \{z \in C: |z-2| \leqslant 2\}$ and prove that, for all $z \in S$, $3 \leqslant |z+2+3j| \leqslant 7$.

9. Show that $|z-6j| \leqslant 3 \Rightarrow 10 \leqslant |z-5+6j| \leqslant 16$.

10. Show that $|z-3-j| < 1 \Rightarrow \sqrt{10}-1 < |z| < \sqrt{10}+1$.

11. Prove, using mathematical induction, the following results:
(i) $|z_1+z_2+\ldots+z_n| \leqslant |z_1|+|z_2|+\ldots+|z_n|$;
(ii) $|z_1z_2z_3\ldots z_n| = |z_1||z_2||z_3|\ldots|z_n|$;
(iii) $\arg(z_1z_2z_3\ldots z_n) = \sum_{i=1}^{n} \arg z_i$;
(iv) $(z_1+z_2+z_3+\ldots+z_n)^* = \sum_{i=1}^{n} z_i^*$;
(v) $(z_1z_2z_3\ldots z_n)^* = z_1^*z_2^*z_3^*\ldots z_n^*$.

12. The function $f: C \to C$ is defined by $f(z) = 2z+j$. Describe the effect of this mapping geometrically and find the value of z which is invariant under the mapping, again interpreting your result geometrically.

13. Answer the same question as in Question 12 (substituting where necessary, *values* for *value*) for the functions defined by: (i) $f(z) = \mathrm{j}z+1$; (ii) $f(z) = z^2+1$; (iii) $f(z) = z^2+1+\mathrm{j}$; (iv) $f(z) = 2z^*+1+\mathrm{j}$.

14. Show that the points representing the complex numbers $3+4\mathrm{j}$, $7+2\mathrm{j}$, $5+8\mathrm{j}$ in the Argand diagram are three vertices of a square and find the fourth vertex.

Show that the interior of the square represents the set

$$\{z \in C: z = 3+4\mathrm{j}+\lambda(2-\mathrm{j})+\mu(1+2\mathrm{j}), \text{ real } \lambda, \mu, 0 < \lambda < 2, 0 < \mu < 2\}.$$

15. Prove that, if $S = \{z \in C: |z-2| < 1\}$, then, for all $z \in S$, $-\frac{1}{6}\pi < \arg z < \frac{1}{6}\pi$.

16. Prove that the modulus of the quotient of two complex numbers is the quotient of the two moduli, and that the difference of the two arguments is an argument of the quotient.

(i) If z is a complex number, and $\left|\dfrac{z-1}{z+1}\right| = 2$, prove that the locus of the point which represents z in the Argand diagram is a circle, and find its centre and radius.

(ii) If

$$\arg\left(\frac{z-1}{z+1}\right) = \tfrac{1}{2}\pi,$$

find the locus of the point which represents z. (O & C)

17. If z is the complex number $\cos\theta+\mathrm{j}\sin\theta$, express $1/(1+z)$ and $1/(1-z)$ in the form $x+y\mathrm{j}$.

If $|z| = 1$, prove that the real part of $2z/(1-z^2)$ is zero. (O & C)

18. If A, B, C, D are the affixes of the complex numbers a, b, c, d prove that $a+c = b+d \Leftrightarrow ABCD$ is a parallelogram.

What can you say about the parallelogram if $a+b\mathrm{j} = c+d\mathrm{j}$?

19. Explain with the help of a sketch, why no complex number z can be found such that $\arg z = \frac{1}{4}\pi$ and $|z-4-\mathrm{j}| = |z-2-3\mathrm{j}|$.

20. a, b are real numbers; show that

$$\left\{z \in C: \operatorname{Im}\left(\frac{z-\mathrm{j}a}{z-b}\right) = 0\right\}$$

is represented by a straight line in the Argand diagram.

Miscellaneous Exercises 17

1. By writing $1 = -\mathrm{j}^2$, express $z = (\sqrt{3}+1)+\mathrm{j}(\sqrt{3}-1)$ as the product of two complex numbers and hence write down the values of $|z|$ and $\arg z$.

2. If $|z-2-\mathrm{j}| < 2$ and $|w-5-5\mathrm{j}| < 1$, find the maximum and minimum values of $|z-w|$.

3. Describe geometrically the relation between the complex numbers $6\mathrm{j}$ and $2+4\mathrm{j}$ and the complex numbers (i) $1+5\mathrm{j}$; (ii) $3+3\mathrm{j}$.

If λ is an arbitrary real number, what can you say about the complex number $(1+\lambda)+\mathrm{j}(5+\lambda)$? (over)

If B, C are the affixes of the complex numbers 6j, 2+4j, find the two complex numbers whose affixes, A, A' are such that the triangles ABC, $A'BC$ are equilateral.

4. z_1, z_2, w_1, w_2 are four complex numbers and

$$\mathbf{A} = \begin{pmatrix} z_1 & z_2 \\ -z_2^* & z_1^* \end{pmatrix}, \quad \mathbf{B} = \begin{pmatrix} w_1 & w_2 \\ -w_2^* & w_1^* \end{pmatrix}.$$

By considering det (**AB**), prove that the product of two sums of four squares is itself the sum of four squares.

5. Write down the product of the complex numbers $x_1+y_1\text{j}$ and $x_2+y_2\text{j}$. Show that this product has modulus r_1r_2 and argument $\theta_1+\theta_2$, where r_1, r_2 are moduli and θ_1, θ_2 the arguments of the given numbers. Solve the equation

$$\frac{z}{3+4\text{j}}+\frac{z-1}{5\text{j}} = \frac{5}{3-4\text{j}}.$$

Show that, if $(z-2)/(z-\text{j})$ is real, the point corresponding to $z = x+y\text{j}$ in the Argand diagram lies on a straight line through the points 2 and j, and find the equation relating x and y when the ratio is purely imaginary. (O & C)

6. If the complex numbers z_1, z_2, z_3 are represented in the Argand diagram by points Z_1, Z_2, Z_3, interpret geometrically (giving a justification) the modulus and argument of the complex number $(z_3-z_1)/(z_2-z_1)$.

If the complex numbers a, b, c, x, y, z are represented by points A, B, C, X, Y, Z and if

$$\frac{x-c}{b-c} = \frac{y-a}{c-a} = \frac{z-b}{a-b},$$

prove that the triangles BCX, CAY, ABZ are similar.

Prove also that the triangles ABC and XYZ have the same centroid.

(O & C)

7. The *cross-ratio* of four complex numbers z_1, z_2, z_3, z_4, written (z_1z_2, z_3z_4), is defined by

$$(z_1z_2, z_3z_4) = \frac{(z_1-z_3)(z_2-z_4)}{(z_1-z_4)(z_2-z_3)}.$$

Show that, if $(z_1z_2, z_3z_4) = \lambda$, then the twenty-four possible permutations of the numbers z_1, z_2, z_3, z_4 give rise to six distinct cross-ratios with values λ, $1-\lambda$, λ^{-1}, $(1-\lambda)^{-1}$, $\lambda(\lambda-1)^{-1}$, $\lambda^{-1}(\lambda-1)$.

If a, b, c, d are four complex numbers, and the complex numbers w_i, z_i $(i = 1, 2, 3, 4)$ are connected by the relation

$$w_i = \frac{az_i+b}{cz_i+d},$$

prove that $(z_1z_2, z_3z_4) = (w_1w_2, w_3w_4)$.

8. A complex number z_r is represented on the Argand diagram by a point Z_r. Prove that arg $[(z_3-z_1)/(z_2-z_1)]$ is equal to the angle $Z_2Z_1Z_3$.

If z_1, z_2, z_3, z_4 are distinct, and the cross-ratio $(z_2-z_1)(z_4-z_3)/(z_4-z_1)(z_2-z_3)$ (that is (z_2z_4, z_1z_3)) is real, prove that, in general, Z_1, Z_2, Z_3, Z_4 lie on a circle.

In the exceptional case, what is the relation between Z_1, Z_2, Z_3, Z_4?

(O & C)

9. The complex numbers a, b, c, x, y, z are represented in the Argand diagram by points A, B, C, X, Y, Z.

(i) If the triangles ABC, XYZ are directly similar (that is, corresponding angles are equal in sense as well as magnitude) prove that

$$\begin{vmatrix} a & b & c \\ x & y & z \\ 1 & 1 & 1 \end{vmatrix} = 0.$$

(ii) If the triangles ABC, XYZ have equal areas, prove that

$$\begin{vmatrix} a & b & c \\ a^* & b^* & c^* \\ 1 & 1 & 1 \end{vmatrix} = \pm \begin{vmatrix} x & y & z \\ x^* & y^* & z^* \\ 1 & 1 & 1 \end{vmatrix}.$$ (O & C)

10. If β and γ are complex numbers whose moduli are both equal to 1, prove that $(1+\beta\gamma)/(\beta+\gamma)$ is real.

Hence, or otherwise, prove that, if a, b, c are complex numbers whose moduli are all equal, then $(a^2+bc)/[a(b+c)]$ is real. (O & C)

11. The points A, B, C are the affixes of the complex numbers a, b, c in the Argand diagram. If the circumcentre of the triangle ABC is at the origin, prove that the orthocentre H, is the affix of the point $a+b+c$. Deduce that O, G, H are collinear, where G is the centroid of the triangle, and that $OH = 3OG$.

12. Points A, B in the Argand diagram represent complex numbers a, b respectively. O is the origin, and P represents one of the values of $\sqrt{(ab)}$. Prove that, if $OA = OB = r$, then also $OP = r$, and OP is perpendicular to AB.

A, B, C lie on a circle with centre O, and represent complex numbers a, b, c respectively. Prove that the point D which represents $-bc/a$ also lies on the circle, and that AD is perpendicular to BC.

The perpendiculars from B, C to CA, AB meet the circle again at E, F respectively. Prove that OA is perpendicular to EF. (O & C)

13. A complex number z is represented by a point Z on the Argand diagram. Prove that multiplication of z by w is represented by taking the point Z' on OZ (where O is the origin) such that $OZ' = OZ|w|$ and rotating OZ' through the angle $\arg w$.

If z is represented by a point on the circle of radius a which touches the y-axis at O and lies in the first and fourth quadrants, prove that

$$z-2a = \mathrm{j}z \tan(\arg z).$$ (O & C)

14. A point P representing the complex number z moves in the Argand diagram so that it lies always in the region defined by

$$|z-1| \leqslant |z-\mathrm{j}| \quad \text{and} \quad |z-2-2\mathrm{j}| \leqslant 1.$$

Indicate on a sketch the region within which P lies.

If P describes the boundary of this region find: (i) the value of $|z|$ when $\arg z$ has its smallest value; (ii) the values of z in the form $x+\mathrm{j}y$ when $\arg(z-1) = \frac{1}{4}\pi$. (London)

15. In the Argand diagram A, B, C and D are four points representing the complex numbers z_1, z_2, z_3 and z_4 respectively. If O is the origin, prove that the triangles OAB and OCD are similar if

$$\frac{z_1}{z_2} = \frac{z_3}{z_4}.$$

The complex number z is given by

$$z = \left(\frac{6j+8}{3j-4}\right)^2.$$

Find, by drawing to scale, the position of the point representing z on the Argand diagram. Check your result by calculation. (London)

16. The complex number $z = x+jy = r(\cos\theta+j\sin\theta)$ is represented in the Argand diagram by the point (x, y). Prove that, if three variable points z_1, z_2, z_3 are such that $z_3 = \lambda z_2+(1-\lambda)z_1$, where λ is a complex constant, then the triangle with vertices z_1, z_2 and z_3 is similar to the triangle with vertices at the points 0, 1 and λ.

ABC is a triangle. On the sides BC, CA, AB triangles BCA', CAB', ABC' are described similar to a given triangle DEF. Prove that the centroids of the triangle ABC and $A'B'C'$ are coincident. (O & C)

18. *Polynomials and partial fractions*

1. THE EVALUATION OF POLYNOMIALS

(Unless specifically stated otherwise, we shall assume that all polynomials have integral coefficients.)

We have seen in Chapter 4 that, given a polynomial $P(x)$ the process of dividing $P(x)$ by $x-a$ leads to the identity

$$P(x) \equiv (x-a)\,Q(x)+R,$$

where $Q(x)$ is a polynomial of degree one less than the degree of $P(x)$. The actual coefficients of $Q(x)$ are obtained by the process of long division, but the process may be shortened by *the method of synthetic division* which consists, essentially, of comparing coefficients on both sides of the above identity, as illustrated in the following example.

Suppose that $P(x) \equiv 2x^4+3x^3-5x+7$ is to be divided by $x+3$. Then the quotient, $Q(x)$ will be a cubic polynomial, and we have

$$2x^4+3x^3+0x^2-5x+7 \equiv (x+3)\,(ax^3+bx^2+cx+d)+R.$$

Comparing coefficients

$$a = 2,\quad 3a+b = 3,\quad 3b+c = 0,\quad 3c+d = -5,\quad 3d+R = 7$$

and, working from the left, we have, by successive substitution

$$a = 2,\quad b = -3,\quad c = 9,\quad d = -32,\quad R = 103.$$

Observing that successive coefficients of $Q(x)$ are obtained by multiplying the preceding coefficient by -3 and adding the result to the corresponding coefficient of $P(x)$, the work may be set out succinctly:

$$\begin{array}{rrrrrl} 2 & 3 & 0 & -5 & 7 & (-3 \\ & -6 & 9 & -27 & 96 & \\ \hline 2 & -3 & 9 & -32 & 103 & \end{array}$$

giving $Q(x) \equiv 2x^3-3x^2+9x-32$ and $R = 103$.

Example 1. *Find the quotient and remainder on dividing* $4x^4-5x^2+7x+2$ *by* $x-5$.

$$\begin{array}{rrrrrl} 4 & 0 & -5 & 7 & 2 & (5 \\ & 20 & 100 & 475 & 2410 & \\ \hline 4 & 20 & 95 & 482 & 2412 & \end{array}$$

Thus $Q(x) \equiv 4x^3+20x^2+95x+482$ and $R = 2412$.

It will be seen that the above procedure constitutes an efficient method for evaluating $R = P(a)$, even if the coefficients of $Q(x)$ are not required explicitly.

Ex. 1. Explain how the method above may be modified for division by $ax+b$, $a \neq 1$, by first dividing $P(x)$ by a. Obtain $Q(x)$ and R when

$$P(x) \equiv 2x^4-15x^3-33x^2-x+14$$

is divided by $2x+3$. (In such cases $Q(x)$ will not, of course, generally have integral coefficients.)

Given a polynomial $P(x)$ of say, degree four,

$$P(x) \equiv a_0x^4+a_1x^3+a_2x^2+a_3x+a_4,$$

the procedure outlined above to calculate $P(b)$ amounts to finding successively

$$a_0,\ a_0b+a_1,\ b(a_0b+a_1)+a_2,\ b[b(a_0b+a_1)+a_2]+a_3$$

and finally, $$P(b) \equiv b\{b[b(a_0b+a_1)+a_2]+a_3\}+a_4.$$

Thus the polynomial $P(x)$ has been *nested* into the form

$$P(x) \equiv x\{x[x(a_0x+a_1)+a_2]+a_3\}+a_4,$$

and $P(b)$ calculated by substituting $x = b$ and working outwards from the middle bracket. The process is often referred to as *nested multiplication*: the reader familiar with the use of hand calculating machines will see that, using the facility of back transfer, the whole computation may be effected without having to record any numbers and clear the machine.

Ex. 2. Use a hand calculator to evaluate $P(37)$ where

$$P(x) \equiv 15x^4-61x^3+46x^2+18x+56.$$

Details of the use of hand-calculating machines in evaluating polynomials and, indeed, in a wide variety of other applications will be found in *Numerical Mathematics* by A. J. Moakes.

It is often useful to express a polynomial $P(x)$, of degree n, in the form

$$P(x) \equiv A_0(x-a)^n+A_1(x-a)^{n-1}+A_2(x-a)^{n-2}+\ldots+A_{n-1}(x-a)+A_n$$

($A_0, A_1, A_2, \ldots, A_n$ integral). Such a reduction may be effected by successive applications of *Horner's method of synthetic division*. (See Example 1.)

Example 2. *Express* $4x^4-5x^2+7x+2$ *in the form*

$$A_0(x-5)^4+A_1(x-5)^3+A_2(x-5)^2+A_3(x-5)+A_4.$$

4	0	−5	7	2	(5
	20	100	475	2410	
4	20	95	482	**2412**	(5
	20	200	1475		
4	40	295	**1957**		(5
	20	300			
4	60	**595**			(5
	20				
4	**80**				

Thus $P(x) \equiv 4(x-5)^4+80(x-5)^3+595(x-5)^2+1957(x-5)+2412.$

Ex. 3. Show that, when $x \approx 5$,

$$P(x) \equiv 4x^4-5x^2+7x+2 \approx 1957x-7373.$$

Find $P(5{\cdot}1)$ correct to 2 significant figures.

Ex. 4. With the notation of Ex. 3, what is the equation of the tangent to the curve $y = P(x)$ at the point (5, 2412)?

Ex. 5. With the notation of Ex. 3, evaluate $P(5)$, $P'(5)$, $P''(5)$.

Ex. 6. Show how, in Example 2, the values of A_4, A_3, A_2, A_1, A_0 enable us to calculate $P(5)$, $P'(5)$, $P''(5)$, $P'''(5)$, $P^{\text{iv}}(5)$.

2. DIFFERENCES

Consider the quartic polynomial

$$f(x) \equiv x^4-5x^3+x+2.$$

Tabulated overleaf are the values of $f(x)$ for $x = -4(1)5$ (that is, for all values of x in steps of 1 from $x = -4$ to $x = 5$). The first and second columns give respectively the values of x and $f(x)$, while each subsequent column gives the successive *differences* between entries in the preceding column.

It is seen from this table that, although the first and second difference columns follow no readily discernible pattern, the third difference column contains the terms of an arithmetic sequence and the fourth differences are constant. This is no coincidence: for any polynomial of degree n, the nth differences are constant (see Ex. 11).

x	f	Δf	$\Delta^2 f$	$\Delta^3 f$	$\Delta^4 f$	$\Delta^5 f$
−4	574					
		−359				
−3	215		200			
		−159		−90		
−2	56		110		24	
		−49		−66		0
−1	7		44		24	
		−5		−42		0
0	2		2		24	
		−3		−18		0
1	−1		−16		24	
		−19		6		0
2	−20		−10		24	
		−29		30		0
3	−49		20		24	
		−9		54		
4	−58		74			
		65				
5	7					

If we are content to assume for the moment that the fourth differences continue constant, it is an easy matter to extend the table working backwards from the constant fourth difference, to supply further values for the polynomial:

x	f	Δf	$\Delta^2 f$	$\Delta^3 f$	$\Delta^4 f$
2	−20				
		−29			
3	−49		20		
		−9		54	
4	−58		74		24
		65		**78**	
5	7		**152**		**24**
		217		**102**	
6	**224**		**254**		
		471			
7	**695**				

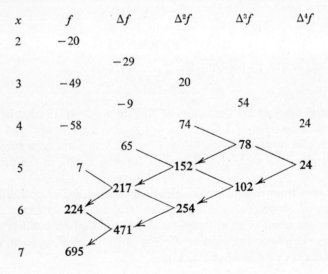

Ex. 7. The cubic polynomial $f(x)$ has the values $f(0) = 2, f(1) = -8, f(2) = -20, f(3) = -28$. Form a difference table and deduce the values of $f(4), f(5)$.

Ex. 8. Explain the following check on a difference table : the sum of any column of differences is equal to the difference between the first and last entries in the preceding difference column.

Differences may be calculated for non-integral values. For example, in the table below, differences are calculated for the polynomial

$$f(x) = x - x^3$$

for values of $x = 0\,(0{\cdot}1)\,0{\cdot}6$. (That is, for values of x from $x = 0$ to $x = 0{\cdot}6$ in steps of 0·1.) Since the values of $f(x)$ are calculated to 3 decimal places all the differences will also be to 3 decimal places and it is unnecessary to enter the decimal point; for example, the constant difference ($\Delta^3 f$), written as -6, is really $-0{\cdot}006$.

x	f	Δf	$\Delta^2 f$	$\Delta^3 f$
0	0			
		99		
0·1	0·099		−6	
		93		−6
0·2	0·192		−12	
		81		−6
0·3	0·273		−18	
		63		−6
0·4	0·336		−24	
		39		−6
0·5	0·375		−30	
		9		
0·6	0·384			

Ex. 9. Extend the table above, back from the constant difference in the $\Delta^3 f$ column, to obtain $f(-0{\cdot}1)$ and $f(0{\cdot}7)$.

If we denote the successive values of the variable x by $x_0, x_1, x_2, \ldots$, *where the step between successive values of x is constant*, the corresponding values of $f(x)$ may be written $f_0, f_1, f_2, \ldots$. Again, we write

$$\Delta f_r = f_{r+1} - f_r$$

and

$$\Delta^2 f_r = \Delta(\Delta f_r) = \Delta f_{r+1} - \Delta f_r, \quad \text{etc.}$$

Similarly, if we work *back* from x_0 through values $x_{-1}, x_{-2}, x_{-3}, \ldots$ to obtain values $f_{-1}, f_{-2}, f_{-3}, \ldots$ of the polynomial, the successive differences are written $\Delta f_{-1}, \Delta f_{-2}, \ldots$, and $\Delta^2 f_{-1}, \Delta^2 f_{-2}, \ldots$, etc.:

$$\begin{array}{ccccc}
 & f & \Delta f & \Delta^2 f & \Delta^3 f \\
x_{-2} & f_{-2} & & & \\
 & & \Delta f_{-2} & & \\
x_{-1} & f_{-1} & & \Delta^2 f_{-2} & \\
 & & \Delta f_{-1} & & \Delta^3 f_{-2} \\
x_0 & f_0 & & \Delta^2 f_{-1} & \quad \ldots. \\
 & & \Delta f_0 & & \Delta^3 f_{-1} \\
x_1 & f_1 & & \Delta^2 f_0 & \\
 & & \Delta f_1 & & \\
x_2 & f_2 & & &
\end{array}$$

If we wish to refer to an explicit value of x, we may write

$$\Delta f(x) = f(x+h) - f(x),$$

$$\Delta^2 f(x) = \Delta f(x+h) - \Delta f(x),$$

where h is the difference between successive values of x for which the polynomial $f(x)$ is being tabulated.

**Ex.* 10. If $f(x) \equiv x(x-1)(x-2) \ldots (x-n+1)$ (and thus $f(x)$ is a polynomial of degree n) prove that, if f is tabulated for *integral* values of x,

(i) $\Delta f(x) = nx(x-1)(x-2)\ldots(x-n+2)$,

(ii) $\Delta^2 f(x) = n(n-1)x(x-1)(x-2)\ldots(x-n+3)$.

Suggest a form for $\Delta^r f(x)$ and prove your conjecture by mathematical induction.

**Ex.* 11. Prove that any polynomial $f(x)$ of degree n may be expressed in the form

$$f(x) \equiv A_1 x(x-1)(x-2) \ldots (x-n+1) + A_2 x(x-1)\ldots(x-n+2) + \ldots + A_n x + A_{n+1},$$

where $A_1, A_2, \ldots, A_{n+1}$ are numbers, and deduce, using Ex. 10, that the nth differences of an nth degree polynomial are constant. Prove furthermore, that $\Delta^n f = A_1 n!$.

**Ex.* 12. Prove, by mathematical induction, the *Gregory–Newton formula*

$$f(n) = f(0) + \binom{n}{1} \Delta f(0) + \ldots + \binom{n}{r} \Delta^r f(0) + \ldots + \Delta^n f(0).$$

Example 3. *The quadratic polynomial* $f(x)$ *has values* $f(2) = 15$, $f(3) = 41$, $f(4) = 81$. *Find an explicit form for* $f(x)$.

Since $f(x)$ is a quadratic polynomial, the second differences will be constant. Thus, we are able to build up the table of differences below to infer the values of $f(0)$ and $\Delta f(0)$.

x	f	Δf	$\Delta^2 f$
0	**5**		
		−2	
1	**3**		**14**
		12	
2	15		**14**
		26	
3	41		14
		40	
4	81		

Thus $f(0) = 5$, $\Delta f(0) = -2$, $\Delta^2 f(0) = 14$, and by the Gregory–Newton formula (Ex. 12)

$$f(n) \equiv f(0)+\binom{n}{1}\Delta f(0)+\binom{n}{2}\Delta^2 f(0)$$
$$= 5-2n+7n(n-1)$$
$$= 7n^2-9n+5.$$

The Gregory–Newton formula was only proved for integral n, but

$$f(x) \equiv 7x^2-9x+5$$

is the unique quadratic polynomial defined by the three values

$$f(2) = 15, \quad f(3) = 41, \quad f(4) = 81.$$

(We may thus be permitted to write formally

$$\binom{x}{1} = x, \quad \binom{x}{2} = \tfrac{1}{2}x(x-1), \quad \binom{x}{3} = \tfrac{1}{6}x(x-1)(x-2), \quad \ldots$$

for all real x.)

Ex. 13. The quadratic polynomial $f(x)$ has values $f(1) = 0$, $f(2) = 1$, $f(3) = 6$; use the Gregory–Newton formula to find the explicit form of $f(x)$.

Exercise 18(a)

1. Use the method of synthetic division to find the quotient and remainder:
(i) when x^3-5x^2+x+16 is divided by $x-2$;
(ii) when $x^4+3x^3-9x^2-23x+14$ is divided by $x+3$;
(iii) when $3x^4+x^3-12x^2-11x-24$ is divided by $x+2$;
(iv) when x^4-5x-5 is divided by $x-4$;
(v) when $7x^5-3x-5$ is divided by $x+5$.

2. Use the method of synthetic division to find the quotient and remainder:
(i) when $2x^3-5x^2-17x$ is divided by $2x+1$;
(ii) when $3x^4+2x^3-3x^2+7x+6$ is divided by $3x+2$;
(iii) when $2x^5-9x^4+2x^3+7x^2+7x-3$ is divided by $2x-1$;
(iv) when $4x^4-9x^3-9x^2+x+2$ is divided by $2x+1$;
(v) when $3x^5-x^4-21x^3+9x^2-14x+12$ is divided by $3x-1$.

3. Use a hand calculating machine to evaluate the following polynomials for the given values of x, giving your answers to 1 decimal place:
(i) $3{\cdot}5x^3+7{\cdot}1x^2+2{\cdot}8x+5{\cdot}9$, $x = 2{\cdot}9$;
(ii) $7{\cdot}7x^3-9{\cdot}6x^2-3{\cdot}8x+17{\cdot}2$, $x = 1{\cdot}97$;
(iii) $8{\cdot}6x^4-9{\cdot}2x^3+7{\cdot}3x^2-5{\cdot}8x+13{\cdot}2$, $x = 2{\cdot}61$.

4. Express the following polynomials in the form

$$a_0(x-h)^n+a_1(x-h)^{n-1}+\ldots+a_{n-1}(x-h)+a_n$$

for the given values of h
(i) x^3+x^2+2x+1, $h = 1$;
(ii) $2x^3-17x^2+30x$, $h = 2$;
(iii) x^3+7x^2-6x-4, $h = -3$;
(iv) x^4-1, $h = -2$.

5. Use the method of synthetic division to find the quotient and remainder:
(i) when $x^4+(1+4\text{j})\,x^2-2x+(1+\text{j})$ is divided by $(x+\text{j})$;
(ii) when $x^4+(1-\text{j})\,x^3-\text{j}x^2+3(1+2\text{j})\,x+(4-3\text{j})$ is divided by $(x-2\text{j})$.

6. By expressing $P(x) \equiv x^3-11x^2+35x-30$ in the form

$$a_0(x-3)^3+a_1(x-3)^2+a_2(x-3)+a_3,$$

estimate (i) $P(3{\cdot}1)$, (ii) $P(3{\cdot}01)$, both correct to 1 decimal place.

7. Express the polynomial $P(x) \equiv x^4+16x^3+100x^2+288x+320$ in the form $\sum_{r=0}^{4} a_r(x+4)^r$.

What does your answer tell you about the shape of the curve

$$y = x^4+16x^3+100x^2+288x+320$$

at the point $(-4, 0)$?

Solve the equation $x^4+16x^3+100x^2+288x+320 = 0$.

8. Show how the method of synthetic division may be modified for division by the polynomial x^2+x+2.

Find the quotient and remainder on dividing

$$x^7+x^6-x^5-3x^4-4x^3+3x^2+5x+3 \quad \text{by} \quad x^2+x+2.$$

9. If x^2+1 is a factor of $x^7+Ax^5+x^3+Bx^2+2x+1$, find the values of A and B.

10. If x^2+4 is a factor of $x^6+Ax^4-x^3+Bx^2-4x-4$, find the values of A and B.

11. A cubic polynomial $P(x)$ has values $P(0) = -1$, $P(1) = 4$, $P(2) = 21$, $P(3) = 62$. Use a table of differences to find $P(-1)$ and $P(4)$.

12. A quartic polynomial $P(x)$ has values $P(-2) = 11$, $P(-1) = 2$, $P(0) = 5$, $P(1) = 8$, $P(2) = 23$, Find $P(3)$ and $P(-3)$.

13. Find the quadratic polynomial $f(x)$ defined by the values

$$f(1) = -15, \quad f(2) = -20, \quad f(3) = -23.$$

14. Find the cubic polynomial $f(x)$ defined by the values

$$f(1) = 3, \quad f(2) = 4, \quad f(3) = 21, \quad f(4) = 66.$$

15. Find an expression for $\sum_{r=1}^{n} r^4$ as a polynomial of degree five in n.

3. PARTIAL FRACTIONS

The process of adding together fractions with polynomials as denominators will be familiar to the reader. For example

$$\begin{aligned} &\frac{2}{x-3}-\frac{3}{x-2} \\ &= \frac{2(x-2)-3(x-3)}{(x-3)(x-2)} \\ &= \frac{5-x}{(x-3)(x-2)}. \end{aligned}$$

The reverse process is called 'resolving a rational function into *partial fractions*'. Such a resolution is of importance in differentiation and integration and also in obtaining power series expansions of rational functions, as we shall see later in this chapter.

Two questions arise—can the resolution always be effected, and, if it can, are the partial fractions obtained unique? Detailed analysis of these questions lies outside the scope of this book.†

Ex. 14. Express $\frac{9}{91}$ in the form $\frac{a}{13}+\frac{b}{7}$, a, b integers, and show that such a resolution into arithmetical partial fractions is *not* unique.

Ex. 15. Express $\frac{73}{105}$ in the form $\frac{a}{7}+\frac{b}{5}+\frac{c}{3}$, a, b, c integers.

Suppose we are given a rational function of the form

$$\frac{P(x)}{Q(x)}$$

where $P(x)$, $Q(x)$ are polynomials in x. As a first step, if the degree of $P(x)$

† See, for example, Ferrar: *Higher Algebra*. We shall convince ourselves, in specific examples, that the answer to both questions is a qualified 'yes'.

is greater than, or the same as, the degree of $Q(x)$, we may divide out until this is no longer the case. For example, we should express

$$\frac{x^4-2x^3-3x^2+x+4}{x^2-1}$$

as

$$x^2-2x-2-\frac{x-2}{x^2-1}.$$

We now suppose that this has been done; that is, that we have a rational function of the form

$$\frac{P(x)}{Q(x)} \quad \text{where } \textit{degree } [P(x)] < \textit{degree } [Q(x)].$$

Let us first consider the case in which $Q(x)$ contains only non-repeated linear factors. Consider, for example, the function

$$\frac{x+18}{(x-3)(2x+1)}.$$

If we can find numerical values A, B such that

$$x+18 \equiv A(2x+1)+B(x-3) \tag{1}$$

then, on division by $(x-3)(2x+1)$, we have

$$\frac{x+18}{(x-3)(2x+1)} = \frac{A}{x-3}+\frac{B}{2x+1}. \tag{2}$$

From (1), suitable values of A and B can be found by equating coefficients:

$$1 = 2A+B, \quad 18 = A-3B,$$

and then solving the resulting simultaneous equations. A preferable method is to put $x = 3$ in (1) to find A, and then to put $x = -\frac{1}{2}$ to find B. The values obtained are $A = 3$, $B = -5$ and

$$\frac{x+18}{(x-3)(2x+1)} = \frac{3}{x-3}-\frac{5}{2x+1}.$$

Ex. 16. Students often object that, since (2) is meaningless for $x = -\frac{1}{2}$ or $x = 3$, it is invalid to substitute these values in (1) to obtain A and B. Expose the fallacy in this argument.

Observe that, in our last example, A and B were also given by

$$A = \frac{3+18}{(2\times 3+1)}, \quad B = \frac{(-\frac{1}{2})+18}{(-\frac{1}{2}-3)};$$

that is, A was found by covering up the term $(x-3)$ in

$$\frac{x+18}{(x-3)(2x+1)}$$

and then putting $x = 3$ in what was left. Similarly, B was found by covering up the term $(2x+1)$ and putting $x = -\frac{1}{2}$ in what was left. This useful technique for obtaining a partial fraction expansion is usually called the *cover-up rule.*

Example 4. *Express in partial fractions*

$$\text{(i)}\ \frac{4x^2-19x+7}{(x-1)(x-2)(x+3)};\quad \text{(ii)}\ \frac{x^3+x}{x^2-4}.$$

(i) By the cover-up rule

$$\frac{4x^2-19x+7}{(x-1)(x-2)(x+3)} = \frac{1}{(x-1)}\left(\frac{4-19+7}{(-1)(4)}\right)+\frac{1}{(x-2)}\left(\frac{16-38+7}{(1)(5)}\right)$$
$$+\frac{1}{(x+3)}\left(\frac{36+57+7}{(-4)(-5)}\right)$$
$$= \frac{2}{x-1}-\frac{3}{x-2}+\frac{5}{x+3}.$$

(Note the check by putting $x = 0$:

$$\tfrac{7}{6} = -2+\tfrac{3}{2}+\tfrac{5}{3}.)$$

(ii) Since x^3+x is a cubic polynomial and x^2-4 is only a quadratic, we must first divide out

$$\frac{x^3+x}{x^2-4} = \frac{x^3-4x+5x}{x^2-4} = x+\frac{5x}{x^2-4}.$$

Now,
$$\frac{5x}{x^2-4} = \frac{5x}{(x-2)(x+2)}$$
$$= \frac{1}{x-2}\left(\frac{10}{4}\right)+\frac{1}{x+2}\left(\frac{-10}{-4}\right),$$

by the cover-up rule. Thus we have

$$\frac{x^3+x}{x^2-4} = x+\frac{5}{2(x-2)}+\frac{5}{2(x+2)}.$$

Ex. 17. Express in partial fractions

$$\text{(i)}\ \frac{x+11}{(x-1)(x+2)};\qquad \text{(ii)}\ \frac{x+6}{x^2-9};$$

$$\text{(iii)}\ \frac{2a}{x^2-a^2}\quad (a \neq 0);\qquad \text{(iv)}\ \frac{2x}{x^2-a^2}\quad (a \neq 0).$$

*Ex. 18. A common application of partial fraction expansions is to the evaluation of certain integrals. Evaluate, by first expressing the integral in partial fractions:

(i) $\int_2^3 \frac{dx}{x^2-1}$; (ii) $\int_0^1 \frac{dx}{(x-2)(x-3)}$.

*Ex. 19. Partial fraction expansions can also frequently be used to simplify the work in differentiating. Differentiate

(i) $\frac{2x}{x^2-5x+6}$; (ii) $\frac{x^2}{x^2-5x-6}$.

It is worthwhile considering the 'cover-up rule' from a rather different standpoint. Let us suppose that we have to express

$$f(x) = \frac{5x+5}{(x-3)(2x-1)}$$

in partial fractions.

We proceed by considering approximations to this rational function. It has singularities at $x = \frac{1}{2}$ and $x = 3$; that is, $f(x)$ is very large when x is near $\frac{1}{2}$ or 3 (but is not defined at $x = \frac{1}{2}$ or $x = 3$). Near $x = 3$, the dominant term in $f(x)$ is $1/(x-3)$: this term is multiplied by a factor with numerical value almost exactly equal to

$$\frac{5\times 3+5}{2\times 3-1} = 4$$

(the 'cover-up rule' again). We may thus say that $f(x)$ is approximately equal to

$$\frac{4}{x-3}$$

near $x = 3$.

The approximate shape of the graph of $f(x)$ is shown in Figure 18.1 while the approximate shape of the graph of $y = 4/(x-3)$ is shown in Figure 18.2.

Near the common asymptote $x = 3$ the two curves are seen to be very similar in shape and position.

Now consider the deviation of $f(x)$ from the approximation:

$$\begin{aligned} f(x)-\frac{4}{x-3} &= \frac{5x+5}{(2x-1)(x-3)}-\frac{4}{x-3} \\ &= \frac{5x+5-8x+4}{(2x-1)(x-3)} \\ &= \frac{-3(x-3)}{(2x-1)(x-3)} \\ &= -\frac{3}{2x-1}. \end{aligned}$$

Thus we have finally
$$f(x) = \frac{4}{x-3} - \frac{3}{2x-1}$$
and our rational function has been expressed in partial fractions.

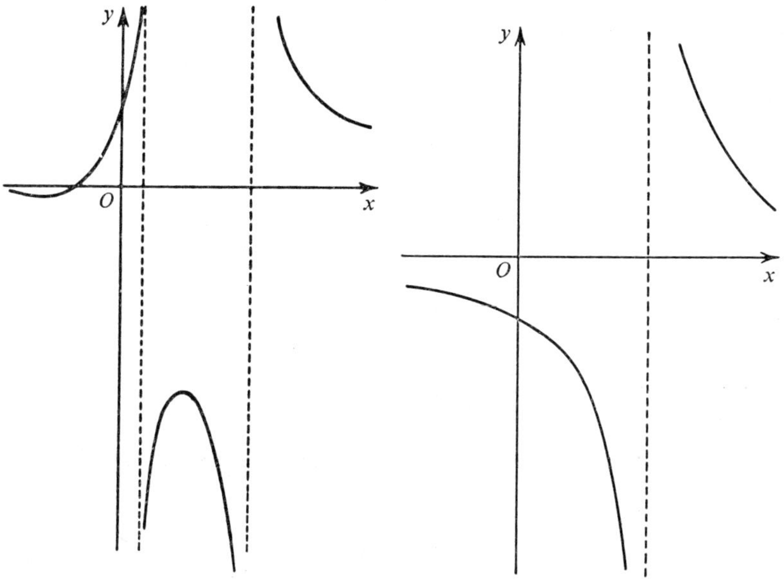

Fig. 18.1 Fig. 18.2

**Ex.* 20. The fact that $(x-3)$ cancelled at the final stage in the example above was no coincidence. Suppose that $f(x)$, $F(x)$ are polynomials and that $F(x)$ has a linear factor $x-\alpha$, that is,
$$F(x) \equiv (x-\alpha)\, g(x), \quad g(\alpha) \neq 0.$$
Show that, near $x = \alpha$, the rational function $f(x)/F(x)$ is approximately
$$\frac{f(\alpha)}{g(\alpha)} \frac{1}{x-\alpha},$$
with deviation
$$\frac{f(x)\, g(\alpha) - g(x)\, f(\alpha)}{(x-\alpha)\, g(x)\, g(\alpha)}.$$
Prove that the numerator of this fraction is a polynomial with a factor $(x-\alpha)$ and deduce that cancelling is always possible.

The method of obtaining partial fractions outlined in Ex. 20 is clearly applicable whenever the denominator contains a (strictly) linear factor. We conclude this section with two worked examples.

Example 5. *Express*

$$f(x) = \frac{4x^2-3x+5}{(x-1)^2(x+2)}$$

in partial fractions.

$f(x)$ has singularities at $x = 1$ and $x = -2$. The results proved in Ex. 20 suggest that we consider the singularity $x = -2$. Near $x = -2$, $f(x)$ is approximately

$$\left(\frac{4(-2)^2-3(-2)+5}{(-2-1)^2}\right)\left(\frac{1}{x+2}\right) = \frac{3}{x+2}.$$

The deviation is

$$\frac{4x^2-3x+5}{(x-1)^2(x+2)} - \frac{3}{x+2} = \frac{4x^2-3x+5-3(x^2-2x+1)}{(x-1)^2(x+2)}$$

$$= \frac{x^2+3x+2}{(x-1)^2(x+2)}$$

$$= \frac{(x+1)(x+2)}{(x-1)^2(x+2)},$$

and thus we have $$f(x) = \frac{3}{x+2} + \frac{x+1}{(x-1)^2}.$$

The reduction can, if desired, be carried one stage further by writing

$$x+1 \equiv (x-1)+2:$$

$$f(x) = \frac{3}{x+2} + \frac{1}{x-1} + \frac{2}{(x-1)^2}.$$

Example 6. *Express*

$$f(x) = \frac{x^2+12}{(x-2)(x^2+4)}$$

in partial fractions.

$f(x)$ has just one singularity, at $x = 2$. Near $x = 2$, $f(x)$ approximates to

$$\left(\frac{2^2+12}{2^2+4}\right)\left(\frac{1}{x-2}\right) = \frac{2}{x-2}.$$

The deviation is

$$\frac{x^2+12}{(x-2)(x^2+4)} - \frac{2}{x-2} = \frac{x^2+12-2(x^2+4)}{(x-2)(x^2+4)}$$

$$= -\frac{(x-2)(x+2)}{(x-2)(x^2+4)}$$

$$= -\frac{x+2}{x^2+4},$$

and thus we have $$f(x) = \frac{2}{x-2} - \frac{x+2}{x^2+4}.$$

Ex. 21. Express in partial fractions

(i) $\dfrac{3x-15}{(x-2)^2(x+1)}$; (ii) $\dfrac{x^2-7x-4}{(x+1)^2(2x+1)}$.

Ex. 22. Express in partial fractions

(i) $\dfrac{x+2}{(x-1)(x^2+2)}$; (ii) $\dfrac{x^2+1}{x^3-1}$.

Ex. 23. By writing $x^2+4 \equiv (x+2j)(x-2j)$, use the 'cover-up' rule to find the partial fraction expansion for

$$\frac{x^2+12}{(x-2)(x^2+4)}.$$

Ex. 24. Express

$$\frac{x^2-x-1}{(x-1)(x^2-2)}$$

(i) as the sum of partial fractions with rational coefficients;
(ii) as the sum of partial fractions with real coefficients.

Exercise 18(b)

1. Express in partial fractions

(i) $\dfrac{11-7x}{3-7x+2x^2}$; (ii) $\dfrac{4-x}{x^2+2x}$; (iii) $\dfrac{9x-22}{2x^2+5x-3}$;

(iv) $\dfrac{2x-1}{1-x^2}$; (v) $\dfrac{2x^2+3}{x^2+x}$; (vi) $\dfrac{13x-19}{(x-1)(x+2)(x-3)}$;

(vii) $\dfrac{x^2-8x+1}{x-x^3}$; (viii) $\dfrac{20-x^3}{x^2-4}$; (ix) $\dfrac{x^4}{x^2-5x+6}$;

(x) $\dfrac{1+3x}{x^2+2x^3}$; (xi) $\dfrac{1+6x^2}{(2+x)(1-2x)^2}$; (xii) $\dfrac{x-2}{(2+x^2)(1+x)}$;

(xiii) $\dfrac{2x}{1+8x^3}$; (xiv) $\dfrac{(x-1)^2}{(x-3)^2(x-2)}$; (xv) $\dfrac{x^3+x^2+x}{(x+1)(x^2+2)}$;

(xvi) $\dfrac{2x^2+3x+2}{x^4(x-2)}$.

2. Find numbers A, B, C, D such that

$$2x^2+8x+3 \equiv A(x-2)^3+B(x-2)^2(x-1)+C(x-2)(x-1)+D(x-1)$$

and deduce the partial fraction expansion of

$$\frac{2x^2+8x+3}{(x-2)^3(x-1)}.$$

3. Adapt the method of Question 2 to express

$$\frac{9x^3-x^2+7x-5}{(x-1)^4(x+4)}$$

in partial fractions.

4. Express

$$\frac{1}{(x+1)(x+2)}$$

in partial fractions.

Hence evaluate

$$\sum_{r=1}^{n} \frac{1}{(r+1)(r+2)}.$$

Evaluate similarly

$$\sum_{r=1}^{n} \frac{1}{(r+1)(r+4)}.$$

5. Express $\dfrac{1}{1+x^3}$

(i) as the sum of two partial fractions with real coefficients;
(ii) as the sum of three partial fractions with complex coefficients.

6. Express $1/(x^2-3x+2)$ in partial fractions and deduce the partial fraction expansion of $y^2/(2y^2-3y+1)$.

Express $x^2/(8x^2-6x+1)$ in partial fractions.

4. POLYNOMIAL APPROXIMATIONS: THE BINOMIAL SERIES

We saw, in Chapter 8, that, for any number x and any positive integer n

$$(1+x)^n \equiv 1+\binom{n}{1}x+\binom{n}{2}x^2+\ldots+\binom{n}{r}x^r+\ldots+x^n.$$

If x is numerically much less than one, so that, say, its rth and higher powers are negligible, the identity above may be used to give a polynomial approximation of degree $(r-1)$ for $(1+x)^n$. Thus, for example, with $n = 10$,

$$(1+x)^{10} \approx 1+10x+45x^2,$$

neglecting terms in x of degree greater than 2 (*terms of order* 3).

Ex. 25. Use the quadratic approximation to $(1+x)^{10}$ obtained above to find $(0{\cdot}998)^{10}$ correct to 4 decimal places.

By using Taylor's series expansion†

$$f(x+a) = f(a)+xf'(a)+\tfrac{1}{2}x^2f''(a)+\ldots,$$

polynomial approximations may be obtained for many other functions. In particular, taking $a = 1$, and $f(x) = x^k$ $(k \in Q)$ we have

$$f'(1) = k,\quad f''(1) = k(k-1),\quad f'''(1) = k(k-1)(k-2),\ \ldots,$$

† See, for example, Siddons, Snell and Morgan: Calculus III.

giving, for sufficiently small x† the *binomial series expansion*

$$(1+x)^k = 1+kx+\frac{k(k-1)}{2!}x^2+\frac{k(k-1)(k-2)}{3!}x^3+\ldots.$$

**Ex.* 26. Verify the following expansions:

(i) $(1+x)^{-1} = 1-x+x^2-x^3+x^4\ldots;$

(ii) $(1+x)^{-2} = 1-2x+3x^2-4x^3+5x^4\ldots.$

(Notice that this result may be obtained by the formal differentiation of (i).)

(iii) $(1+x)^{-3} = 1-\frac{2.3}{2}x+\frac{3.4}{2}x^2-\frac{4.5}{2}x^3+\frac{5.6}{2}x^4\ldots;$

(iv) $(1+x)^{1/2} = 1+(\frac{1}{2}x)-\frac{1}{2!}(\frac{1}{2}x)^2+\frac{1.3}{3!}(\frac{1}{2}x)^3-\left(\frac{1.3.5}{4!}\right)(\frac{1}{2}x)^4\ldots;$

(v) $(1+x)^{-1/2} = 1-(\frac{1}{2}x)+\frac{1.3}{2!}(\frac{1}{2}x)^2-\frac{1.3.5}{3!}(\frac{1}{2}x)^3+\frac{1.3.5.7}{4!}(\frac{1}{2}x)^4\ldots.$

Ex. 27. Suppose, when x is very small, that $(1+x)^{\frac{1}{2}}$ can be approximated by $1+ax+bx^2$. Then $1+x$ and $(1+ax+bx^2)^2$ are nearly equal. Deduce that, for a 'best fit', $2a = 1$ and $a^2+2b = 0$, giving the first three terms of expansion (iii) of Ex. 26.

Deduce similarly a quadratic approximation to $(1+x)^{-\frac{1}{2}}$ if x is very small.

Example 7. *If x is sufficiently small for x^4 and higher powers to be neglected, find cubic polynomial approximations for*

(i) $\dfrac{1}{(1-2x)(2-x)}$; (ii) $\dfrac{1}{(1-x)\sqrt{(1+x)}}$.

(i) By the 'cover-up' rule

$$\frac{1}{(1-2x)(2-x)} = \frac{2}{3(1-2x)}-\frac{1}{3(2-x)}$$
$$= \frac{2}{3}(1-2x)^{-1}-\frac{1}{6}\left(1-\frac{x}{2}\right)^{-1}$$
$$\approx \frac{2}{3}(1+2x+4x^2+8x^3)-\frac{1}{6}\left(1+\frac{x}{2}+\frac{x^2}{4}+\frac{x^3}{8}\right)$$
$$\equiv \tfrac{1}{2}+\tfrac{5}{4}x+\tfrac{21}{8}x^2+\tfrac{85}{16}x^3.$$

† The decision as to what constitutes a 'sufficiently small' x depends, of course, upon the accuracy demanded of the answer. In numerical approximations, answers are usually required to a certain number of significant figures or places of decimals and the point at which to truncate the binomial series can be found by inspection.

It can be shown that

$$\lim_{N\to\infty}\sum_{r=1}^{N}\frac{(n-1)(n-2)\ldots(n-r+1)}{r!}x^r = (1+x)^n-1$$

for *any real n*, provided $|x| < 1$. The result is also true for $x = 1$, provided $n > -1$ and for $x = -1$ provided $n > 0$. It is not true for any $|x| > 1$, unless n is a positive integer or zero.

(ii) $$\frac{1}{(1-x)\sqrt{(1+x)}} = (1-x)^{-1}(1+x)^{-\frac{1}{2}}$$
$$\approx (1+x+x^2+x^3)(1-\tfrac{1}{2}x+\tfrac{3}{8}x^2-\tfrac{5}{16}x^3)$$
$$\approx 1+x(1-\tfrac{1}{2})+x^2(1-\tfrac{1}{2}+\tfrac{3}{8})+x^3(1-\tfrac{1}{2}+\tfrac{3}{8}-\tfrac{5}{16})$$
$$\equiv 1+\tfrac{1}{2}x+\tfrac{7}{8}x^2+\tfrac{9}{16}x^3.$$

Alternatively, we may write

$$\frac{1}{(1-x)\sqrt{(1+x)}} \approx 1+ax+bx^2+cx^3.$$

Then, using Ex. 26 (v), we have

$$1-\tfrac{1}{2}x+\tfrac{3}{8}x^2-\tfrac{5}{16}x^3 \approx (1-x)(1+ax+bx^2+cx^3),$$

giving $$a-1 = -\tfrac{1}{2}, \quad b-a = \tfrac{3}{8}, \quad c-b = -\tfrac{5}{16}$$

or $$a = \tfrac{1}{2}, \quad b = \tfrac{7}{8}, \quad c = \tfrac{9}{16}$$

as before.

Ex. 28. Solve Example 7 (i) by the method of Example 7(ii).

Example 8. *By using the binomial expansion for* $(1-2x)^{-\frac{1}{2}}$, *estimate* $\sqrt{5}$ *and calculate an upper limit for the error in your approximation.*

Since $(1-2\times\frac{1}{10})^{-\frac{1}{2}} = 1/\sqrt{(\frac{4}{5})} = \frac{1}{2}\sqrt{5}$ we have (see Ex. 26 (v))

$$\tfrac{1}{2}\sqrt{5} = 1-(-\tfrac{1}{10})+\frac{1.3}{2!}(-\tfrac{1}{10})^2-\frac{1.3.5}{3!}(-\tfrac{1}{10})^3$$
$$+\frac{1.3.5.7}{4!}(-\tfrac{1}{10})^4-\frac{1.3.5.7.9}{5!}(-\tfrac{1}{10})^5+\ldots,$$

$$1+0{\cdot}1+0{\cdot}015+0{\cdot}0025+0{\cdot}00044+0{\cdot}00008.$$

Thus $\frac{1}{2}\sqrt{5} \approx 1{\cdot}11802$ and $\sqrt{5} \approx 2{\cdot}23604$.

To estimate the accuracy of our answer we have to obtain an upper limit for the sum of the remaining terms in the series. Now the coefficient, u_r, of $(\frac{1}{10})^r$ in the expansion above is given by

$$u_r = \frac{1.3.5.\ldots.(2r-1)}{r!}$$

and we have $$\frac{u_r}{u_{r-1}} = \frac{2r-1}{r} = 2-\frac{1}{r} < 2.$$

Thus $$u_r < 2u_{r-1}$$

and the sum of the remaining terms in the binomial series above is certainly less than

$$\frac{1.3.5.7.9.11}{6!}(-\tfrac{1}{10})^6\,[1+\tfrac{2}{10}+\tfrac{4}{100}+\tfrac{8}{1000}+\ldots]$$

$$=\frac{1.3.5.7.9.11}{6!}(\tfrac{1}{10})^6\left[\frac{1}{1-\frac{1}{5}}\right] \quad \text{(sum of infinite geometric series)}$$

$$\approx 1{\cdot}805\times 10^{-5}.$$

Since $4{\cdot}4\times 10^{-4}$ and 8×10^{-5} are overestimates of the fifth and sixth terms of the series above we may safely conclude that the error in taking $1{\cdot}11802$ as $\frac{1}{2}\sqrt{5}$ is less than $1{\cdot}805\times 10^{-5}$ and so our answer of $2{\cdot}23604$ for $\sqrt{5}$ has an error of less than $3{\cdot}61\times 10^{-5}$.

Ex. 29. In fact, $\sqrt{5} = 2{\cdot}236068$ (6 d.p.), and our limit for the error in Example 8 appears to have been rather crude. Explain why this is so. How could a more accurate estimate of the error be obtained?

Ex. 30. To 4 d.p., $\sqrt{5} = 2{\cdot}2361$, while in Example 8 above, we obtain $\sqrt{5} = 2{\cdot}2360$ (to 4 d.p.) although we subsequently show our error to be less than 5×10^{-5}. Explain.

The binomial series gives a polynomial approximation to $(1+x)^n$ valid in the neighbourhood of $x = 0$. If we require a polynomial approximation valid in some other neighbourhood, it is necessary to shift the origin, as shown in our next example.

Example 9. *Obtain a quadratic approximation to* $(1+2x)^{-1}$ *in the neighbourhood of* $x = 3$.

$$(1+2x)^{-1} = [7+2(x-3)]^{-1}$$

$$= \frac{1}{7}\left[1+\frac{2y}{7}\right]^{-1}, \quad \text{where} \quad y = x-3$$

$$\approx \frac{1}{7}\left(1-\frac{2y}{7}+\frac{4y^2}{49}\right), \quad \text{since } y \text{ is approximately zero}$$

$$= \tfrac{1}{7}-\tfrac{2}{49}(x-3)+\tfrac{4}{343}(x-3)^2.$$

The quadratic approximation is best left in this form but it may be reduced to

$$\tfrac{127}{343}-\tfrac{38}{343}x+\tfrac{4}{343}x^2.$$

Notice that this is quite different from the quadratic approximation to $(1+2x)^{-1}$ in the neighbourhood of $x = 0$:

$$1-2x+4x^2.$$

Exercise 18(c)

1. Obtain cubic polynomial approximations for the following expressions in the neighbourhood of $x = 0$:

(i) $\dfrac{1}{1-3x}$; (ii) $\dfrac{1}{3-x}$; (iii) $\dfrac{1}{(1-2x)^2}$; (iv) $\dfrac{1}{(1-x^2)^2}$; (v) $\dfrac{1}{(2-x)^3}$.

2. Obtain cubic polynomial approximations for the following expressions in the neighbourhood of $x = 0$.

(i) $\dfrac{1}{\sqrt{(1-2x)}}$; (ii) $\sqrt{(1-4x)}$; (iii) $\sqrt{(4-x)}$; (iv) $\dfrac{1}{\sqrt{(4-x)}}$; (v) $\dfrac{1}{\sqrt{(2-x)}}$.

3. Obtain cubic polynomial approximations for the following expressions, in the neighbourhood of $x = 0$:
(i) $(1+x)^{\frac{1}{3}}$; (ii) $(1+2x)^{-4}$; (iii) $(1-x)^{-\frac{2}{3}}$; (iv) $(2-x)^{-5}$; (v) $(8-x)^{\frac{1}{3}}$.

4. Obtain cubic polynomial approximations for the following expressions, in the neighbourhood of $x = 0$:

(i) $\dfrac{1}{(1-x)(1-2x)}$; (ii) $\dfrac{1}{(x-2)(x-3)}$; (iii) $\dfrac{1}{(1+x)^2(2+x)}$;

(iv) $\dfrac{x(x+3)}{(1+x^2)(1+2x)}$; (v) $\dfrac{1+4x}{(1-2x)^2(1+2x^2)}$.

5. Obtain quadratic polynomial approximations for the following expressions in the neighbourhood of $x = 0$:

(i) $\dfrac{1}{(1+x)\sqrt{(1+x)}}$; (ii) $\dfrac{1}{(1+x^2)\sqrt{(1-2x)}}$; (iii) $\dfrac{1}{\sqrt{(1-5x+6x^2)}}$;

(iv) $\dfrac{\sqrt{(1-x)}}{1+x}$; (v) $\dfrac{1}{(1+x)^2\sqrt{(1+2x)}}$.

6. Obtain quadratic polynomial approximations for:
(i) $(1+x)^{-1}$ in the neighbourhood of $x = 1$;
(ii) $(2-x)^{-1}$ in the neighbourhood of $x = -1$;
(iii) $\sqrt{(3+2x)}$ in the neighbourhood of $x = -1$;
(iv) $(2+3x)^{\frac{1}{3}}$ in the neighbourhood of $x = 2$;
(v) $\dfrac{\sqrt{(3-x)}}{x}$ in the neighbourhood of $x = -1$.

7. Prove that

$$\frac{1}{\sqrt{(1+x^2)}+1} = \frac{\sqrt{(1+x^2)}-1}{x^2}$$

and deduce that, if x is very small,

$$\frac{1}{\sqrt{(1+x^2)}+1} \approx \tfrac{1}{2}-\tfrac{1}{8}x^2.$$

8. Find, correct to 5 significant figures, the values of (i) $(0{\cdot}998)^{1/3}$; (ii) $(1{\cdot}02)^{1/2}$; (iii) $(4{\cdot}01)^{1/2}$; (iv) $(0{\cdot}799)^{1/3}$.

9. By substituting $x = 1/1000$ in the expansion of $(1-x)^{1/3}$, find $(999)^{1/3}$ correct to 4 significant figures.

10. Prove that $19{\cdot}97 < \sqrt{399} < 19{\cdot}98$.

11. Prove that, if

$$E = \frac{x^2}{2-x-2\sqrt{(1-x)}},$$

then $E = 2-x+2\sqrt{(1-x)}$.

Deduce that, if x is small, then E is approximately equal to $\frac{1}{4}x^2$.

12. If x is so small that x^3 and higher powers of x may be neglected, express the function

$$\frac{\sqrt{(4+x)}}{1-2x+(1+3x)^{2/3}}$$

in the form, $a+bx+cx^2$. (O & C)

13. Express the function E given by

$$E = \frac{x+3}{(2x+1)(1+x^2)}$$

in partial fractions.

Hence prove that, if x is so large that x^{-4} can be neglected, then

$$E = \frac{5+2x}{4x^3}. \qquad \text{(O \& C)}$$

14. Use the binomial expansion to calculate the value of $(16{\cdot}32)^{1/4}$ correct to six places of decimals. (O & C)

Miscellaneous Exercise 18

1. The polynomial $x^6+Ax^5+Bx^4-Ax^3+Ax^2+Bx+C$ is exactly divisible by x^3+1; find A, B, C.

2. If $(1+x+x^2)^n \equiv a_0+a_1x+a_2x^2+\ldots+a_{2n}x^{2n}$, write down the values of a_0 and a_{2n} and prove that $a_r = a_{2n-r}$.

Show that $\sum_{r=1}^{2n} a_r = 3^n$ and find $\sum_{r=1}^{2n} (-1)^r a_r$.

If n is an even number, prove that $\sum_{r=1}^{n} (-1)^{r+1} a_{2r-1} = 0$.

3. Find the value of the constant A for which the expression

$$2x+3-A(x^2+x+1)$$

has $(x-2)$ as a factor and find the remaining factor.

Hence, or otherwise, put into partial fractions the expression

$$\frac{2x+3}{(x^2+x+1)(x-2)} \qquad \text{(O \& C)}$$

4. Prove that
$$x^2+2bx+1 \equiv (x-a)(x+2b+a)+a^2+2ab+1.$$

Put
$$\frac{1}{(x-2)(x^2+2bx+1)}$$

into partial fractions
(i) when $b = \frac{1}{2}$,
(ii) when $b = 1$. (O & C)

5. Find A, B such that, for all values of x other than $\frac{1}{2}$, $-\frac{1}{2}$, $-\frac{3}{2}$,
$$\frac{x+1}{(2x-1)(2x+1)(2x+3)} = \frac{Ax+B}{(2x-1)(2x+1)} - \frac{A(x+1)+B}{(2x+1)(2x+3)}.$$

Find the sum to n terms of the series whose rth term is
$$\frac{r+1}{(2r-1)(2r+1)(2r+3)}.$$ (O & C)

6. Prove that, if $-\frac{1}{2} < x \leqslant \frac{1}{2}$,
$$1-x+\frac{1.3}{1.2}x^2-\frac{1.3.5}{1.2.3}x^3+\frac{1.3.5.7}{1.2.3.4}x^4+\ldots = (1+2x)^{-1/2}.$$

Deduce that
$$1+\frac{1}{4}+\frac{1.3}{4.8}+\frac{1.3.5}{4.8.12}+\frac{1.3.5.7}{4.8.12.16}+\ldots = \sqrt{2}.$$

7. Sum to infinity the series:

(i) $1+\frac{1}{3}+\frac{1.3}{3.6}+\frac{1.3.5}{3.6.9}+\frac{1.3.5.7}{3.6.9.12}+\ldots;$

(ii) $1+\frac{1}{4}+\frac{1.4}{4.8}+\frac{1.4.7}{4.8.12}+\frac{1.4.7.10}{4.8.12.16}+\ldots;$

(iii) $1-\frac{3}{8}+\frac{3.9}{8.16}-\frac{3.9.15}{8.16.24}+\frac{3.9.15.21}{8.16.24.32}+\ldots;$

(iv) $1+\frac{1}{6}+\frac{1.5}{6.12}+\frac{1.5.9}{6.12.18}+\frac{1.5.9.13}{6.12.18.24}+\ldots.$

8. The cubic polynomial $P(x)$ assumes the values 2, 3, 2, 11 respectively for $x = 1, 2, 3, 4$. By writing
$$P(x) \equiv A+B(x-1)+C(x-1)(x-2)+D(x-1)(x-2)(x-3) \quad \text{find } P(x).$$

Show how to fit a quadratic polynomial to coincide with the values of
$$y = \sin x \quad \text{at} \quad x = 0, \ \tfrac{1}{2}\pi, \ \pi.$$

9. Find a linear approximation to $\sqrt{(3-x)}$ in the neighbourhood of $x = -1$, and interpret your result graphically.
Show that $\sqrt{3{\cdot}9} \approx 1{\cdot}975$.

10. Find the sixth term and the rth term of the series whose first five terms are
$$4, \quad 1, \quad 0, \quad 13, \quad 76$$
on the assumption that the rth term is a polynomial in r of as low a degree as possible.

19. *Complex numbers (2)*

1. INTEGRAL POWERS OF COMPLEX NUMBERS

In Chapter 17 we saw that the multiplication of complex numbers was best expressed in terms of their moduli and arguments: if $z = r(\cos\theta + j\sin\theta)$ and $w = s(\cos\phi + j\sin\phi)$ then $zw = rs[\cos(\theta+\phi) + j\sin(\theta+\phi)]$. In particular, if $z = \cos\theta + j\sin\theta$ then $z^2 = \cos 2\theta + j\sin 2\theta$, from which it follows that $z^3 = \cos 3\theta + j\sin 3\theta$ and so on; the general result,

$$z^n = \cos n\theta + j\sin n\theta \quad (n \in Z^+),$$

may be proved by mathematical induction.

Theorem 19.1. (*De Moivre's theorem for positive integral exponent.*) *For any positive integer n*

$$(\cos\theta + j\sin\theta)^n = \cos n\theta + j\sin n\theta.$$

Proof.

$$(\cos\theta + j\sin\theta)^{n-1} = \cos(n-1)\theta + j\sin(n-1)\theta$$

$$\Rightarrow (\cos\theta + j\sin\theta)^n = [\cos(n-1)\theta + j\sin(n-1)\theta](\cos\theta + j\sin\theta)$$

$$\Rightarrow (\cos\theta + j\sin\theta)^n = \cos n\theta + j\sin n\theta$$

But

$$(\cos\theta + j\sin\theta)^1 = \cos\theta + j\sin\theta$$

and the result holds for all positive integral n, by induction.

Ex. 1. Illustrate de Moivre's theorem using the Argand diagram for the cases $n = 2, 3, 4$. What can you say about θ if $(\cos\theta + j\sin\theta)^n = 1$ (n a positive integer)?

Ex. 2. Express $\sqrt{3}+j$ in modulus-argument form and deduce the value of $(\sqrt{3}+j)^9$.

De Moivre's theorem may be extended without difficulty to include negative integral exponents. If we define z^0 to be 1, this enables us to say that $(\cos\theta + j\sin\theta)^n = \cos n\theta + j\sin n\theta$ for all integral n.

Theorem 19.2. (*De Moivre's theorem for negative integral exponents.*) *For any negative integer n,*

$$(\cos\theta + j\sin\theta)^n = \cos n\theta + j\sin n\theta.$$

Proof. Write $n = -m$; then m is a positive integer and

$$(\cos\theta + \mathrm{j}\sin\theta)^m = \cos m\theta + \mathrm{j}\sin m\theta,$$

by Theorem 19.1. Thus

$$\begin{aligned}(\cos\theta+\mathrm{j}\sin\theta)^n &= \frac{1}{\cos m\theta+\mathrm{j}\sin m\theta}\\ &= (\cos m\theta - \mathrm{j}\sin m\theta), \quad \text{since } |\cos m\theta + \mathrm{j}\sin m\theta| = 1,\\ &= \cos n\theta + \mathrm{j}\sin n\theta.\end{aligned}$$

Ex. 3. Illustrate de Moivre's theorem, using the Argand diagram, for the cases $n = -2, -3, -4$. What can you say about (i) $z^n + z^{-n}$, (ii) $z^n - z^{-n}$ where $|z| = 1$?

Ex. 4. Evaluate $(\sqrt{3}+\mathrm{j})^{-9}$.

The results proved in Theorems 19.1 and 19.2 are frequently useful in deriving further results. We shall illustrate, in the next examples, some of the techniques most commonly employed.

Example 1. *Express* $\cos 6\theta$ *in terms of* $\cos\theta$ *and* $\sin\theta$.

Since $\cos 6\theta + \mathrm{j}\sin 6\theta = (\cos\theta + \mathrm{j}\sin\theta)^6$ we have, using the Binomial Theorem,

$$\cos 6\theta + \mathrm{j}\sin 6\theta = c^6 + 6\mathrm{j}c^5s - 15c^4s^2 - 20\mathrm{j}c^3s^3 + 15c^2s^4 + 6\mathrm{j}cs^5 - s^6,$$

where $c = \cos\theta$, $s = \sin\theta$. Comparing the real parts of both sides,

$$\cos 6\theta = \cos^6\theta - 15\cos^4\theta\sin^2\theta + 15\cos^2\theta\sin^4\theta - \sin^6\theta.$$

Notice that, in Example 1,

(i) by comparing imaginary parts we immediately derive an expression for $\sin 6\theta$ in terms of $\cos\theta$ and $\sin\theta$;

(ii) the expression for $\cos 6\theta$ (but not $\sin 6\theta$) may be written as a polynomial in either $\cos\theta$ or $\sin\theta$.

Ex. 5. Show that

$$\cos 6\theta = 32\cos^6\theta - 48\cos^4\theta + 18\cos^2\theta - 1.$$

Ex. 6. Express $\tan 6\theta$ as a rational function of $\tan\theta$.

Example 2. *Express* $\cos^6\theta$ *in terms of multiple angles.*

Writing $z = \cos\theta + \mathrm{j}\sin\theta$, we have $z^{-1} = \cos\theta - \mathrm{j}\sin\theta$ and thus

$$2\cos\theta = z + z^{-1}.$$

Then

$$\begin{aligned}64\cos^6\theta &= (z+z^{-1})^6\\ &= (z^6+z^{-6}) + 6(z^4+z^{-4}) + 15(z^2+z^{-2}) + 20.\end{aligned}$$

But

$$z^6+z^{-6} = (\cos 6\theta+\mathrm{j}\sin 6\theta)+(\cos 6\theta-\mathrm{j}\sin 6\theta), \text{ by de Moivre's theorem}$$
$$= 2\cos 6\theta,$$

and similar results hold for z^4+z^{-4} and z^2+z^{-2}. Thus

$$32\cos^6\theta = \cos 6\theta+6\cos 4\theta+15\cos 2\theta+10.$$

**Ex.* 7. Suggest a quick check on the accuracy of the answer to Example 2.

**Ex.* 8. By writing $2\mathrm{j}\sin\theta = z-z^{-1}$, express $\sin^6\theta$ in terms of multiple angles.

Ex. 9. Evaluate $\int_0^{\frac{1}{2}\pi}\cos^6\theta\,\mathrm{d}\theta$.

Example 3. *Find the sum of the series*

$$x\sin\theta+x^2\sin 2\theta+x^3\sin 3\theta+\ldots+x^{n-1}\sin(n-1)\theta,$$

where x is a real number.

The given series is reminiscent of a geometric series, but we have sines of multiple angles, $\sin r\theta$, rather than powers of sines, $\sin^r\theta$. De Moivre's theorem suggests a possible way of changing from the multiple angle form into an exponent form. Consider the two series:

$$C = 1+x\cos\theta+x^2\cos 2\theta+\ldots+x^{n-1}\cos(n-1)\theta,$$

$$S = \quad x\sin\theta+x^2\sin 2\theta+\ldots+x^{n-1}\sin(n-1)\theta.$$

Multiplying the second series by j and adding, this gives

$$C+\mathrm{j}S = 1+x(\cos\theta+\mathrm{j}\sin\theta)+x^2(\cos 2\theta+\mathrm{j}\sin 2\theta)+\ldots$$
$$+x^{n-1}(\cos(n-1)\theta+\mathrm{j}\sin(n-1)\theta).$$

Thus, writing $z = \cos\theta+\mathrm{j}\sin\theta$,

$$C+\mathrm{j}S = 1+xz+x^2z^2+\ldots+x^{n-1}z^{n-1}, \quad \text{by de Moivre's theorem}$$

$$= \frac{1-x^nz^n}{1-xz}, \quad \text{from the formula for the sum of a geometric series}$$

$$= \frac{1-x^nz^n}{1-xz}\cdot\frac{(1-xz^*)}{(1-xz^*)}$$

$$= \frac{1-xz^*-x^nz^n+x^{n+1}z^{n-1}|z|^2}{1-x(z+z^*)+x^2|z|^2}$$

$$= \frac{1-x(\cos\theta-\mathrm{j}\sin\theta)-x^n(\cos n\theta+\mathrm{j}\sin n\theta)+x^{n+1}(\cos(n-1)\theta+\mathrm{j}\sin(n-1)\theta)}{1-2x\cos\theta+x^2},$$

since $|z| = 1$.

Comparing the imaginary parts of both sides we then have

$$S = \frac{x \sin \theta - x^n \sin n\theta + x^{n+1} \sin (n-1)\theta}{1-2x\cos\theta + x^2}.$$

Ex. 10. What is the value of C in Example 3?

Ex. 11. If $|x| < 1$, find $\sum_{r=1}^{\infty} x^r \sin r\theta$.

Exercise 19(a)

1. Simplify

(i) $(\cos\frac{1}{4}\pi + \mathrm{j}\sin\frac{1}{4}\pi)^4$; (ii) $(\cos\frac{1}{6}\pi + \mathrm{j}\sin\frac{1}{6}\pi)^{-3}$;
(iii) $(\cos\frac{1}{4}\pi - \mathrm{j}\sin\frac{1}{4}\pi)^6$; (iv) $(\sin\frac{5}{36}\pi + \mathrm{j}\cos\frac{5}{36}\pi)^6$;
(v) $(\cos\frac{1}{3}\pi + \mathrm{j}\sin\frac{1}{3}\pi)^3(\cos\frac{1}{4}\pi + \mathrm{j}\sin\frac{1}{4}\pi)^4$;
(vi) $(1+\mathrm{j}\tan\frac{1}{6}\pi)^5$; (vii) $(1-\mathrm{j}\cot\frac{1}{6}\pi)^{-4}$;
(viii) $(1+\cos 2\theta + \mathrm{j}\sin 2\theta)^{-4}$.

2. Express in standard form

(i) $(\cos\frac{1}{4}\pi + \mathrm{j}\sin\frac{1}{4}\pi)^{12}$; (ii) $(1+\mathrm{j})^8$; (iii) $(\sqrt{3}-\mathrm{j})^4(1+\mathrm{j}\sqrt{3})^6$;
(iv) $(\cos\theta + \mathrm{j}\sin\theta)^n/(\cos\phi - \mathrm{j}\sin\phi)^m$.

3. Express in terms of $\sin\theta$: (i) $\cos 4\theta$; (ii) $\sin 5\theta$.

4. Express in terms of $\cos\theta$: (i) $\cos 4\theta$; (ii) $\cos 5\theta$; (iii) $\dfrac{\sin 6\theta}{\sin\theta}$.

5. Express $\tan 4\theta$ in terms of $\tan\theta$.

6. Prove that $\cos 7\theta = 64\cos^7\theta - 112\cos^5\theta + 56\cos^3\theta - 7\cos\theta$.
Write down the seven roots of the equation

$$64x^7 - 112x^5 + 56x^3 - 7x = 0,$$

and also the seven roots of the equation

$$64x^7 - 112x^5 + 56x^3 - 7x - 1 = 0.$$

7. Express, in terms of cosines of multiple angles:
(i) $\cos^5\theta$; (ii) $\sin^4\theta$; (iii) $\cos^7\theta$.

8. Express, in terms of sines of multiple angles:
(i) $\sin^5\theta$; (ii) $\sin^7\theta$; (iii) $\sin^3\theta\cos\theta$.

9. Express $\sin^5\theta\cos^4\theta$ in terms of sines of multiple angles.

10. Evaluate $\int_0^{\frac{1}{2}\pi} \cos^4\theta\, d\theta$ and $\int_0^{\frac{1}{2}\pi} \cos^4\theta \sin^6\theta\, d\theta$.

11. Sum to n terms the series:
(i) $\cos x + \cos 2x + \ldots + \cos nx$; (ii) $\sin x + \sin 2x + \ldots + \sin nx$.

12. Sum to n terms the series

$$\cos x + 2\cos 2x + 4\cos 3x + \ldots + 2^{n-1}\cos nx.$$

13. Sum the series

$$1+\binom{n}{1}\cos\theta+\binom{n}{2}\cos 2\theta+\ldots+\cos n\theta.$$

14. Find $\sum_{r=0}^{n-1}(r+1)\sin r\theta$.

15. Find $\sum_{r=1}^{n}\sin^r\theta\sin r\theta$.

16. By considering $\cos(\theta_1+\theta_2+\theta_3+\theta_4)+\mathrm{j}\sin(\theta_1+\theta_2+\theta_3+\theta_4)$, express

$$\tan(\theta_1+\theta_2+\theta_3+\theta_4) \quad \text{in terms of} \quad \tan\theta_1, \tan\theta_2, \tan\theta_3, \tan\theta_4.$$

17. Find $\sum_{r=1}^{n}\sin(2r+1)\theta.\sin^{2r+1}\theta$.

18. If $z = \cos\theta+\mathrm{j}\sin\theta$, show that

$$z-1/z = 2\mathrm{j}\sin\theta \quad \text{and} \quad z^n-1/z^n = 2\mathrm{j}\sin n\theta.$$

Express $\sin^5\theta$ in the form $a\sin 5\theta+b\sin 3\theta+c\sin\theta$ and hence solve completely the equation $16\sin^5\theta = \sin 5\theta$. (O & C)

2. RATIONAL POWERS OF COMPLEX NUMBERS

In this section we shall denote a general rational number by p/q, where p, q are integers and $q > 0$.

Given a *real* number $r > 0$, there is just one positive qth root of r^p, that is, there is just one $a > 0$ such that $r^p = a^q$. We write

$$a = r^{p/q} \quad \text{or} \quad a = \sqrt[q]{(r^p)}.$$

**Ex.* 12. Prove that, if p, q are integral and $r > 0$, there is only one positive real number a such that $r^p = a^q$. (Show that, for $b > 0$, $r^p = b^q \Rightarrow b = a$.)

**Ex.* 13. Prove that, if q is odd there is just one real number a such that, for real r, $r^p = a^q$ and that, if q is even, there are either two or no such numbers.

Now let us consider the problem of finding a complex number w such that $z^p = w^q$, where z is the complex number $r(\cos\theta+\mathrm{j}\sin\theta)$, $r > 0$. Suppose $w = s(\cos\phi+\mathrm{j}\sin\phi)$; then

$$r^p(\cos p\theta+\mathrm{j}\sin p\theta) = s^q(\cos q\phi+\mathrm{j}\sin q\phi).$$

(p, q being integers, we may apply de Moivre's theorem to both sides.) Now two non-zero complex numbers can be equal only if they have the same moduli, and arguments differing by an integral multiple of 2π. Thus

$$r^p = s^q,$$

$$p\theta+2k\pi = q\phi.$$

Whether q is even or odd, we may take $s = r^{p/q}$ (see Ex. 13) and

$$\phi = \frac{p+2k\pi}{q},$$

giving $$w = r^{p/q}\left\{\cos\left(\frac{p\theta+2k\pi}{q}\right)+\mathrm{j}\sin\left(\frac{p\theta+2k\pi}{q}\right)\right\}.$$

In this expression, k can take any integral value, positive or negative. Since values of k differing by q or any multiple of q give rise to the same w, there are precisely q different values of w given, for example, by taking

$$k = 0, 1, 2, ..., (q-1);$$

in other words, a *complex number has exactly* q qth *roots*. Since the real numbers form a subset of the complex numbers, every real number has q qth roots; by Ex. 13 at least $(q-2)$ of these will be complex.

Ex. 14. What are the four fourth roots of: (i) 1; (ii) 4; (iii) -1?

If $z = r(\cos\theta+\mathrm{j}\sin\theta)$, where $\theta = \arg z$, we shall define $z^{p/q}$ to be

$$r^{p/q}\left(\cos\frac{p\theta}{q}+\mathrm{j}\sin\frac{p\theta}{q}\right).$$

It is important to realise that $z^{p/q}$ is only one of the qth roots of z^p. We shall sometimes use the notation $\sqrt[q]{z^p}$ for $z^{p/q}$; in particular, $\sqrt{z} = z^{\frac{1}{2}}$.

**Ex.* 15. Under what circumstances do $(z^p)^{1/q}$ and $z^{p/q}$ represent the same complex number?

Ex. 16. Verify that the definition given above for $z^{p/q}$ yields the correct value for $r^{p/q}$, where $r > 0$ is a real number.

Example 4. *Find* $(-1+\mathrm{j})^{\frac{1}{5}}$ *and the other fifth roots of the complex number* $(-1+\mathrm{j})$.

Writing $z = -1+\mathrm{j}$

$= \sqrt{2}(\cos\frac{3}{4}\pi+\mathrm{j}\sin\frac{3}{4}\pi)$

and $w^5 = z$,

we have $$w = 2^{\frac{1}{10}}\left\{\cos\frac{\frac{3}{4}\pi+2k\pi}{5}+\mathrm{j}\sin\frac{\frac{3}{4}\pi+2k\pi}{5}\right\}.$$

Distinct values for w are given by $k = 0, 1, 2, 3, 4$; $k = 0$ gives $(-1+\mathrm{j})^{\frac{1}{5}}$. Thus

$$(-1+\mathrm{j})^{\frac{1}{5}} = 2^{\frac{1}{10}}(\cos\tfrac{3}{20}\pi+\mathrm{j}\sin\tfrac{3}{20}\pi).$$

The remaining four fifth roots of $(-1+\mathrm{j})$ are $2^{\frac{1}{10}}(\cos\frac{11}{20}\pi+\mathrm{j}\sin\frac{11}{20}\pi)$, $2^{\frac{1}{10}}(\cos\frac{19}{20}\pi+\mathrm{j}\sin\frac{19}{20}\pi)$, $2^{\frac{1}{10}}(\cos\frac{27}{20}\pi+\mathrm{j}\sin\frac{27}{20}\pi)$, and $2^{\frac{1}{10}}(\cos\frac{7}{4}\pi+\mathrm{j}\sin\frac{7}{4}\pi)$.

Using tables we have

$$(-1+\mathrm{j})^{\frac{1}{5}} \approx 0{\cdot}955+0{\cdot}487\mathrm{j},$$

and similar approximations may be found for each of the other fifth roots of $(-1+\mathrm{j})$.

Ex. 17. Plot the positions, in the Argand diagram, of the affixes of the complex number $-\frac{1}{\sqrt{2}}+\frac{\mathrm{j}}{\sqrt{2}}$ and of its five fifth roots.

Ex. 18. Find the three cube roots of j, and verify that their sum is zero. Plot their positions in the Argand diagram.

3. THE *n*TH ROOTS OF UNITY

The equation $z^n = 1$ has, by the results proved in the last section, precisely n roots, for

$$1 = \cos 2k\pi+\mathrm{j}\sin 2k\pi,$$

and thus $$z = \cos\frac{2k\pi}{n}+\mathrm{j}\sin\frac{2k\pi}{n} \quad (k = 0, 1, 2, \ldots (n-1)).$$

These n complex numbers are the nth *roots of unity*. Writing

$$\omega = \cos\frac{2\pi}{n}+\mathrm{j}\sin\frac{2\pi}{n}$$

and applying de Moivre's theorem

$$\left(\omega^k = \cos\frac{2k\pi}{n}+\mathrm{j}\sin\frac{2k\pi}{n}\right)$$

the nth roots of unity may be written $1, \omega, \omega^2, \omega^3, \ldots, \omega^{n-1}$.

**Ex.* 19. Show that the nth roots of unity are represented by the vertices of a regular n-sided polygon inscribed in the unit circle $|z| = 1$.

Since $\omega^n-1 = 0$ and $\omega \neq 1$, we have, by summing the geometric series in the usual way, the following important result:

$$1+\omega^1+\omega^2+\omega^3+\ldots+\omega^{n-1} = 0;$$

that is, *the sum of the n* nth *roots of unity is zero.*

**Ex.* 20. If p is a prime number, and if ξ is any pth root of unity other than 1, show that the complete set of pth roots of unity may be written as $1, \xi, \xi^2, \ldots, \xi^{p-1}$. Discuss possible generalizations of this result for the case where p is not prime. (If you have difficulty in proving the general result, consider the particular cases $p = 3, p = 4$.)

Example 5. *Prove that* $(1+z)^n = z^n \Rightarrow \mathrm{Re}(z) = -\frac{1}{2}$.

$$\begin{aligned} (1+z)^n &= z^n \\ \Rightarrow \quad (1+z) &= z\xi, \quad \text{where } \xi \text{ is an } n\text{th root of unity other than 1,} \\ \Rightarrow \quad (1+z^*) &= z^*\xi^* \\ \Rightarrow \quad (1+z)(1+z^*) &= zz^*\xi\xi^* \\ \Rightarrow 1+(z+z^*)+zz^* &= zz^* \text{ (since } \xi\xi^* = 1) \\ \Rightarrow \quad \mathrm{Re}(z) &= -\tfrac{1}{2}. \end{aligned}$$

(See p. 383 Ex. 14 (v).)

Exercise 19(b)

1. Find in standard form the three cube roots of:
(i) 8; (ii) -1; (iii) $-\mathrm{j}$; (iv) $(1+\mathrm{j})^3$.

2. Express in the form $a+b\mathrm{j}$, giving a, b to 2 significant figures in (ii)–(iv)
(i) $\sqrt{(3-4\mathrm{j})}$; (ii) $\sqrt{(1+2\mathrm{j})}$; (iii) $\sqrt{(-3-\mathrm{j})}$; (iv) $\sqrt{(1-3\mathrm{j})}$.

3. If ω is a complex cube root of unity, show that the cube roots of z^3 are z, $z\omega$, $z\omega^2$.
Find the three cube roots of $-2+2\mathrm{j}$ and deduce surd expressions for $\cos(\pi/12)$ and $\sin(\pi/12)$.

4. Find, correct to 2 significant figures, the real and imaginary parts of $(2-\mathrm{j})^{1/6}$.

5. Simplify:
(i) $\sqrt{(\cos\theta-\mathrm{j}\sin\theta)^5} \div \sqrt{(\cos\theta+\mathrm{j}\sin\theta)}$;
(ii) $(\sin\theta-\mathrm{j}\cos\theta)^{1/3}$;
(iii) $\sqrt{\left(\dfrac{\cos\theta-\mathrm{j}\sin\theta}{\cos 3\theta+\mathrm{j}\sin 3\theta}\right)}$;
(iv) $(1+\mathrm{j}\cot\theta)^{1/4}$;
(v) $\{(\cos\theta+\mathrm{j}\sin\theta)(\sin\theta-\mathrm{j}\cos\theta)\}^{1/4}$.

6. Plot in an Argand diagram the four fourth roots of 16 and, on a separate diagram the six sixth roots of 64.

7. Solve the equation $z^4-z^2+1 = 0$.

8. Solve the equation $1+z+z^2+z^3+z^4+z^5 = 0$.

9. Solve the equation $z^3-(\mathrm{j}-z)^3 = 0$.

10. Solve the equation $(1+\mathrm{j}z)^5-(1-\mathrm{j}z)^5 = 0$.

11. Show that, if $(\mathrm{j}-z)^n = (\mathrm{j}z-1)^n$, then z must be a real number, and find all the real numbers satisfying this equation.

12. If 1, ω, ω^2 are the cube roots of unity, prove that:
(i) $(a+\omega-\omega^2)(a-\omega+\omega^2) = a^2+3$;
(ii) $(1+j\omega-\omega^2)(1-\omega+j\omega^2) = 2$;
(iii) $(a+b)(a+b\omega)(a+b\omega^2) = a^3+b^3$;
(iv) $(a+b+c)(a+b\omega+c\omega^2)(a+b\omega^2+c\omega) = a^3+b^3+c^3-3abc$.

13. Describe geometrically the effect of multiplying a general complex number z by ω, a complex cube root of unity. Deduce geometrically that, if $|z| = 1$, then $|z+\omega z| = 1$.

14. 1, ξ, ξ^2, ξ^3, ξ^4 are the five fifth roots of unity and a, b two given complex numbers, and if A_1, A_2, A_3, A_4, A_5 are the affixes of the complex numbers $a+b$, $a+b\xi$, $a+b\xi^2$, $a+b\xi^3$, $a+b\xi^4$ show that $A_1A_2A_3A_4A_5$ is a regular pentagon.

15. If ω is a complex cube root of unity, prove that $a+b+c$ is a factor of the determinant

$$\Delta = \begin{vmatrix} a & b & c \\ c & a & b \\ b & c & a \end{vmatrix}.$$

Show also that

$$\Delta = \begin{vmatrix} a & \omega b & \omega^2 c \\ \omega^2 c & a & \omega b \\ \omega b & \omega^2 c & a \end{vmatrix}.$$

Hence express Δ as a product of linear factors and also as a product of a real linear and a real quadratic factor.

Factorize into four linear factors the determinant

$$\begin{vmatrix} a & b & c & d \\ d & a & b & c \\ c & d & a & b \\ b & c & d & a \end{vmatrix}.$$

4. COMPLEX POWERS OF COMPLEX NUMBERS

(*This section may be omitted at a first reading*)

If e is the base of the natural logarithms, and if y is a real number, it seems plausible to assume that, provided e^{jy} has a meaning,

$$\frac{d}{dy}(e^{jy}) = j\,e^{jy}.$$

On this assumption, $z = e^{jy}$ satisfies the differential equation

$$\frac{dz}{dy} = jz.$$

But $w = \cos y + \mathrm{j}\sin y$ also satisfies this differential equation (by direct verification). Also, when $y = 0$, $z = 1 = w$ and thus $z = w$ for all real y:

$$\mathrm{e}^{\mathrm{j}y} = \cos y + \mathrm{j}\sin y.$$

Since $$\mathrm{e}^{x+\mathrm{j}y} = \mathrm{e}^{x}.\mathrm{e}^{\mathrm{j}y},$$

it follows that $$\mathrm{e}^{x+\mathrm{j}y} = \mathrm{e}^{x}(\cos y + \mathrm{j}\sin y). \tag{1}$$

**Ex.* 21. Show that $e^{-\mathrm{j}y} = \cos y - \mathrm{j}\sin y$ and deduce that $\cos y = \frac{1}{2}(\mathrm{e}^{\mathrm{j}y} + \mathrm{e}^{-\mathrm{j}y})$, $\sin y = \dfrac{1}{2\mathrm{j}}(\mathrm{e}^{\mathrm{j}y} - \mathrm{e}^{-\mathrm{j}y})$. Deduce from these expressions and the infinite series for e^{x} the series expansions for $\cos y$ and $\sin y$.

Ex. 22. Show that $\cos y = \cosh \mathrm{j}y$ and that $\mathrm{j}\sin y = \sinh \mathrm{j}y$.

**Ex.* 23. Prove that $\mathrm{e}^{z_1} = \mathrm{e}^{z_2} \Leftrightarrow z_1 - z_2 = 2k\pi\mathrm{j}$, k an integer.

**Ex.* 24. Discuss de Moivre's theorem in the light of the expression of a complex number in the form $r\mathrm{e}^{\mathrm{j}\theta}$.

Now consider the equation
$$z = \mathrm{e}^{w}.$$

Given any non-zero z we can certainly find a w which satisfies this equation. For example, if $w = \ln|z| + \mathrm{j}\arg z$,

$$\begin{aligned}\mathrm{e}^{w} &= \mathrm{e}^{\ln|z| + \mathrm{j}\arg z}\\ &= \mathrm{e}^{\ln|z|}\,\mathrm{e}^{\mathrm{j}\arg z}\\ &= |z|\,(\cos\arg z + \mathrm{j}\sin\arg z)\\ &= z.\end{aligned}$$

By Ex. 23, any other solution of the equation may be written in the form $w = \ln|z| + \mathrm{j}(\arg z + 2k\pi)$. Such an expression is called a *logarithm of* z; the particular expression with $k = 0$ is called the *principal logarithm of* z and is written $\ln z$:
$$\ln z = \ln|z| + \mathrm{j}\arg z.$$

With this definition of the logarithm of a complex number we are in a position to define a complex power of a complex number:

$$z^{w} = \mathrm{e}^{w\ln z} \quad (z \neq 0).$$

For example,
$$\begin{aligned}\mathrm{j}^{\mathrm{j}} &= \mathrm{e}^{\mathrm{j}\ln\mathrm{j}}\\ &= \mathrm{e}^{\mathrm{j}(\frac{1}{2}\pi\mathrm{j})}\\ &= \mathrm{e}^{-\frac{1}{2}\pi}\\ &\approx 0{\cdot}208,\end{aligned}$$

while $$\begin{aligned}(1+j)^{1-j} &= e^{(1-j)\ln(1+j)}\\ &= e^{(1-j)\{\ln\sqrt{2}+\frac{1}{4}\pi j\}}\\ &= e^{(\ln\sqrt{2}+\frac{1}{4}\pi)+j(\frac{1}{4}\pi-\ln\sqrt{2})}\\ &= \sqrt{2}e^{\frac{1}{4}\pi}(\cos(\tfrac{1}{4}\pi-\ln\sqrt{2})+j\sin(\tfrac{1}{4}\pi-\ln\sqrt{2}))\\ &\approx 2{\cdot}8+1{\cdot}3j.\end{aligned}$$

Ex. 25. Find expressions for:
(i) $\ln(-1)$; (ii) $\ln(-j)$, (iii) $(-1)^j$; (iv) $|(1+j)^{1/j}|$.

Ex. 26. Verify that the definition given in Section 2 for $z^{p/q}$ (p/q rational) is in accordance with the definition given here for a complex power of a complex number.

Exercise 19(c)

1. If a, b, r, s are real numbers and
$$u = re^{j\theta}, \quad v = se^{j\theta},$$
find:
(i) $|au+bv|$;
(ii) an argument of the complex number $au+bv$.

2. Express in the form $a+bj$:
$$\text{(i) } e^{1+\frac{1}{4}\pi j}; \quad \text{(ii) } e^{(n\pi+\pi j)}.$$

3. Express in the form $a+bj$;
$$\text{(i) } \ln(\sqrt{3}-j); \quad \text{(ii) } (\sqrt{3}-j)^j.$$

4. If z moves once anticlockwise around the unit circle in the Argand diagram, starting at the point -1, describe the motion of the point representing z^j.

5. If z is a complex number and $\sin z$, $\cos z$ are defined by
$$\sin z = \frac{1}{2j}(e^{jz}-e^{-jz}), \quad \cos z = \frac{1}{2}(e^{jz}+e^{-jz}),$$
prove that:
(i) $\sin^2 z+\cos^2 z = 1$; (ii) $\sin 2z = 2\sin z\cos z$;
(iii) $\cos(\frac{1}{2}\pi-z) = \sin z$; (iv) $\cos(z_1+z_2) = \cos z_1\cos z_2-\sin z_1\sin z_2$.

6. Show that:
(i) $\cos(x+yj) = \cos x\cosh y-j\sin x\sinh y$;
(ii) $\sin(x+yj) = \sin x\cosh y+j\cos x\sinh y$.

7. Show that:
$$\tan\tfrac{1}{2}(u+jv) = \frac{\sin u+j\sinh v}{\cos u+\cosh v}.$$

If $x+jy = c\tan\frac{1}{2}(u+jv)$, express $x^2+y^2+c^2$ in terms of u, v and c. If v and c are positive constants show that the locus of the point (x, y) referred to Cartesian axes is a circle of radius c cosech v. ($\tan z = \sin z/\cos z$.) (London)

8. Give a sketch of the representation in the Argand diagram of the two sets:

$$A = \{z \in C: \ |z| = 1, \ -\tfrac{1}{2}\pi < \operatorname{Im} z < \tfrac{1}{2}\pi\};$$
$$B = \{w \in C: \ z \in A, \ e^z w - e^z + w + 1 = 0\}.$$

Miscellaneous Exercise 19

1. Solve the equation $z^8 - z^4 + 1 = 0$ and mark the positions of the roots in the Argand diagram.

2. Prove that, if n is a positive integer,

$$(\cos\theta + \mathrm{j}\sin\theta)^n = \cos n\theta + \mathrm{j}\sin n\theta.$$

By putting n equal to 5 in this formula, or otherwise, prove that

$$\sin\frac{\pi}{10} = \frac{\sqrt{5}-1}{4}.$$ (O & C)

3. Prove that $\cos 7\theta = \cos^7\theta(1 - 21\tan^2\theta + 35\tan^4\theta - 7\tan^6\theta)$.
Find the real part of

$$\frac{(1+\cos\theta+\mathrm{j}\sin\theta)^n}{(1+\cos\theta-\mathrm{j}\sin\theta)^n}.$$ (London)

4. Express each of the complex numbers $z_1 = (1+\mathrm{j})\sqrt{2}$, $z_2 = 4(-1+\mathrm{j})\sqrt{2}$ in the form $r(\cos\theta+\mathrm{j}\sin\theta)$, where r is positive. Prove that $z_1^3 = z_2$, and find the other cube roots of z_2 in the form $r(\cos\theta+\mathrm{j}\sin\theta)$. (O & C)

5. Solve the equation $z^4 + 2z^2 + 3 = 0$, giving the real and imaginary parts of each root correct to 2 significant figures.

6. If $\alpha + \mathrm{j}\beta = \sqrt{\{(a+\mathrm{j}b)(c+\mathrm{j}d)\}}$ where a, b, c, d, α and β are real, find the value of α^2 in terms of a, b, c and d. (London)

7. Prove that the roots of the equation $(z+1)^n - (z-1)^n = 0$ $(n \geqslant 3)$ all lie on the imaginary axis. Illustrate the result geometrically in the case $n = 3$.

8. Given that

$$D = \begin{vmatrix} 1 & \omega & \omega^2 & 1 \\ \omega & \omega^2 & 1 & 1 \\ \omega^2 & 1 & 1 & \omega \\ 1 & 1 & \omega & \omega^2 \end{vmatrix},$$

where ω is a complex cube root of unity, prove that $D^2 = -27$. (O & C)

9. Evaluate $\sum_{r=0}^{n-1} \cos(\alpha + r\beta)$ and $\sum_{r=0}^{n-1} \sin(\alpha + r\beta)$.

10. If $z = \cos\theta + \mathrm{j}\sin\theta$, show that $z + 1/z = 2\cos\theta$ and find the corresponding result for $z - 1/z$.

Prove that

$$\cos^8\theta = \tfrac{1}{128}[\cos 8\theta + 8\cos 6\theta + 28\cos 4\theta + 56\cos 2\theta + 35].$$

Evaluate $$\int_{\frac{1}{4}\pi}^{\frac{1}{2}\pi} (\cos^8\theta + \sin^8\theta)\,d\theta.$$ (London)

11. If $|z| = 1$ and $\arg z = \theta \neq 0$, express in modulus argument form:

(i) $1+z$, (ii) $1-z$; (iii) $\sqrt{\left(\dfrac{1+z}{1-z}\right)}$.

Prove that
$$w = \sqrt{\left(\frac{1+z}{1-z}\right)}$$
lies on a fixed straight line, whatever the value of θ.

12. Demonstrate the following results geometrically, where ω is a complex cube root of unity:

(i) $(1+\omega)(1+\omega^2) = 1$; (ii) $\dfrac{1+\omega}{1+\omega^2} = \omega$; (iii) $\mathrm{Re}(1+2\omega) = 0$.

13. Find the roots of the equation $z^5+1 = 0$ in the form
$$\cos\phi + \mathrm{j}\sin\phi,$$
where ϕ is to be determined.

Deduce, or prove otherwise, that the roots of
$$16x^5-20x^3+5x+1 = 0$$
are $\cos\frac{1}{5}[(2r+1)\pi]$ $(r = 0, 1, 2, 3, 4)$. (London)

14. Solve the equation
$$\mathrm{j}(1-x\mathrm{j})^n = (1+x\mathrm{j})^n,$$
and verify your solution by setting: (i) $n = 1$; (ii) $n = 2$.

15. Find $|\omega+1+\mathrm{j}|$ and $\arg(\omega+1+\mathrm{j})$ geometrically, where ω is that complex cube root of unity with positive imaginary part.

Hence, or otherwise, find surd expressions for $\sin\frac{5}{12}\pi$ and $\cos\frac{5}{12}\pi$.

16. Prove that:
$$(x+y+z)(x+\omega y+\omega^2 z)(x+\omega^2 y+\omega z) \equiv x^3+y^3+z^3-3xyz,$$
where ω is a complex cube root of unity.

Hence, or otherwise, solve the following problems:

(i) Prove that the product
$$(x^3+y^3+z^3-3xyz)(a^3+b^3+c^3-3abc)$$
is expressible in the form
$$A^3+B^3+C^3-3ABC,$$
where $A = ax+by+cz$, $B = ay+bz+cx$, $C = az+bx+cy$.

(ii) Solve the equation
$$x^3-9x+12 = 0.$$
(O & C)

17. In the determinant
$$\Delta = \begin{vmatrix} a & b & c & d \\ d & a & b & c \\ c & d & a & b \\ b & c & d & a \end{vmatrix}$$
the cofactors of a, b, c, d are denoted by A, B, C, D respectively (the expansion of Δ by its first column being $aA+bB+cC+dD$).

By considering the product $\Delta\Omega$, where

$$\Omega = \begin{vmatrix} 1 & 0 & 0 & 0 \\ \omega & 1 & 0 & 0 \\ \omega^2 & 0 & 1 & 0 \\ \omega^3 & 0 & 0 & 1 \end{vmatrix}$$

prove that, if ω is any root of the equation $x^4-1=0$, then $a+b\omega+c\omega^2+d\omega^3$ divides into Δ, the quotient being

$$A+B\omega^3+C\omega^2+D\omega.$$

Hence show that

$$A+B+C+D = (a+c-b-d)\{(a-c)^2+(b-d)^2\}. \qquad \text{(O \& C)}$$

18. P, Q are the affixes of the complex numbers p, q and $\xi = \cos 2\pi/n + \text{j} \sin 2\pi/n$. Locate the affix of the complex numbers $(p-q)\,\xi$.

If a, b are two complex numbers, prove that the affixes of the complex numbers $z_1, z_2, \ldots, z_n$ are the vertices of a regular n-sided polygon if and only if

$$(z_r-a)^n = b^n \quad (r = 1, 2, 3, \ldots, n).$$

19. If Z_1, Z_2, Z_3 are the affixes of the complex numbers z_1, z_2, z_3 prove that a necessary and sufficient condition for the triangle $Z_1Z_2Z_3$ to be equilateral is that

$$z_1^2+z_2^2+z_3^2-z_2z_3-z_3z_1-z_1z_2 = 0.$$

20. Prove that, if n is a positive integer ($n > 1$) and

$$\omega = \cos 2\pi/n + \text{j} \sin 2\pi/n,$$

then $1+\omega+\omega^2+\ldots+\omega^{n-1} = 0$.

In the Argand diagram the points $A_1, A_2, \ldots, A_n$ are the vertices of a regular polygon inscribed in a circle of radius a with its centre at the origin. The complex numbers represented by the points $A_1, A_2, \ldots, A_n$ are $z_1, z_2, \ldots, z_n$. Prove that

$$z_1^2+z_2^2+\ldots+z_n^2 = 0.$$

The perpendicular distances of the points $A_1, A_2, \ldots, A_n$ from any given line through the centre are $d_1, d_2, \ldots, d_n$. Prove that

$$d_1^2+d_2^2+\ldots\, d_n^2 = \tfrac{1}{2}na^2. \qquad \text{(O \& C)}$$

21. If $x+\text{j}y = \tanh(u+\text{j}v)$ where u, v, x, y are real, find x and y in terms of u and v.

Prove that

$$x^2+y^2-2x \coth 2u+1 = 0$$

and

$$x^2+y^2+2y \cot 2v-1 = 0.$$

If u and y are regarded as variable parameters, and x and y as Cartesian co-ordinates, describe the relationship between the two families of circles.

22. Prove that

$$\sinh(\theta+\text{j}\phi) = \sinh\theta\cos\phi+\text{j}\cosh\theta\sin\phi.$$

For all real or complex values of z the sum of the infinite series

$$a_0+a_1z+a_2z^2+\ldots$$

is $f(z)$. Prove that if ω is a root of the equation $\omega^2+\omega+1 = 0$, then

$$a_0+a_3x^3+a_6x^6+\ldots+ = \tfrac{1}{3}\{f(x)+f(\omega x)+f(\omega^2x)\}.$$

By considering the series for sinh x, prove that

$$\frac{x^3}{3!}+\frac{x^9}{9!}+\frac{x^{15}}{15!}+\ldots = \frac{2}{3}\sinh\frac{x}{2}\left\{\cosh\frac{x}{2}-\cos\frac{x\sqrt{3}}{2}\right\}. \qquad \text{(O \& C)}$$

23. A sequence $u_1, u_2 \ldots$ is defined by $u_1 = 1$ and

$$u_{n+1} = \alpha u_n + n + 1 \quad (n \geqslant 1),$$

where α is independent of n. Find an expression for u_n when $\alpha = 1$.

By induction or otherwise show that, when $\alpha \neq 1$, u_n is of the form

$$u_n = A\alpha^n + Bn + C,$$

where A, B, C are independent of n, and find A, B and C.

If now α is a complex mth root of unity and n is a multiple of m, determine the real part of u_n. (O & C)

24. Prove that

$$x^{2n}-1 \equiv (x-1)(x+1)\prod_{k=1}^{n-1}\left(x^2-2x\cos\frac{k\pi}{n}+1\right)$$

and devise a similar expression for $x^{2n+1}-1$.

25. Prove that

$$x^{2n}-2x^n\cos n\theta+1 = \prod_{k=0}^{n-1}\left\{x^2-2x\cos\left(\theta+\frac{2k\pi}{n}\right)+1\right\}.$$

Deduce the following results:

(i) $\sin n\alpha = 2^{n-1}\prod_{k=0}^{n-1}\sin\left(\alpha+\frac{k\pi}{n}\right)$;

(ii) $\cos n\alpha-\cos n\beta = 2^{n-1}\prod_{k=0}^{n-1}\left\{\cos\alpha-\cos\left(\beta+\frac{2k\pi}{n}\right)\right\}$.

From (i) and (ii) deduce, by logarithmic differentiation, the further results:

(iii) $\cot n\alpha = \frac{1}{n}\sum_{k=0}^{n-1}\cot\left(\alpha+\frac{k\pi}{n}\right), \quad \alpha \neq \frac{r\pi}{n}$;

(iv) $\operatorname{cosec}^2 n\theta = \frac{1}{n^2}\sum_{k=0}^{n-1}\operatorname{cosec}^2\left(\theta+\frac{k\pi}{n}\right), \quad \theta \neq \frac{r\pi}{n}$.

26. Prove that $x^2+y^2+z^2-yz-zx-xy$ has a linear factor $x+\omega y+\omega^2 z$, where ω is a complex cube root of unity. Deduce that, if $x \neq 3$, $x^2+y^2+z^2-yz-zx-xy$ is a factor of $(y-z)^n+(z-x)^n+(x-y)^n$.

20. *Mappings in the Argand diagram*

The *equation* of a curve in the Argand diagram is the condition imposed upon the complex number z whose affix, P, is any point of the curve. For example, the equation of a straight line through O and containing the point B (represented by the complex number b) is

$$z = \lambda b, \quad \lambda \text{ real.}$$

A straight line not containing O is uniquely defined by the foot, A, of the perpendicular from O to the line (see Figure 20.1). Suppose A is the affix of the complex number a and that $P(z)$ is any point on the given line.

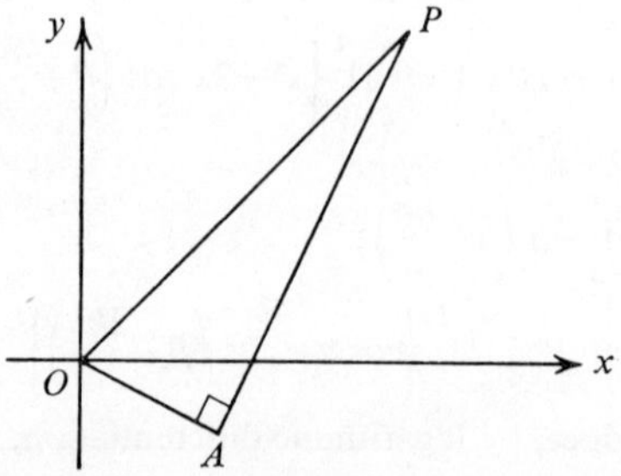

Fig. 20.1

Then **OP** = **OA**+**AP** or, in complex number notation,

$$z = a + \mathrm{j}\lambda a, \quad \lambda \text{ real.} \tag{1}$$

(Recall that multiplication by j rotates the vector representing a complex number through $\frac{1}{2}\pi$.) Taking complex conjugates of both sides of (1)

$$z^* = a^* - \mathrm{j}\lambda a^*.$$

Eliminating λ this gives

$$a^*(z-a) + a(z^*-a^*) = 0,$$

or

$$a^*z + az^* = c, \tag{2}$$

where $c\ (= 2|a|^2)$ is a real number, as the equation of the line through A and perpendicular to OA.

**Ex.* 1. By writing $a = h+k\mathrm{j}$ and $z = x+y\mathrm{j}$, prove conversely that any equation of type (2) above represents a straight line.

Ex. 2. What is the equation of the perpendicular bisector of the line joining the points which represent the complex numbers 0 and $2-3j$?

**Ex.* 3. Show that

$$a^*z+az^* = 2k|a|^2, \quad k \text{ real},$$

represents a straight line and describe its relationship to the line

$$a^*z+az^* = 2|a|^2.$$

Interpret geometrically the quantity Re (a^*z).

Ex. 4. Where does the line $(1+2j)\,z+(1-2j)\,z^* = 12$ cut
(i) the real axis; (ii) the imaginary axis?

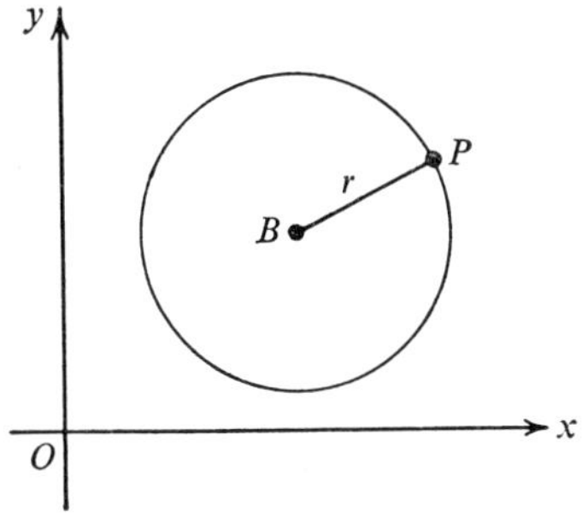

Fig. 20.2

The equation of a circle in the Argand diagram is also easily found. If the centre of the circle is B, corresponding to the complex number b, and if the radius is r, then, for any point $P(z)$ of the circumference,

$$|z-b| = r,$$

or

$$(z-b)\,(z^*-b^*) = r^2,$$

which gives, on writing the *real* number $|b|^2-r^2$ as d,

$$zz^*-b^*z-bz^*+d = 0 \tag{3}$$

as the equation of the circle centre B and radius $[|b|^2-d]^{\frac{1}{2}}$.

Ex. 5. Show conversely, that for real $a \neq 0$ and real c, the equation

$$azz^*+b^*z+bz^*+c = 0$$

represents a circle, provided $ac < |b|^2$.

Ex. 6. What is the equation of the circle, centre $1-2j$ and radius 3 in the form (3)?

Ex. 7. Show that $2zz^*+(3-j)\,z+(3+j)\,z^*+1 = 0$ is the equation of a circle, centre $-\frac{1}{2}(3+j)$ and radius $\sqrt{2}$. Find the centre and radius of the circle with equation:

(i) $zz^*-z(1-3j)-z^*(1+3j)+6 = 0$;
(ii) $4zz^*-z(2+4j)-z^*(2-4j)+1 = 0$.

We now consider the images of straight lines and circles under three simple functions f, g, h from the set C into the set C given by

(I) $f(z) = w$, where $w = z+\beta$;
(II) $g(z) = w$, where $w = \alpha z$;
(III) $h(z) = w$, where $w = z^{-1}$;

α, β being complex constants.

(I) $w = z+\beta$.

This is clearly a translation of the whole plane by an amount equivalent to the position vector corresponding to the complex number β (see Figure 20.3).

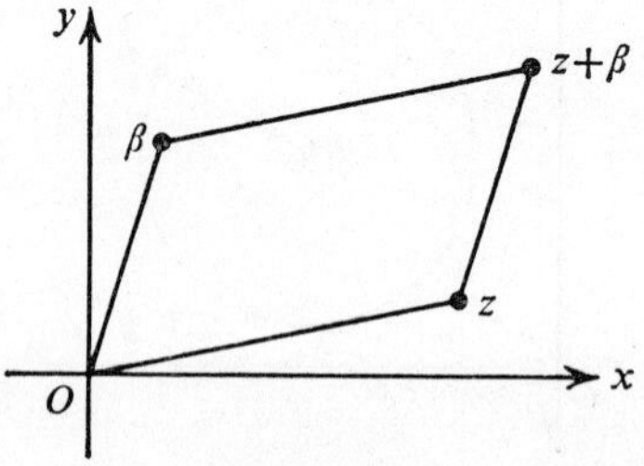

Fig. 20.3

Geometrically it is obvious that this function maps a straight line into a parallel straight line, and a circle into an equal circle with its centre translated an amount β.

Analytically,

$$a^*z+az^* = c$$

$$\Rightarrow a^*(w-\beta)+a(w^*-\beta^*) = c$$

$$\Rightarrow \quad a^*w+aw^* = c+a^*\beta+a\beta^*.$$

But $a^*\beta+a\beta^* = 2\,\mathrm{Re}\,(a^*\beta)$; thus the right-hand side is real and the equation represents a straight line.

Again, for a circle we have

$$|z-b| = r \quad (r \text{ real and positive})$$

$$\Rightarrow |w-(b+\beta)| = r,$$

which represents a circle of the same radius but with its centre translated by an amount β.

Ex. 8. Explain why the straight lines $a^*z+az^* = c$ and

$$a^*w+aw^* = c+a^*\beta+a\beta^*$$

are parallel.

(II) $w = \alpha z$.

As shown in Chapter 17 this maps $P \to Q$ where $OQ = |\alpha| OP$ and $\angle POQ = \arg \alpha$; that is, it represents an extension by the factor $|\alpha|$ followed by an anticlockwise rotation of magnitude $\arg \alpha$ (Figure 20.4).

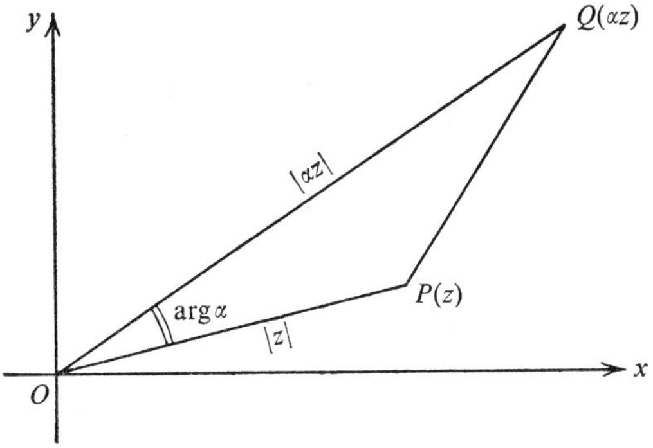

Fig. 20.4

The effect of this mapping upon straight lines and circles may be deduced analytically as follows:

$$a^*z + az^* = c$$

$$\Rightarrow a^*\left(\frac{w}{\alpha}\right) + a\left(\frac{w}{\alpha}\right)^* = c$$

$$\Rightarrow (\alpha a)^* w + (\alpha a) w^* = c|\alpha|^2$$

and

$$|z-b| = r$$

$$\Rightarrow \quad \left|\frac{w}{\alpha} - b\right| = r$$

$$\Rightarrow \quad |w - b\alpha| = r|\alpha|.$$

Thus straight lines are mapped into straight lines and circles are mapped into circles under the transformation $w = \alpha z$.

**Ex.* 9 Show that the centre of the z circle maps into the centre of the w circle under this transformation and that the ratio of the two radii is $|\alpha| : 1$.

Ex. 10. Illustrate geometrically the effect of the transformation $w = (1 + \mathrm{j})z$ upon the circle $|z - 1| = 1$.

(III) $w = z^{-1}$.

We must restrict the domain of this function to the whole of C with the number 0 deleted. To describe the transformation geometrically it is useful to define the *inverse* of a point P with respect to a circle. Given a circle, centre O and radius r, the inverse of the point P is the point P' on OP such that $OP.OP' = r^2$.

**Ex.* 11. Show that the transformation $w = z^{-1}$ maps the point P into the reflection in the real axis of the inverse of P with respect to the unit circle $|z| = 1$ (see Figure 20.5).

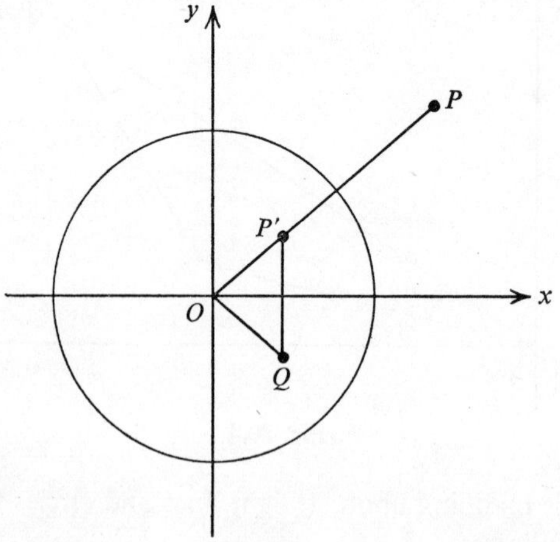

Fig. 20.5

Now consider the effect of the transformation $w = z^{-1}$ upon a general straight line and circle. The straight line

$$a^*z + az^* = c$$

maps into
$$a^*\left(\frac{1}{w}\right) + a\left(\frac{1}{w}\right)^* = c,$$

that is, into
$$cww^* - a^*w^* - aw = 0.$$

This represents a circle through O, if $c \neq 0$, and a straight line through O if $c = 0$. Thus, a straight line not containing O maps into a circle through O, while a straight line through O maps into another straight line through O.

**Ex.* 12 Show that a straight line through O maps into its reflection in the real axis under the transformation $w = z^{-1}$. What lines map into themselves under this transformation?

**Ex.* 13. Show that, under the transformation $w = z^{-1}$, the straight line through A and perpendicular to OA maps into a circle with its centre on the line OA^*, where A, A^* are the affixes of the conjugate complex numbers a, a^*.

Now consider the effect of the transformation $w = z^{-1}$ upon the circle

$$zz^* - b^*z - bz^* + d = 0.$$

The image is the set of points defined by

$$\left(\frac{1}{w}\right)\left(\frac{1}{w}\right)^* - b^*\left(\frac{1}{w}\right) - b\left(\frac{1}{w}\right)^* + d = 0,$$

that is, by $$dww^* - b^*w^* - bw + 1 = 0.$$

If $d = 0$ this represents a straight line, otherwise it represents another circle. Thus, a circle passing through O maps into a straight line not through O, while a circle not through O maps into another circle not through O.

**Ex.* 14. Show that, under the transformation $w = z^{-1}$, a circle, centre B, which passes through the origin maps into a straight line perpendicular to OB^*, where B, B^* are the affixes of the conjugate complex numbers b, b^*.

**Ex.* 15. Show that, under the transformation $w = z^{-1}$, a circle, centre B, which does not pass through the origin maps into a circle with its centre lying on the line OB^*, where B, B^* are the affixes of the conjugate complex numbers b, b^*.

**Ex.* 16. Given a circle, centre B, not passing through O, show that the transformation $w = z^{-1}$ does not in general map B into the centre of the image circle.

**Ex.* 17. Show that a diameter of a circle is mapped onto a diameter of the image circle under the transformations $w = z+\beta$ and $w = \alpha z$. Show also that, under the transformation $w = z^{-1}$, a diameter *through the origin* is mapped onto a diameter of the image circle.

To summarize the effect of the mapping $w = z^{-1}$ upon straight lines and circles we have (reading $\rightarrow$ as 'maps into'):

(i) straight line through $O \rightarrow$ straight line through O.
(ii) straight line not through $O \rightarrow$ circle through O;
(iii) circle through $O \rightarrow$ straight line not through O;
(iv) circle not through $O \rightarrow$ circle not through O.

Example 1. *If the point z lies on the circle* $|z| = 1$, *find the locus of the point w where* $w = \mathrm{j}/(z+\mathrm{j})$.

Method (i)

Consider the sequence of transformations:

$$f_1: z \rightarrow u = z + \mathrm{j},$$

$$f_2: u \rightarrow v = \frac{1}{u},$$

$$f_3: v \rightarrow w = \mathrm{j}v,$$

upon the circle $|z| = 1$. Their effects are shown in the following sequence of diagrams (Figure 20.6). (For f_2, recall the result of Ex. 14.)

The locus is seen to be the line $w = \tfrac{1}{2} + \lambda\mathrm{j}$ or $w + w^* = 1$.

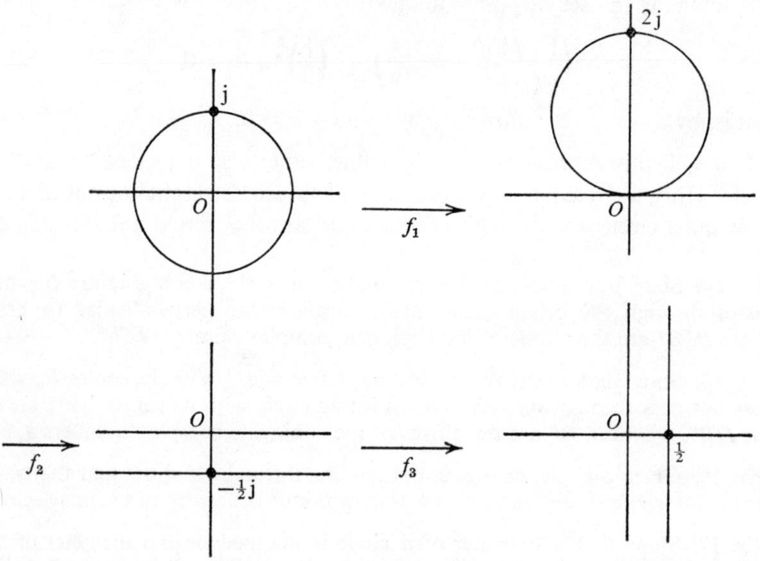

Fig. 20.6

Method (ii)

$$w(z+\mathrm{j}) = \mathrm{j}$$

$$\Rightarrow \quad z = \frac{\mathrm{j}(1-w)}{w}.$$

Thus

$$|z| = 1$$

$$\Rightarrow \quad |w| = |\mathrm{j}(1-w)|$$

$$\Rightarrow \quad |w| = |(1-w)|$$

$$\Rightarrow \quad ww^* = (1-w)(1-w^*)$$

$$\Rightarrow w+w^* = 1, \quad \text{with the same conclusion as before.}$$

Ex. 18. If z moves anticlockwise around the circle $|z| = 1$, starting at $z = \mathrm{j}$, how does w move along the line $w+w^* = 1$?

Example 2. *Show that the transformation*

$$w = \frac{2z-3+\mathrm{j}}{\mathrm{j}z-2}$$

maps the unit circle $|z| = 1$ *into a circle with centre on the real axis and radius* $\sqrt{2}$.

Method (i)

Write
$$w = \frac{2z-3+j}{jz-2} = \frac{2(z+2j)-3-3j}{j(z+2j)} = -2j - \frac{3-3j}{z+2j},$$
and consider the effect of the successive transformations

$$f_1: z \to t = z+2j,$$
$$f_2: t \to u = 1/t,$$
$$f_3: u \to v = -3(1-j)\,u = 3\sqrt{2}\left(\cos\frac{3\pi}{4} + j\sin\frac{3\pi}{4}\right)u,$$
$$f_4: v \to w = v-2j$$

upon the circle $|z| = 1$. (Recall the result of Ex. 15 for deducing the effect

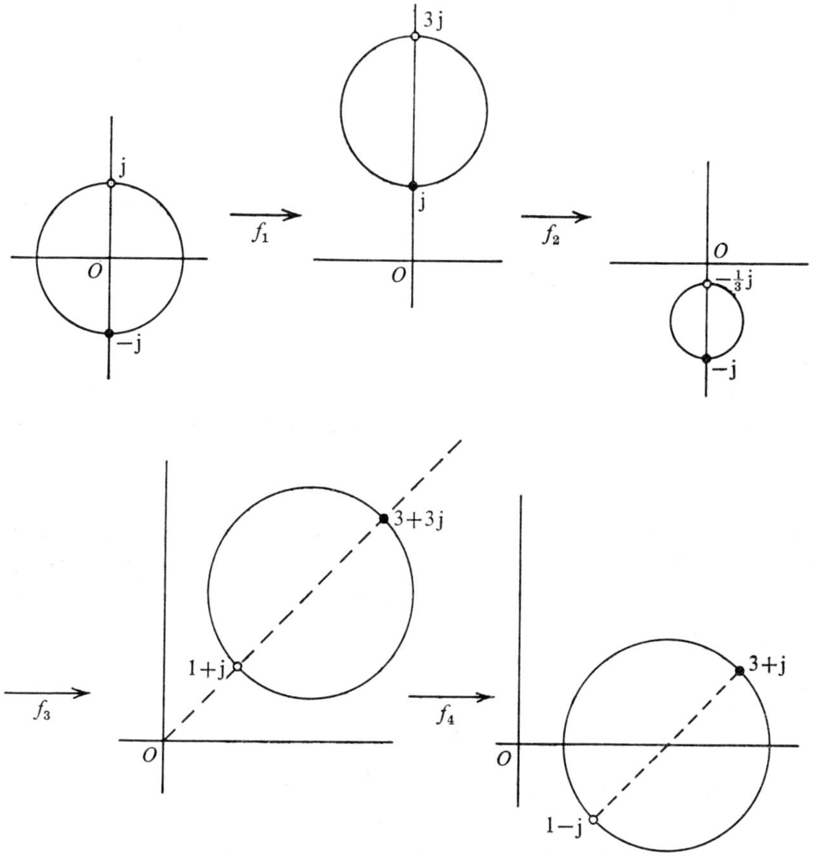

Fig. 20.7

of f_2. Notice that, by Ex. 17, the line joining the successive images of $\pm$j remains a diameter of the corresponding circle.)

Thus, the image is a circle, centre $\frac{1}{2}[(1-\mathrm{j})+(3+\mathrm{j})] = 2$ and radius $\frac{1}{2}|(3+\mathrm{j})-(1-\mathrm{j})| = \frac{1}{2}|2+2\mathrm{j}| = \sqrt{2}$.

Method (ii)

$$w = \frac{2z-3+\mathrm{j}}{\mathrm{j}z-2},$$

$$\Rightarrow \qquad (\mathrm{j}z-2)\,w = 2z-3+\mathrm{j}$$

$$\Rightarrow \qquad z = \frac{2w-3+\mathrm{j}}{\mathrm{j}w-2}.$$

Thus $\qquad |z| = 1$

$$\Rightarrow \qquad |\mathrm{j}w-2| = |2w-3+j|$$

$$\Rightarrow \quad (\mathrm{j}w-2)(-\mathrm{j}w^*-2) = (2w-3+\mathrm{j})(2w^*-3-\mathrm{j})$$

$$\Rightarrow ww^*-2\mathrm{j}(w-w^*)+4 = 4ww^*-6(w+w^*)-2\mathrm{j}(w-w^*)+10$$

$$\Rightarrow \quad ww^*-2(w+w^*)+2 = 0.$$

This equation certainly represents a circle. Furthermore, since the interchange of w and w^* does not affect the equation, it is symmetrical about the real axis and thus its centre lies on the real axis. It cuts the real axis at points given by $w = w^*$,

i.e. $\qquad x^2-4x+2 = 0,$

i.e. $\qquad x = 2\pm\sqrt{2},$

giving a radius of $\sqrt{2}$, as before.

Ex. 19. With the notation of Example 2, if z moves anticlockwise round the circle $|z| = 1$, starting at $z = \mathrm{j}$, how does w move around the circle

$$ww^*-2(w+w^*)+2 = 0?$$

Example 3. *The function f: $C \to C$ is defined by $f(z) = (z-1)^2$. Prove that the image of the unit circle $|z| = 1$ under this mapping is a closed curve, consisting of all points w with the property that $r = 2(1+\cos\theta)$, where $r = |w|$ and $\theta = \arg w$.*

Since $|z| = 1$, we may write $z = \cos\phi+\mathrm{j}\sin\phi$; then

$$\begin{aligned} z-1 &= \cos\phi-1+\mathrm{j}\sin\phi \\ &= -2\sin^2\tfrac{1}{2}\phi+2\mathrm{j}\sin\tfrac{1}{2}\phi\cos\tfrac{1}{2}\phi \\ &= 2\sin\tfrac{1}{2}\phi(-\sin\tfrac{1}{2}\phi+\mathrm{j}\cos\tfrac{1}{2}\phi) \\ &= 2\sin\tfrac{1}{2}\phi\{\cos(\tfrac{1}{2}\pi+\tfrac{1}{2}\phi)+\mathrm{j}\sin(\tfrac{1}{2}\pi+\tfrac{1}{2}\phi)\}. \end{aligned}$$

Thus $w = (z-1)^2 = 4\sin^2\frac{1}{2}\phi\{\cos(\pi+\phi)+\mathrm{j}\sin(\pi+\phi)\}$ and it follows that

$r = 4\sin^2\frac{1}{2}\phi$, $\theta = \pi+\phi-2k\pi$, where $k = -1, 0, 1$, according to which value is required to bring θ in the range $-\pi < \theta \leqslant \pi$. Eliminating ϕ gives

$$r = 4\cos^2\tfrac{1}{2}\theta = 2(1+\cos\theta).$$

In the equation $r = 2(1+\cos\theta)$, any value of θ in the range $0 \leqslant \theta \leqslant \pi$ determines a unique value of r and thus a unique complex number $r(\cos\theta+\mathrm{j}\sin\theta)$. As θ varies the corresponding complex number moves around a curve Γ which is clearly closed since r returns to its original value after θ has turned through an angle of 2π. $r = 2(1+\cos\theta)$ is called the *polar equation of* Γ; the shape of the curve, known as a *cardioid*, is shown in Figure 20.8.

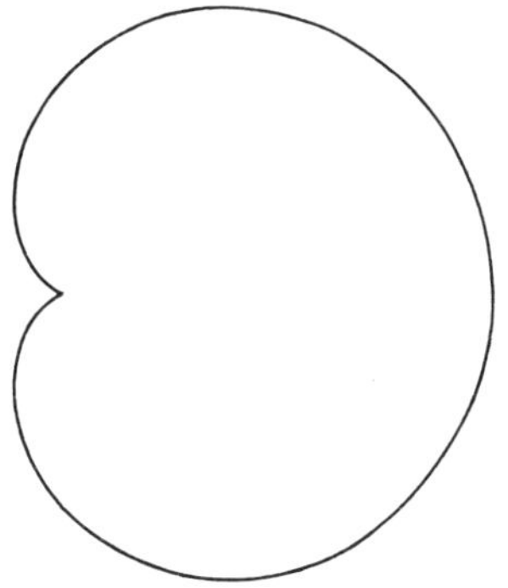

Fig. 20.8

Ex. 20. Prove the first part of Example 3 geometrically by constructing the unit circle and the affixes of numbers $z-1$, $(z-1)^2$.

Exercise 20

1. The function f has domain $\{z \in C: |z| = 1\}$. Find the range of f in each of the following cases:
(i) $f(z) = z+3$; (ii) $f(z) = z+\mathrm{j}$; (iii) $f(z) = 2z$; (iv) $f(z) = \mathrm{j}z$;
(v) $f(z) = (1+\mathrm{j})z$.

2. In Question 1, if the point z is regarded as moving anticlockwise around the circle $|z| = 1$, starting at $z = 1$, describe the motion of the image point $w = f(z)$ in each of the cases (i)–(v).

3. The function $f: C \to C$ is defined by $f(z) = z^3+z^2+2z+1$. Find the values of z which remain invariant under this mapping and illustrate your answer by reference to the Argand diagram.

4. The function $f: C \to C$ is defined by

$$f(z) = az^2+bz+c,$$

where $a, b, c, \in C$. If $f(1) = 0$, $f(0) = 2\mathrm{j}$, $f(\mathrm{j}) = 1+\mathrm{j}$, show that just one z remains invariant under this mapping, and find its value.

What two complex numbers map into $3+\mathrm{j}$?

5. As z moves once anticlockwise around the unit circle $|z| = 1$, starting at $z = 1$, describe the motion of the point w where:
(i) $w = z^2$, (ii) $w = z^3$; (iii) $w = \mathrm{j}z^4$, (iv) $w = z^{-1}$.

6. Show that the affixes of the complex numbers 1, $\frac{1}{2}(-1+\sqrt{3}\mathrm{j})$, $\frac{1}{2}(-1-\sqrt{3}\mathrm{j})$ form the vertices of an equilateral triangle. Describe the effect of the transformation $w = jz+1+j$ upon this triangle, illustrating your result by a sketch of the Argand diagram.

7. Answer the same question as in Question 6 for the transformation

$$w = (1+j)\,z+1+j.$$

8. Show that, under the mapping defined by $w = j/(z-1)$, the interior of the circle $|z-1| = 1$ is mapped into the exterior of the circle $|w| = 1$. Show that two points remain invariant under this transformation and locate their approximate positions on an Argand diagram.

9. Show that, under the mapping defined by $w = (j-jz)/(1+z)$, the interior of the circle $|z| = 1$ is mapped into the *half-plane* $\text{Im}(w) > 0$.

10. Show that, under the mapping defined by $w = (j+2jz)/(1-z)$, the image of the set $\{z \in C: |z| < 1\}$ is the set $\{w \in C: \text{Im}(w) > -\frac{1}{2}\}$.

11. Find the equation, in the form $a^*z+az^* = b$, of the line joining the points $z = 1$ and $z = j$.

Find the image, under the mappings defined by $w = z/(z+1)$, of the set of points represented by the interior of the triangle with vertices $z = 0, z = 1, z = j$.

In nos. **12–17** find the image of the circle $|z| = 1$ under the given transformation. If the image is a circle, find its centre and radius.

12. $w = \dfrac{2}{j-2z}$. **13.** $w = \dfrac{2+z}{j-z}$. **14.** $w = \dfrac{1+z}{1-jz}$.

15. $w = \dfrac{(1+j)\,z-1}{(1-j)\,z+j}$. **16.** $w = \dfrac{jz+1-j}{(1+j)\,z-1}$.

17. $w = \dfrac{2z-j}{1+z}$.

18. Prove that, if $(z-8j)/(z+6)$ is purely imaginary, the locus of z in the Argand diagram is a circle with centre at the point $4j-3$ and radius 5. (O & C)

19. Show that the set of points $\{z\}$ satisfying

$$\arg\left(\frac{z-a}{z-b}\right) = \theta,$$

where θ is a real number in the interval $-\pi < \theta \leqslant \pi$ and a, b are complex numbers, is represented by the arc of a circle through the affixes A, B of a, b.

What is the condition for a point z to lie on the other arc of this circle?

20. Show that, if k is a real number not equal to 1, the set of points $\{z\}$ satisfying

$$\left|\frac{z-a}{z-b}\right| = k$$

is a circle, with its centre the point C on AB such that $CA/CB = k^2$, and that A and B are inverses with respect to this circle.

What can you say about this circle (i) if $k = 0$; (ii) if k is very large?

Draw on the same diagram the system of circles obtained for various values of k by taking $a = 1$, $b = -1$.

What happens if $k = 1$?

21. Show that the transformation defined by $w = 1/(3+\mathrm{j}-z)$ maps the circle $|z-2-\mathrm{j}| = 1$ into the straight line $\mathrm{Re}\,(w) = \frac{1}{2}$. What transformation maps the line $\mathrm{Re}(z) = \frac{1}{2}$ into the circle $|w-2-\mathrm{j}| = 1$?

22. Show that the transformation defined by $w = 1+z^2$ maps the unit circle $|z| = 1$ into the unit circle $|w-1| = 1$. Draw a sketch of the Argand diagram, construct the point $1+z^2$ and deduce the above result geometrically.

23. Show that the transformation defined by $w = 1/(1-z)^2$ maps the unit circle $|z| = 1$ into the curve with polar equation $2r(1+\cos\theta) = 1$. Draw a sketch of this curve, indicating on it the images of the points 1, j, -1, $-\mathrm{j}$.

24. Prove that if

$$x+\mathrm{j}y = \frac{1}{\lambda+\mathrm{j}\mu},$$

then the points on the Argand diagram defined by making λ constant lie on a circle, and the points defined by making μ constant lie on a circle.

Prove also that, whatever be the values of the constants, the centres of the two systems of circles obtained lie on two fixed perpendicular lines.

25. In the transformation defined by $w = 1/(z-\mathrm{j})$, describe the motion of the image point w if z moves anticlockwise around the rectangle with vertices 0, b, $b+a\mathrm{j}$, $a\mathrm{j}$, starting at the origin (where a, b are positive real numbers).

26. The points P and Q represent, in the Argand diagram, the complex numbers z and $1/(z^2+2)$. The point P describes a quadrant of a circle, from the origin along the real axis to the point $z = a$, round the arc of the circle $|z| = a$ to the point $z = \mathrm{j}a$, and back along the imaginary axis to the origin. Describe the path traced out by Q (i) when $a = 1$, (ii) when $a = \sqrt{2}$.

27. Find the transformation of the form

$$w = \frac{az+b}{cz+d},$$

which maps 0 into j, j into 0 and 1 into -1.

Show, that, if this particular transformation maps the complex number ξ into the complex number η, then it also maps η into ξ.

Can you generalize these results in any way?

28. Show that, if the image of the complex number z under the transformation $w = z+z^{-1}$ is real, then $|z| = 1$.

Deduce a geometrical construction for determining the roots of the quadratic equation $z^2+az+1 = 0$ where a is a real number in the interval $-2 < a < 2$.

Extend your results to deal with the quadratic equation $z^2+az+b = 0$, where a, b are real numbers such that $a^2 < 4b$, by considering the transformation $w = z+bz^{-1}$.

21. *Quadratic equations and quadratic functions*

1. THE QUADRATIC EQUATION

We have shown in Chapter 19 that it is always possible to extract the mth root of a complex number p; that is, it is always possible to solve the equation

$$w^m - p = 0.$$

In particular, if $m = 2$, the quadratic equation

$$w^2 - p = 0 \tag{1}$$

has roots $+\sqrt{p}$ and $-\sqrt{p}$, and the real and imaginary parts of these two numbers may be calculated to any required degree of accuracy.

The more general quadratic equation

$$az^2 + bz + c = 0 \quad (a, b, c \in C, a \neq 0) \tag{2}$$

may be reduced to form (1) by the transformation

$$w = z + \frac{b}{2a},$$

a process usually known as 'completing the square':

$$az^2 + bz + c = 0$$

$$\Rightarrow \quad z^2 + \frac{b}{a}z = -\frac{c}{a} \quad (a \neq 0)$$

$$\Rightarrow \left(z + \frac{b}{2a}\right)^2 = \frac{b^2 - 4ac}{4a^2}$$

$$\Rightarrow \quad z = \frac{-b \pm \sqrt{(b^2 - 4ac)}}{2a}.$$

If $b^2 = 4ac$, only one root is obtained, otherwise equation (2) has two distinct roots. For uniformity it is useful, in the case $b^2 = 4ac$, to say that (2) has *two coincident roots*, or a *repeated root*; with this convention, *every quadratic equation has two roots.*

Ex. 1. Solve the equation $(3+j)\,z^2 - 8jz - 6 + 2j = 0$.

If the two roots of equation (2) are denoted by α and β then az^2+bz+c has linear factors $(z-\alpha)$ and $(z-\beta)$ and we may write

$$az^2+bz+c \equiv a(z-\alpha)(z-\beta),$$

or

$$az^2+bz+c \equiv az^2-a(\alpha+\beta)z+a\alpha\beta.$$

Thus

$$\alpha+\beta = -b/a \tag{3}$$

and

$$\alpha\beta = c/a. \tag{4}$$

Relations (3) and (4) frequently enable us to avoid the explicit solution of a quadratic equation, especially in those cases where we are concerned with symmetrical relations between the roots. (Particularly important cases of this arise in analytical geometry. See Chapter 22.)

Example 1. *If the roots of the equation* $(1+j)z^2-2jz+1-j = 0$ *are* α, β *find:*

(i) $\dfrac{1}{\alpha}+\dfrac{1}{\beta}$; (ii) $\alpha^2+\beta^2$;

(iii) $\alpha\beta^3+\alpha^3\beta$; (iv) $\alpha^3+\beta^3$;

(v) $\alpha^5+\beta^5$; (vi) $|\alpha-\beta|$.

From (3) and (4),

$$\alpha+\beta = \frac{2j}{1+j} = 1+j,$$

$$\alpha\beta = \frac{1-j}{1+j} = -j.$$

(i) $\dfrac{1}{\alpha}+\dfrac{1}{\beta} = \dfrac{\alpha+\beta}{\alpha\beta} = \dfrac{1+j}{-j} = -1+j.$

(ii) Since α, β both satisfy the given equation:

$$(1+j)\alpha^2-2j\alpha+1-j = 0,$$

$$(1+j)\beta^2-2j\beta+1-j = 0,$$

whence, by addition,

$$(1+j)(\alpha^2+\beta^2)-2j(\alpha+\beta)+2-2j = 0.$$

But $\alpha+\beta = 1+j$; thus

$$(1+j)(\alpha^2+\beta^2) = 2j(1+j)-2+2j$$
$$= -4+4j$$

giving

$$\alpha^2+\beta^2 = 4j.$$

[Alternatively we could write $\alpha^2+\beta^2 = (\alpha+\beta)^2-2\alpha\beta$, but this approach is less easy to generalize to higher powers—see (iv).]

(iii) $$\alpha\beta^3+\beta\alpha^3 = \alpha\beta(\alpha^2+\beta^2)$$
$$= -j(4j), \quad \text{by (ii)},$$
$$= 4.$$

(iv) Since $$(1+j)\,\alpha^2-2j\alpha+1-j = 0$$

we have $$(1+j)\,\alpha^3-2j\alpha^2+(1-j)\,\alpha = 0;$$

similarly, $$(1+j)\,\beta^3-2j\beta^2+(1-j)\,\beta = 0$$
whence, by addition,
$$(1+j)\,(\alpha^3+\beta^3)-2j(\alpha^2+\beta^2)+(1-j)\,(\alpha+\beta) = 0.$$

Thus $$(1+j)\,(\alpha^3+\beta^3) = 2j(4j)-(1-j)\,(1+j), \quad \text{by (ii)},$$
$$= -10,$$

giving $\alpha^3+\beta^3 = -5+5j$, on multiplying both sides by $\frac{1}{2}(1-j)$.

[Alternatively, $$\alpha^3+\beta^3 = (\alpha+\beta)\,(\alpha^2-\alpha\beta+\beta)$$
$$= (\alpha+\beta)\,\{(\alpha+\beta)^2-3\alpha\beta\}, \text{ etc.]}$$

(v) As in (iv)
$$(1+j)\,(\alpha^4+\beta^4) = 2j(\alpha^3+\beta^3)-(1-j)\,(\alpha^2+\beta^2)$$
$$= 2j(-5+5j)-(1-j)\,(4j)$$
$$= -14-14j$$

giving $$\alpha^4+\beta^4 = -14.$$

Thus, $$(1+j)\,(\alpha^5+\beta^5) = 2j(\alpha^4+\beta^4)-(1-j)\,(\alpha^3+\beta^3)$$
$$= 2j(-14)-(1-j)\,(-5+5j)$$
$$= -38j$$

giving $$\alpha^5+\beta^5 = -19-19j.$$

[Alternatively,
$$(\alpha^2+\beta^2)\,(\alpha^3+\beta^3) = \alpha^5+\beta^5+\alpha^2\beta^2(\alpha+\beta), \text{ etc.]}$$

(vi) $$(\alpha-\beta)^2 = (\alpha+\beta)^2-4\alpha\beta$$
$$= 6j,$$
$$|\alpha-\beta|^2 = 6,$$
$$|\alpha-\beta| = \sqrt{6}.$$

Example 2. The roots of the equation $2z^2-3z+5 = 0$ are α and β. *Form the equations:* (i) *with roots* $\alpha+\mu, \beta+\mu$; (ii) *with roots* $\lambda\alpha, \lambda\beta$; (iii) *with roots* α^2, β^2.

(i) If $w = z+\mu$, then, when $2z^2-3z+5 = 0$, $w = \alpha+\mu$ or $\beta+\mu$. Thus, $\alpha+\mu$ and $\beta+\mu$ are the roots of the equation

$$2(w-\mu)^2-3(w-\mu)+5 = 0,$$

i.e.
$$2w^2-w(3+4\mu)+2\mu^2+3\mu+5 = 0.$$

Alternatively, $\alpha+\mu$ and $\beta+\mu$ are the roots of the equation

$$w^2+Aw+B = 0,$$

where $\quad A = -(\alpha+\beta+2\mu), \quad B = (\alpha+\mu)(\beta+\mu) = \alpha\beta+\mu(\alpha+\beta)+\mu^2.$

Thus
$$A = \tfrac{3}{2}-2\mu, \quad B = \tfrac{5}{2}+\tfrac{3}{2}\mu+\mu^2,$$

leading to the same equation as before.

(ii) With reasoning similar to that of (i), we make the transformation $z \to (1/\lambda)\,w$: the equation

$$2w^2-3\lambda w+5\lambda^2 = 0$$

has roots $\lambda\alpha$, $\lambda\beta$.
Alternatively, the equation is

$$w^2+Cw+D = 0,$$

where $\quad C = -\lambda(\alpha+\beta) = -\tfrac{3}{2}\lambda; \quad D = \lambda^2\alpha\beta = \tfrac{5}{2}\lambda^2,$ etc.

(iii) If $w = z^2$, then either $z = w^{\frac{1}{2}}$ or $z = -w^{\frac{1}{2}}$. Thus we have

$$\begin{aligned}
&2z^2-3z+5 = 0\\
&\Leftrightarrow 2z^2+5 = 3z\\
&\Leftrightarrow \text{either } 2w+5 = 3w^{\frac{1}{2}} \quad \text{or} \quad 2w+5 = -3w^{\frac{1}{2}}\\
&\Leftrightarrow \{2w+5-3w^{\frac{1}{2}}\}\{2w+5+3w^{\frac{1}{2}}\} = 0\\
&\Leftrightarrow 4w^2-11w+25 = 0,
\end{aligned}$$

on multiplying out and rearranging terms; and this last equation has roots α^2, β^2.

Alternatively, the required equation is

$$w^2-Ew+F = 0,$$

where
$$E = \alpha^2+\beta^2 = (\alpha+\beta)^2-2\alpha\beta, \quad F = \alpha^2\beta^2 = (\alpha\beta)^2,$$

giving
$$E = (\tfrac{3}{2})^2-5, \quad F = (\tfrac{5}{2})^2, \quad \text{etc.}$$

Ex. 2. If α, β are the roots of the equation $2z^2-3z+7 = 0$, find:

$$\text{(i) } \alpha^2+\beta^2; \quad \text{(ii) } \alpha^3+\beta^3; \quad \text{(iii) } |\alpha-\beta|; \quad \text{(iv) } \frac{1}{\alpha^2\beta}+\frac{1}{\alpha\beta^2}.$$

Ex. 3. If α, β are the roots of the equation $3z^2-5z+3 = 0$ form the equation (i) with roots $2\alpha, 2\beta$; (ii) with roots $\alpha-1, \beta-1$; (iii) with roots α^2, β^2.

2. QUADRATIC EQUATIONS WITH REAL COEFFICIENTS

If in the equation

$$az^2+bz+c = 0,$$

a, b, c are all real (and a is non-zero) all the previous results naturally still hold but the additional restriction placed upon the coefficients enables us to deduce further results. In this section it will be assumed, unless explicitly stated otherwise, that a, b, c are real numbers and $a \neq 0$.

**Ex.* 4. If $f(x) \equiv ax^2+bx+c$ and if $f(\alpha) > 0$, $f(\beta) < 0$ (α, β real) prove that the equation $ax^2+bx+c = 0$ has a root lying between α and β. Why is there no corresponding theorem if a, b, c are complex?

Theorem 21.1. *If the equation*

$$az^2+bz+c = 0 \quad (a, b, c \in R, a \neq 0)$$

has a complex root α, $\text{Im}(\alpha) \neq 0$, *then its other root is the complex conjugate* α^*.

Since $a\alpha^2+b\alpha+c = 0$, we have by taking the complex conjugate of both sides,

$$a^*(\alpha^*)^2+b^*\alpha^*+c^* = 0^*.$$

But $a^* = a, b^* = b, c^* = c, 0^* = 0$ and thus

$$a(\alpha^*)^2+b\alpha^*+c = 0$$

and the result follows.

**Ex.* 5. Explain why the proof of Theorem 21.1 breaks down if a, b, c are not restricted to the set R. Explain also why the condition $\text{Im}(\alpha) \neq 0$ was added in the enunciation of the theorem.

**Ex.* 6. Prove that a quadratic equation with real coefficients either has no non-real root or two distinct non-real roots.

Ex. 7. If α, β are two non-real numbers with the property that

$$\text{Im}(\alpha+\beta) = \text{Im}(\alpha\beta) = 0,$$

prove that α, β are complex conjugates.

There is an analogous result to Theorem 21.1 for irrational roots of a quadratic equation with rational coefficients.

Theorem 21.2. *If the equation*

$$az^2+bz+c = 0 \quad (a, b, c \in Q, a \neq 0)$$

has a root $p+q\sqrt{r}$ *where* p, q, r *are rational* ($q \neq 0$) *and* $\sqrt{r}$ *is irrational, then the other root of the equation is* $p-q\sqrt{r}$.

(Notice carefully the change in conditions for this theorem: a, b, c are now restricted to be rational numbers.)

$a(p+q\sqrt{r})^2+b(p+q\sqrt{r})+c$ must be of the form $P+Q\sqrt{r}$, where P and Q are rational. Thus

$$P+Q\sqrt{r} = 0,$$

since $p+q\sqrt{r}$ is a root of the given equation.

But this is possible only if $P = Q = 0$ since $\sqrt{r}$ is irrational; thus

$$P-Q\sqrt{r} = 0$$

and it follows that $p-q\sqrt{r}$ is a root of the given equation.

Ex. 8. Explain why the proof of Theorem 21.2 breaks down if a, b, c are not restricted to the set Q. Explain also why the condition $q \neq 0$ was added in the enunciation of the theorem, and where the fact that $\sqrt{r}$ is irrational is used.

Exercise 21(a)

1. Write down the sum and product of the roots for each of the following equations:
(i) $z^2-3z-7 = 0$; (ii) $2z^2-4z+11 = 0$; (iii) $(1+j)z^2-z+(1-j)=0$; (iv) $(z+j)^2 = (4-j)z$; (v) $(z-1+j)^2+z^2 = (2jz-1)^2$.

2. Write down the equations with roots:
(i) $-2, 3$; (ii) $-\frac{1}{2}, \frac{2}{3}$; (iii) $3-\sqrt{5}, 3+\sqrt{5}$; (iv) $2-\sqrt{2}, 3+2\sqrt{2}$; (v) $3-4j, 3+4j$; (vi) $\frac{1}{2}(1-j\sqrt{3}), \frac{1}{2}(1+j\sqrt{3})$; (vii) $1+2j, 1-j$; (viii) $\sqrt{2}-j, 1-\sqrt{2}j$.

3. Solve the following equations, giving the real and imaginary parts of the roots correct to 2 significant figures. Check your solutions by calculating the approximate product of the roots (using a slide rule):
(i) $2z^2-3z-7 = 0$; (ii) $2z^2-3z+7 = 0$; (iii) $z^2+(1+j)z+(1-2j) = 0$.

4. Find $\alpha+\beta$, $\alpha\beta$, $\alpha^2+\beta^2$, $(1/\alpha)+(1/\beta)$ and $\alpha^4\beta+\alpha\beta^4$:
(i) when α, β are the roots of the equation $z^2-3z-9 = 0$;
(ii) when α, β are the roots of the equation $3z^2-z-5 = 0$;
(iii) when α, β are the roots of the equation $z^2-jz-(1-j) = 0$.

5. α, β are the roots of the equation $2z^2-9z-4 = 0$. Find the values of
(i) $(1/\alpha)+(1/\beta)$; (ii) $(\alpha/\beta^2)+(\beta/\alpha^2)$; (iii) $|\alpha-\beta|$; (iv) $|\alpha^2-\beta^2|$.

6. α, β are the roots of the equation $3z^2-2z-7 = 0$. Find the values of:
(i) $(\alpha+k)(\beta+k)$; (ii) $\alpha^3+\beta^3$; (iii) $(1+\alpha)^{-1}+(1+\beta)^{-1}$.

7. α, β are the roots of the equation $(z-a)^2 = 4zb$. Find:
(i) $\alpha^2+\beta^2$; (ii) $(1/\alpha)+(1/\beta)$; (iii) $|\alpha-\beta|$.

8. The equations

$$\begin{cases} b^2x^2+a^2y^2 = a^2b^2, \\ y = mx+c \end{cases}$$

are solved simultaneously to give two pairs of solutions $x = x_1, y = y_1$ and $x = x_2, y = y_2$. Find $\frac{1}{2}(x_1+x_2)$ and $\frac{1}{2}(y_1+y_2)$.

9. If α, β are the roots of the equation $2z^2-z-7 = 0$, form the equations with roots:
(i) $\alpha-1, \beta-1$; (ii) $10\alpha, 10\beta$; (iii) $(1/\alpha), (1/\beta)$; (iv) α^2, β^2.

10. If α, β are the roots of the equation $(1+\mathrm{j})\,z^2-2\mathrm{j}z+(2-3\mathrm{j}) = 0$, form the equations with roots:
(i) $\alpha-2\mathrm{j}, \beta-2\mathrm{j}$; (ii) $(1+\mathrm{j})\,\alpha, (1+\mathrm{j})\,\beta$; (iii) α^2, β^2.

11. If α, β are the roots of the equation $2z^2-5z+4 = 0$, form the equations with roots:
(i) $\alpha-\beta, \beta-\alpha$; (ii) $\alpha+\mathrm{j}\beta, \beta+\mathrm{j}\alpha$; (iii) $\alpha^2+\alpha, \beta^2+\beta$.

12. The roots of the equation $4z^2+az-37 = 0$ differ by 1; find the possible values of a.

13. The roots of the equation $z^2-az+9\mathrm{j} = 0$ are λ and $\mathrm{j}\lambda$. Find the possible values of a.

14. The roots of the equation $(2-\mathrm{j})\,z^2+(3+\mathrm{j})\,z-4-5\mathrm{j} = 0$ are α and β. Form the equations with roots (i) $(1/\alpha), (1/\beta)$; (ii) α^*, β^*.

15. If the roots of the equation $az^2+bz+c = 0$ are α, β prove that

$$-aw^2+b\mathrm{j}w+c \equiv a(\mathrm{j}w-\alpha)\,(\mathrm{j}w-\beta).$$

16. If a, b are real and non-zero and $z^2+az+b = 0$, where z is non-real, prove that

$$|z| = 1 \Rightarrow b = 1 \quad\text{and}\quad |a| \leqslant 2.$$

17. Discuss the application of the method of completing the square to the vector quadratic equation

$$a\mathbf{r}.\mathbf{r}+\mathbf{b}.\mathbf{r}+c = 0 \quad (a \neq 0).$$

18. The two equations

$$z^2+az+b = 0,$$

$$z^2+Az+B = 0$$

have a common root. Prove that

$$b(a-A)^2-a(a-A)\,(b-B)+(b-B)^2 = 0.$$

19. If one root of the equation $az^2+bz+c = 0$ is the square of the other, find a relation connecting a, b, c.

20. If one root of the equation $az^2+bz+c = 0$ is j times the other, find a relation connecting a, b, c.

21. Prove that, if one of the roots of the equation $z^2+za+1 = 0$ has unit modulus then a is real.

3. THE QUADRATIC FUNCTION

A function $f: R \to R$ defined by

$$f(x) = ax^2+bx+c \quad (a, b, c \in R)$$

is called a (*real*) *quadratic function.* Since in this section we shall be primarily concerned with inequalities, we confine our attention to real quadratic functions and the coefficients a, b, c as well as the variable x, will be assumed to be real.

We now derive a necessary and sufficient condition for $f(x)$ to be positive for *all* real values of x.

Theorem 21.3. *If a is not zero, then*

$$ax^2+bx+c > 0 \quad \textit{for all real } x \Leftrightarrow a > 0 \quad \textit{and} \quad b^2-4ac < 0.$$

Proof. Write $\qquad y = ax^2+bx+c.$

Then
$$y = a\left\{x^2+\frac{b}{a}x+\frac{b^2}{4a^2}+\frac{c}{a}-\frac{b^2}{4a^2}\right\}$$

$$= a\left\{\left(x+\frac{b}{2a}\right)^2+\frac{4ac-b^2}{4a^2}\right\}. \tag{1}$$

(i) If $a > 0$ and $b^2 < 4ac$, then $y > 0$ from (1), since

$$\left(x+\frac{b}{2a}\right)^2 \geqslant 0 \text{ for all } x.$$

(ii) If $ax^2+bx+c > 0$ for all x, then certainly $a > 0$, for we may choose x such that

$$\left(x+\frac{b}{2a}\right)^2+\frac{4ac-b^2}{4a^2} > 0.$$

But, if $b^2 > 4ac$, the equation $ax^2+bx+c = 0$ has real roots, α, β say, and

$$ax^2+bx+c = a(x-\alpha)(x-\beta),$$

from which it follows that $y < 0$ if $\alpha < x < \beta$. But we are given that y is always positive; thus $b^2 < 4ac$.

**Ex.* 9. Give a geometrical demonstration of the truth of Theorem 21.3.

**Ex.* 10. Prove that

$$ax^2+bx+c < 0 \quad \text{for all real } x \Leftrightarrow a < 0 \quad \text{and} \quad b^2-4ac < 0.$$

The quantity $\Delta = b^2-4ac$ occupies a position of central importance when considering the quadratic expression ax^2+bx+c or the quadratic equation $ax^2+bx+c = 0$; Δ is called the *discriminant* of the expression ax^2+bx+c.

**Ex.* 11. If the roots of the quadratic equation $ax^2+bx+c = 0$ are α, β, prove that $a^2(\alpha-\beta)^2 = \Delta$. If the coefficients are real, what does this tell us about the reality of the roots of the quadratic equation?

Example 3. *Determine the range of the function f: $R \to R$ defined by*

$$f(x) = \frac{x+a}{x^2+ax+1}$$

for different values of a.

If $x^2+ax+1 = 0$, $f(x)$ is undefined; excluding this case, if y is a value assumed by the given expression then

$$y(x^2+ax+1) = x+a,$$

or

$$yx^2+(ay-1)\,x+(y-a) = 0.$$

Thu sy is a possible image of the function f if this quadratic† in x has real roots; that is, if

$$(ay-1)^2 \geqslant 4y(y-a)$$

or

$$E \equiv (a^2-4)\,y^2+2ay+1 \geqslant 0.$$

To determine what values of y satisfy this inequality we have, by Theorem 21.3, to consider the discriminant of E and the sign of the coefficient of y^2. The discriminant of E is

$$\Delta = 4a^2-4(a^2-4) = 16$$

and thus, since $\Delta > 0$, the equation $E = 0$ has real roots α, β (where, in fact, $\alpha = (a+2)^{-1}$ and $\beta = (a-2)^{-1}$). Thus we have

$$E \equiv (a^2-4)\,(y-\alpha)\,(y-\beta),$$

provided $|a| \neq 2$.

The coefficient of y^2 in E shows us that the critical values of a are ± 2.

(i) $|a| < 2$:

For E to be greater than or equal to zero, y must lie between α and β; the range is thus

$$\{y \in R: \alpha \leqslant y \leqslant \beta\}.$$

(ii) $|a| > 2$:

By a similar argument, the range is

$$\{y \in R: y \leqslant \alpha \quad \text{or} \quad y \geqslant \beta\}.$$

† If $y = 0$, the equation is no longer a quadratic, but the subsequent inequality is still satisfied for all a.

(iii) $a = 2$:

$E = 4y+1$ and thus the condition $E \geqslant 0$ gives us the range

$$\{y \in R: y \geqslant -\tfrac{1}{4}\}.$$

(iv) $a = -2$:

By a similar argument, the range is

$$\{y \in R: y \leqslant \tfrac{1}{4}\}.$$

Ex. 12. Illustrate the solution of Example 3, by drawing rough sketches of the curve $y = (x+a)/(x^2+ax+1)$ in the separate cases $a = -4$, $a = -2$, $a = 0$, $a = 2$, $a = 4$. [Express y^{-1} in the form $z = y^{-1} = x+(x+a)^{-1}$ and deduce that the graph of z has turning points at $x+a = \pm 1$; hence sketch the y curve.]

Miscellaneous Exercises 21

1. If α, β are the roots of the equation $z^2-5z-11 = 0$ form the equation (i) with roots $\alpha+(\beta/\alpha)$, $\beta+(\alpha/\beta)$; (ii) with roots $(\alpha+j\beta, j\alpha+\beta)$.

2. If α, β are the roots of the equation $az^2+bz+c = 0$, find the roots of the equation

$$a^2z^2+(2ca-b^2)z+c^2 = 0 \quad \text{in terms of} \quad \alpha, \beta.$$

3. If a, b, c, d are real numbers, prove that the roots of the equation

$$\begin{vmatrix} a-z & b \\ c & d-z \end{vmatrix} = 0$$

can be complex only if b and c are of opposite signs.

4. Show that, if x is real,

$$\tfrac{7}{3} \leqslant \frac{3x^2+x+3}{x^2+x+1} \leqslant 5.$$

5. Show that, if x is real, the expression

$$\frac{2x^2-13x-7}{x^2-3x}$$

cannot assume any value between 25/9 and 9.

6. Show that, if x is real, the expression

$$\frac{x^2-4x+3}{x^2-6x+8}$$

can attain all real values.

7. If α, β are the roots of the equation $ax^2+bx+c = 0$ express the roots of the equation

$$ac(x^2+1)-(b^2-2ac)x = 0$$

in terms of α and β.

Prove that, if x_1, x_2 are roots of the equation

$$(x^2+1)(a^2+1)-nax(ax+1) = 0,$$

then

$$(x_1^2+1)(x_2^2+1) = nax_1x_2(x_1+x_2).$$

(O & C)

8. x, y are real numbers.

(i) Show that, if x and y are connected by the equation

$$x^2+y^2-2x+4y-4 = 0,$$

then $-2 \leqslant x \leqslant 4$ and $-5 \leqslant y \leqslant 1$.

(ii) Show that, if x and y are connected by the equation

$$x^2+xy-2y^2-3x+3y+9 = 0,$$

then x may have *any* value, but y cannot lie between -1 and 3.

9. Show that, when

$$y(x-1)(x-3) = x-2,$$

x is real for every real value of y.

Sketch the graph of y as a function of x, indicating the behaviour of the graph when x or y is large. (O & C)

10. Two quadratic equations have roots α, β and $\alpha+\lambda$, $\beta+\lambda$ respectively; prove that their discriminants are equal, provided they are written with their leading coefficients as unity.

11. If a, b, c, p, q are all real and if the solutions of the simultaneous equations

$$px+qy = 1, \quad ax^2+by^2 = 1$$

are all real, prove that $ab \leqslant aq^2+bp^2$.

12. If a, b, h, λ are all real, discuss the reality of the solution of the simultaneous equations

$$ax^2+2hxy+by^2 = \lambda,$$

$$x^2+y^2 = 1.$$

by putting $x = \cos\theta$, $y = \sin\theta$ and considering a quadratic equation in $\tan\theta$.

13. Show that, if

$$y(x-2)(x-3) = x-1,$$

then x is not real when y has any value between $-3-2\sqrt{2}$ and $-3+2\sqrt{2}$. (O & C)

14. Show that, if p, q, r are not all equal and if

$$y = \frac{x-p}{(x-q)(x-r)},$$

then x is real for every real value of y provided that

$$(p-q)(p-r) \leqslant 0.$$

(O & C)

15. If α, β are the roots of

$$x^2-2ax+ab+c = 0$$

and γ, δ are the roots of

$$x^2-2bx+ab-c = 0,$$

prove that

$$\tfrac{1}{2}(\alpha-\beta)(\gamma-\delta) = (\alpha-\delta)(\gamma-\beta) = (\alpha-\gamma)(\beta-\delta).$$

Hence, or otherwise, prove that, if the quadratic equations have a common root, it is a repeated root of one of them. (O & C)

22. *The parabola and rectangular hyperbola*

1. SETS OF POINTS IN A PLANE DEFINED BY PARAMETERS

The reader has already met (Chapter 3) the parametric representation of a line in two and three dimensions. For example, given coordinate axes Ox, Oy in a plane, the set of points

$$S = \{(x, y): x = 2t-1, y = t+2; t \text{ real}\}$$

represents a line. The set of points may be written in the alternative form

$$S = \{(x, y): x-2y+5 = 0\},$$

for

$$\frac{x+1}{2} = y-2 = t,$$

$$\Leftrightarrow x-2y+5 = 0$$

(Notice that, to show that the parametric equations

$$x = 2t-1, \quad y = t+2$$

represent the line

$$x-2y+5 = 0,$$

it is insufficient merely to verify that the point $(2t-1, t+2)$ lies on the line: it is necessary also to show that *any* point of the line has coordinates of this form. See Ex. 29 below.)

Ex. 1. Show that, if both x and y are given as linear expressions in a parameter t, then the equation connecting x and y is linear and thus represents a straight line.

Ex. 2. Find the Cartesian equation of the line given by the parametric form

$$x = 3t-2, \quad y = 2t+3;$$

and find a parametric form for the line with Cartesian equation

$$3x-5y-7 = 0.$$

If the expressions for x and y involve non-linear expressions in a parameter t, the set of points so defined generally constitute a curve. For example,

$$\Gamma = \{(x, y): x = t, y = t^2; t \text{ real}\}$$

represents a curve, since the expression for y in terms of x is not linear.

Ex. 3. Show that the Cartesian equations of the curve Γ defined above is

$$y = x^2.$$

Distinguish between the curve Γ and the curve Γ' defined by

$$\Gamma' = \{(x, y): x = t^2, y = t^4; t \text{ real}\}.$$

Sketch Γ and Γ' for $-3 < x < 3$.

Ex. 4. Show that the set of points

$$\{(x, y) \in R: x = t^2, y = t^2 - 1; t \text{ real}\}$$

constitute a line segment. (Only in such artificial cases does a non-linear parametric expression yield a straight line.)

The expressions for x and y need not be algebraic: the circle with centre (0, 0) and radius 2 may be expressed in the parametric form

$$x = 2\cos\theta, \quad y = 2\sin\theta.$$

Ex. 5. Find the equation of the curve with parametric representation $x = a\sec\theta$, $y = b\tan\theta$.

The use of a parametric representation greatly facilitates the study of the geometry of curves with known equations. This is because the position of a point on the curve may be specified by one variable (the parameter) rather than two variables (x and y) connected by an equation (the equation of the curve).

2. TRANSLATION OF COORDINATE AXES

The equation of a given curve naturally depends upon the choice of origin and axes. Equations can often be much simplified by retaining the directions of the x and y axes (defined by the unit vectors $\mathbf{i}$ and $\mathbf{j}$) but shifting the origin to another point O'. Such a change in coordinate system is referred to as a *translation of coordinate axes.*

Suppose we are given a pair of perpendicular unit vectors $\mathbf{i}$ and $\mathbf{j}$ and a point P with coordinates (x, y) referred to the origin O and axes Ox, Oy in the directions $\mathbf{i}, \mathbf{j}$. Then $\mathbf{r} = \mathbf{OP} = x\mathbf{i}+y\mathbf{j}$ is the position vector of P with respect to O. Now let O' be the point (h, k) referred to the axes Ox, Oy (Figure 22.1). Then

$$\mathbf{a} = \mathbf{OO'} = h\mathbf{i}+k\mathbf{j}$$

is the position vector of O' with respect to O. It follows (Figure 22.1) that

$$\mathbf{s} = \mathbf{O'P} = \mathbf{r}-\mathbf{a} = (x-h)\,\mathbf{i}+(y-k)\,\mathbf{j}$$

is the position vector of P with respect to O'. Put another way, if O' is taken as the origin, with axes $O'x'$, $O'y'$ in directions defined by $\mathbf{i}, \mathbf{j}$ then the coordinates (x', y') of P with respect to the new axes and origin O' are given by

$$x' = x-h, \quad y' = y-k.$$

Ex. 6 The coordinates of A, B, C, D referred to perpendicular axes through O are respectively (1, 2), (−2, 3), (0, 4) and (−1, −5). What are their coordinates referred to a pair of parallel axes through O' (2, −1)?

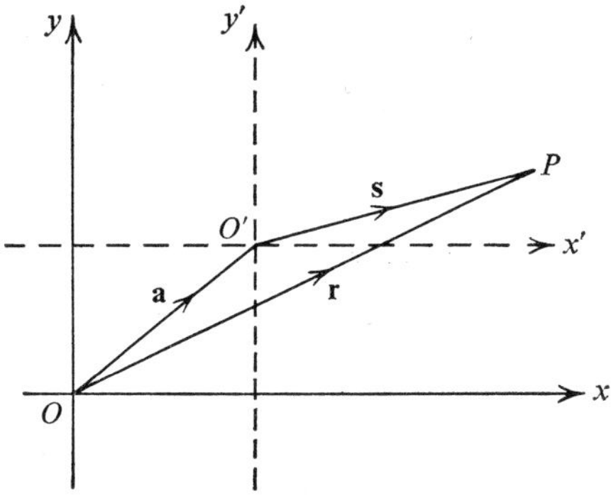

Fig. 22.1

Example 1. *Show, by a translation of coordinates to a new origin* O', *whose coordinates in the original system were* (2, −3), *that the equation*

$$x^2+xy-2y^2-x-14y-20 = 0$$

represents a pair of straight lines.

Denoting new coordinates by dashes we have

$$x' = x-2, \quad y' = y+3,$$

$$\Rightarrow x = x'+2, \quad y = y'-3,$$

and the given equation becomes

$$(x'+2)^2+(x'+2)(y'-3)-2(y'-3)^2-(x'+2)-14(y'-3)-20 = 0,$$

which simplifies to

$$x'^2+x'y'-2y'^2 = 0.$$

But the expression on the left of this equation factorizes:

$$(x'+2y')(x'-y') = 0;$$

and thus the equation represents the pair of straight lines

$$x'+2y' = 0, \quad x'-y' = 0$$

through O'.

Referred to Ox, Oy, the equations are

$$(x-2)+2(y+3) = 0 \quad \text{and} \quad (x-2)-(y+3) = 0,$$

or

$$x+2y+4 = 0 \quad \text{and} \quad x-y-5 = 0.$$

Exercise 22(a)

Find the equations of the curves given parametrically in Questions 1–6.

1. $x = \cos\theta$, $y = 2\sin\theta$.

2. $x = t^2$, $y = t^3$.

3. $x = 1-t$, $y = 1+t+t^2$.

4. $x = \cos\theta-\sin\theta$, $y = \cos\theta+\sin\theta$.

5. $x = \dfrac{at^2}{1+t^3}$, $y = \dfrac{at}{1+t^3}$.

6. $x = t+1/t$, $y = 1+t^2$.

7. Show that the parametric form

$$x = \frac{1-t^2}{1+t^2}, \quad y = \frac{2t}{1+t^2}$$

represents the curve

$$x^2+y^2 = 1,$$

less the point $(-1, 0)$.

8. If the axes are translated to $(1, -4)$ as new origin, find the new coordinates of the points:
(i) $(0, 0)$; (ii) $(3, -4)$; (iii) $(3, 4)$; (iv) $(5, -5)$.

9. By translating the axes to pass through $(1, 1)$ as new origin prove that the equation

$$x^2-4y^2-2x+8y-3 = 0$$

represents a pair of straight lines, and find their gradients.

10. Prove that the equation

$$2x^2+3xy-2y^2-17x+6y+8 = 0$$

represents a pair of straight lines intersecting in the point $(2, 3)$.

11. Show, by a translation of axes to pass through (p, q) as new origin, that the equation

$$ax^2+2hxy+by^2+2gx+2fy+c = 0$$

is transformed into the equation

$$ax'^2+2hx'y'+by'^2+d = 0$$

if p, q are chosen so that

$$ap+hq+g = 0,$$
$$hp+bq+f = 0.$$

Deduce that the equation

$$ax^2+2hxy+by^2+2gx+2fy+c = 0$$

represents a pair of straight lines only if det **A** $= 0$ where

$$\mathbf{A} = \begin{pmatrix} a & h & g \\ h & b & f \\ g & f & c \end{pmatrix}.$$

12. Use the results of Question 11 to show that the equation

$$6x^2+5xy+y^2-x-1 = 0$$

represents a pair of straight lines, and find their point of intersection.

3. THE PARABOLA

A parabola is defined as the set of points in a plane whose distances from a fixed point, S, are equal to their distances from a fixed line l not containing S: see Figure 22.2. For the two typical points P_1 and P_2,

$$SP_1 = P_1M_1, \quad SP_2 = P_2M_2.$$

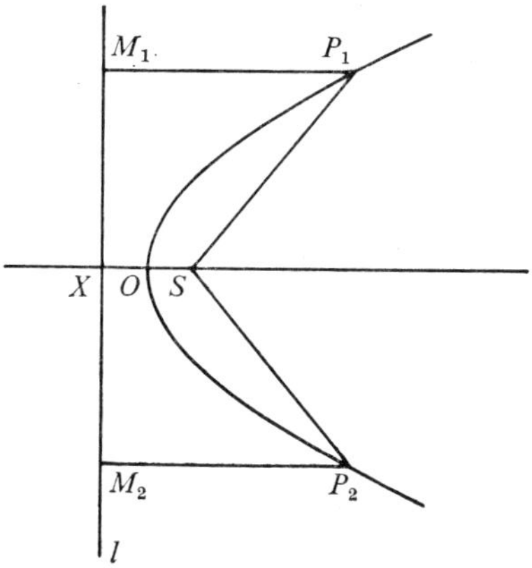

Fig. 22.2

The curve is clearly symmetrical about the line through S perpendicular to l; this line is called the *axis* of the parabola. The axis cuts the parabola at O, the *vertex* of the parabola.

Ex. 7. In Figure 22.2 why does $SO = OX$?

The point S referred to in the definition is the *focus* of the parabola; the line l is the *directrix*. From the symmetry of the figure about the axis, the tangent at the vertex is parallel to the directrix.

Ex. 8. If the chord PQ of a parabola, when produced, cuts the directrix at K, prove that SK bisects the exterior angle between SP and SQ. What can you say about the angle PSZ, where Z is the point at which the tangent at P cuts the directrix? What does this tell you about the angle SPM, where M is the foot of the perpendicular from P on to the directrix?

Ex. 8 shows that a number of properties of the parabola may be derived very quickly from the definition by pure geometry, but the further study of the curve is best made by analytical methods which we shall now develop.

As has been pointed out already, the equation of a parabola depends not only upon the relative positions of the point S and the line l but also upon the choice of coordinate axes. Let us suppose (Figure 22.3) that

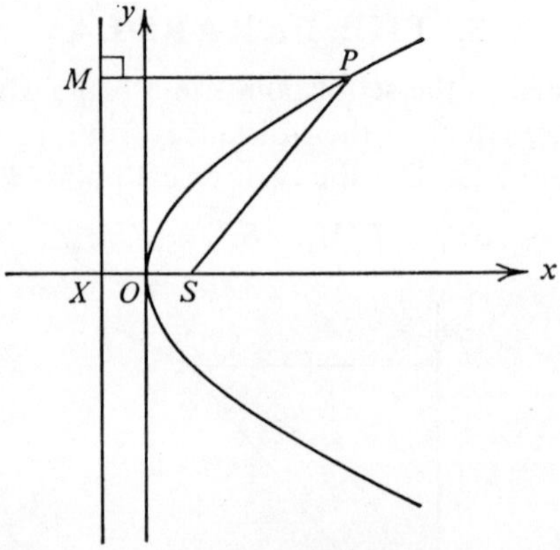

Fig. 22.3

$SX = 2a$. The fact that SX is an axis of symmetry suggests strongly that this line should be taken as one of the axes of coordinates. The absence of a perpendicular axis of symmetry leaves us without an immediately obvious choice for second axis; however, if we take the origin to be a point of the curve, we shall avoid a constant term in the equation. Thus we shall take O as the origin, OS as the x axis and the perpendicular through O as the y axis. Now suppose that P is any point on the parabola, Γ, and that the coordinates of P referred to our axes of coordinates are (x, y); let M be the foot of the perpendicular from P on to l.

Since $XO = OS$, S is the point $(a, 0)$ and l has equation $x+a = 0$. Thus

$$SP^2 = (x-a)^2+y^2,$$

$$PM^2 = (x+a)^2,$$

and

$$\Gamma = \{(x, y): (x-a)^2+y^2 = (x+a)^2\}$$

$$= \{(x, y): y^2 = 4ax\}.$$

Thus, *referred to its axis as the x axis, its vertex as the origin and S as the point* $(a, 0)$, *the equation of a parabola is*

$$y^2 = 4ax. \tag{1}$$

**Ex.* 9. Show that $x = at^2, y = 2at$ is a parametric form for the parabola $y^2 = 4ax$. Need a be positive in this equation?

From the equation derived above it may be seen that all parabolas have a similar shape but differ in size (depending upon the distance of the focus from the directrix). The chord of the parabola through S and perpendicular to the axis is called the *latus rectum*; the length of the latus rectum determines the size of the parabola.

**Ex.* 10. Show that the length of the latus rectum is $2SX$.

**Ex.* 11. What is the length of the latus rectum of the parabola $y^2 = 4ax$? Answer the same question for the parabolas $y^2 = -ax$ and $x^2 = 3ay$.

Ex. 12. Using the same axes, plot, on graph paper, the parabola $y^2 = 4ax$, for $a = 2, 1, \frac{1}{2}, \frac{1}{10}, -1, -2$.

**Ex.* 13. By writing the equation

$$y^2 - 4ax - 2ay - 3a^2 = 0$$

in the form

$$(y-a)^2 = 4a(x+a)$$

and translating the axes to the new origin $(-a, a)$, show that it represents a parabola of latus rectum $4a$. Show also that, with the original coordinate system, the vertex is the point $(-a, a)$, the focus $(0, a)$ and the directrix $x+2a = 0$.

Ex. 14. Show that the equation

$$y^2 - 2ax - 2ay + 5a^2 = 0$$

represents a parabola and find its focus and directrix. Answer the same question for the equations:

(i) $y^2 + 3ax + 6a^2 = 0$;   (ii) $x^2 + 2ax - 8ay - 7a^2 = 0$.

The chord joining the points $P_1(at_1^2, 2at_1)$, $P_2(at_2^2, 2at_2)$ of the parabola is easily obtained. Its gradient, m, is given by

$$m = \frac{2at_2 - 2at_1}{at_2^2 - at_1^2} = \frac{2}{t_1+t_2},$$

and thus its equation is

$$y - 2at_1 = \frac{2}{t_1+t_2}(x - at_1^2)$$

which reduces to

$$2x - (t_1+t_2)y + 2at_1t_2 = 0. \tag{2}$$

By letting $t_2 \to t_1$ this leads immediately to the equation of the tangent at $P_1(at_1^2, 2at_1)$:

$$x - t_1 y + at_1^2 = 0. \tag{3}$$

**Ex.* 15. Interpret the parameter t_1 geometrically.

**Ex.* 16. P_1 is the point $(at_1^2, 2at_1)$, P_2 is the point $(at_2^2, 2at_2)$. Show that P_1P_2 passes through the focus, S, of the parabola if and only if $t_1 t_2 = -1$. (A chord which passes through the focus is called a *focal chord*.)

**Ex.* 17. Obtain the gradient of the tangent to the parabola $y^2 = 4ax$ at the point $(at^2, 2at)$ by differentiating the equation $y^2 = 4ax$.

**Ex.* 18. If $lx + my + na = 0$ cuts the parabola at the points $P_1(at_1^2, 2at_1)$, $P_2(at_2^2, 2at_2)$, show that t_1, t_2 are the roots of the quadratic equation

$$lt^2 + 2mt + n = 0.$$

Write down the values of m/l, n/l in terms of t_1 and t_2 and deduce the equation of the chord P_1P_2. How may this method be modified to obtain the equation of the tangent at $P(at^2, 2at)$?

**Ex.* 19. Show that the line

$$(x - x_1)/l = (y - y_1)/m = \lambda$$

meets the parabola $y^2 = 4ax$ at points with parameters λ_1, λ_2 which are roots of the quadratic equation

$$m^2\lambda^2 + (2my_1 - 4al)\lambda + (y_1^2 - 4ax_1) = 0.$$

Deduce that, if (x_1, y_1) lies on the parabola and the given line is a tangent, then

$$2my_1 - 4al = 0.$$

Hence show that the tangent at (x_1, y_1) has equation

$$(x - x_1)/y_1 = (y - y_1)/2a.$$

The equation of the *normal* to the parabola $y^2 = 4ax$ at the point $P(at^2, 2at)$ (that is, the line through P perpendicular to the tangent) is now easily found. For, since the gradient of the tangent is $1/t$, the gradient of the normal must be $-t$ and thus its equation is

$$y - 2at = -t(x - at^2),$$

which reduces to

$$tx + y = 2at + at^3. \tag{4}$$

Ex. 20. Show that, if the normal to the parabola $y^2 = 4ax$ at the point

$$P(ap^2, 2ap)$$

cuts the parabola again at the point $Q(aq^2, 2aq)$, then q is a root of the quadratic equation in λ

$$p\lambda^2 + 2\lambda - 2p - p^3 = 0.$$

What is the other root of this equation? Find q in terms of p.

The parabola is rich in geometrical properties which may be easily derived analytically from the results already found. We shall content ourselves here by proving two famous results; the reader will find further examples in Exercise 22(*b*).

In proving the following two properties, we shall take the parabola in the standard form $y^2 = 4ax$ by making the vertex the origin, the axis of the parabola the x axis, the tangent at the vertex the y axis, and writing the length OS as a.

(i) *Tangents at the extremities of a focal chord meet at right-angles on the directrix.*

The chord joining $P_1(at_1^2, 2at_1)$ and $P_2(at_2^2, 2at_2)$ is, from equation (2),

$$2x-(t_1+t_2)\,y+2at_1t_2 = 0.$$

If P_1P_2 is a focal chord, $(a, 0)$ lies on this line and thus

$$t_1t_2 = -1.$$

From equation (3), the gradient of the tangents at P_1 and P_2 are $1/t_1$ and $1/t_2$ and thus, since $t_1t_2 = -1$ for a focal chord, tangents at the extremities of such a chord certainly meet at right-angles.

On solving the equations

$$x-t_1y+at_1^2 = 0,$$
$$x-t_2y+at_2^2 = 0,$$

we obtain the point of intersection R of the two tangents in the form $(at_1t_2, a(t_1+t_2))$. For a focal chord P_1P_2, this reduces to $(-a, a(t_1+t_2))$ and thus R lies on the directrix. (See Ex. 22 for an alternative method of obtaining the coordinates of R.)

(ii) *If P is any point on the parabola, focus S, and M is the foot of the perpendicular from P on to the directrix, then the angle SPM is bisected by the tangent at P.*

The reader who has successfully completed Ex. 8 will have already proved this property by pure geometry.

In Figure 22.4 we have

$$\begin{aligned}\tan\alpha &= \text{gradient of tangent at } P\,(at^2, 2at)\\ &= 1/t, \text{ from equation (2)},\\ \tan\beta &= \text{gradient of } SP\\ &= \frac{2at}{at^2-a}\\ &= \frac{2t}{t^2-1}\\ &= \frac{2(1/t)}{1-(1/t)^2}.\end{aligned}$$

Thus
$$\tan\beta = \frac{2\tan\alpha}{1-\tan^2\alpha}$$
$$= \tan 2\alpha$$
and the result follows.

Property (ii) is usually known as the *parabolic reflection property* since it shows that a ray of light emanating from the focus will be reflected by a parabolic mirror along a line parallel to the axis.

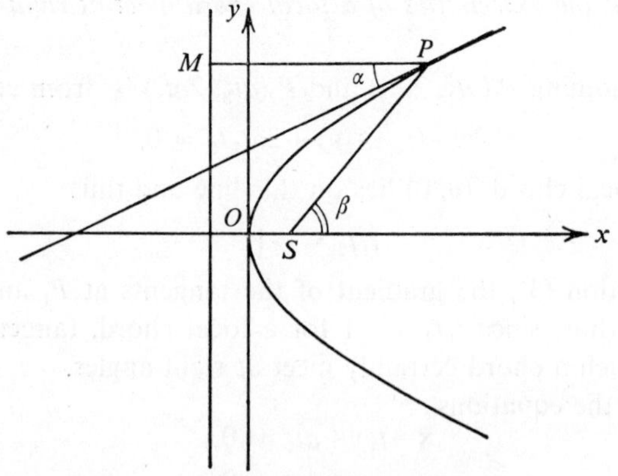

Fig. 22.4

Ex. 21. Explain the use of parabolic reflection in car headlights and reflecting telescopes. Explain, with the aid of diagrams, the 'headlights on' and 'headlights dipped' positions.

Parabolas frequently occur both in nature and in design. For example, a first approximation to the path of a projectile moving under gravity is a parabola. Again, in architecture, it is becoming quite common to find roofs of buildings designed so that certain cross-sections are parabolic. The curves of suspension bridges are also approximately arcs of a parabola.

We conclude this section with an example illustrating the method of determining a locus associated with a parabola.

Example 2. *PQ is a focal chord of a parabola, focus S. Show that the locus of the mid-point of PQ is another parabola and locate its focus and directrix.*

Set up coordinate axes with the vertex, O, as origin, the axis of the parabola as x axis, the tangent at the vertex as y axis and call the length of OS, a. The equation of the parabola is then

$$y^2 = 4ax.$$

Any line through the focus, $S(a, 0)$, other than $x-a=0$, has the form

$$y = m(x-a).$$

(Notice that m is a parameter: different values of m give different lines through the focus and all such lines, with the single exception mentioned above, may be put in this form.)

The line $y = m(x-a)$ cuts the parabola $y^2 = 4ax$ in points with y coordinates y_1 and y_2 given by the roots of the equation

$$y^2 = 4a\left(\frac{y}{m}+a\right),$$

that is, by $$my^2-4ay-4a^2m = 0.$$

Let (h, k) be the mid-point of this focal chord; then

$$k = \tfrac{1}{2}(y_1+y_2) = 2a/m$$

by a property of the roots of a quadratic equation.

Again, since (h, k) lies on the chord,

$$k = m(h-a)$$

or $$h = a+(2a/m^2),$$

on substituting for k.

The locus is thus given parametrically by

$$x = a+(2a/m^2), \quad y = (2a/m).$$

Elimination of m gives this locus as the curve

$$y^2 = 2a(x-a).$$

(Strictly speaking, to complete the question we should also prove that any point of the curve $y^2 = 2a(x-a)$ is a point of the locus. The reader may care to supply the details.)

By translating the axes to the new origin $(a, 0)$ it is easily seen that this equation represents a parabola of latus rectum $2a$. In the original coordinate system, the focus is $S'(\frac{3}{2}a, 0)$ and the directrix, l', $x = \frac{1}{2}a$. The directrix of the locus cuts the common axis at A where $OA = AS = SS'$ (see Figure 22.5).

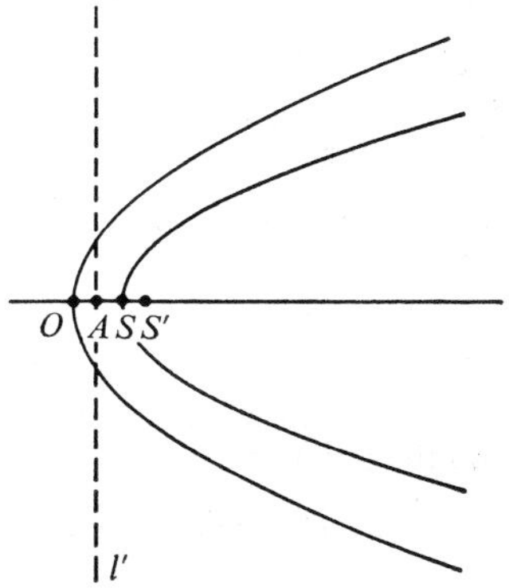

Fig. 22.5

Notes on Example 2.

(i) The use of symmetric functions of the roots of a quadratic equation is frequently applicable to problems involving the parabola and has the advantage of retaining symmetry in the calculation.

(ii) An alternative method of solving the question would have been to write P as the point $(at_1^2, 2at_1)$ and Q as the point $(at_2^2, 2at_2)$. The coordinates of R could then be found in terms of t_1 and t_2, again giving the locus of R parametrically.

(iii) If LL' is the latus rectum of the given parabola, the mid-point of the particular chord LL' is S which must therefore lie on the required locus, giving us a check on our working. Again, the symmetry of the locus about the x axis was to be expected from the symmetry of the parabola about its axis.

Ex. 22. Show that if the tangent at the point $P(at^2, 2at)$ passes through the point $R(h, k)$ then t is a root of the quadratic equation

$$a\lambda^2 - k\lambda + h = 0.$$

Deduce that the coordinates of the meet of tangents at $P_1(at_1^2, 2at_1)$ and $P_2(at_2^2, 2at_2)$ is $R(at_1 t_2, a(t_1+t_2))$.

Exercise 22(b)

1. Find the latus rectum, vertex, focus, and directrix of each of the following parabolas:

(i) $(y-2)^2 = 4(x-3)$; (ii) $(x-3)^2 = 2(y+1)$;
(iii) $(y+1)^2 = -(x-1)$; (iv) $(x+2)^2 = 3(y+3)$;
(v) $(y+3)^2 = 4(x-5)$; (vi) $(x-2)^2 = 8(y-2)$;
(vii) $(y-4)^2+2x = 0$.

2. Find the equations of the parabolas having the given points as foci and the given lines as directrices:

(i) $(2, 1)$, $x = 0$; (ii) $(-3, 2)$, $y = 3$.

3. Find the equations of the parabolas: (i) with focus $(4, 3)$ and vertex $(1, 3)$; (ii) with focus $(-1, -7/4)$ and vertex $(-1, -1)$.

4. Find the equations of the tangents to the following parabolas at the points given:

(i) $y^2 = 2(x+1)$, $(7, -4)$; (ii) $(y+2)^2 = x$, $(9, 1)$;
(iii) $y^2-3y-5x = 0$, $(2, 5)$; (iv) $x^2-2x-3y-21 = 0$, $(-3, -2)$.

5. Show that the line $x+y-2 = 0$ is a tangent to the parabola $x^2+8y = 0$ and find the point of contact.

6. Show that the line $5x+2y+8 = 0$ is a tangent to the parabola

$$x^2-2x-4y = 0,$$

and find the point of contact.

7. Any element P of the set S of points in a plane has a position vector of the form

$$\mathbf{r} = (t-1)\,\mathbf{i}+t^2\mathbf{j},$$

where t is a scalar parameter. Show that S defines a parabola and determine the position vectors of its vertex and focus.

8. A ball is thrown obliquely into the air in such a way that its position vector at time t is given by

$$\mathbf{r} = 10t\mathbf{i}+(10t-16t^2)\,\mathbf{k},$$

where $\mathbf{i}$ is a horizontal unit vector and $\mathbf{k}$ is a unit vector pointing vertically upwards. Show that the path of the ball is a parabola, and determine the distance of its directrix from the origin.

9. Any element P of the set S of points in a plane has a position vector of the form

$$\mathbf{r} = \cos 2\theta.\mathbf{i}-(1+\sin\theta).\mathbf{j},$$

where θ is a scalar parameter. Show that S forms part of a parabola with its axis parallel to $\mathbf{i}$ and sketch the set of points S.

Questions 10–15 refer to Figure 22.6. O is the vertex, S the focus of the parabola, AM is the directrix, $ANPM$ is a rectangle. The tangent at P cuts the axis at T and the normal at P cuts it at G. PT and SM intersect at Z. You may use any method which you regard as suitable.

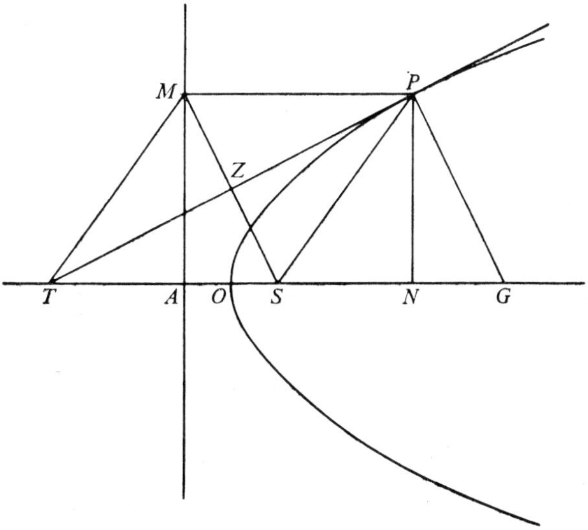

Fig. 22.6

10. Prove that SZ and PT are perpendicular and that Z lies on the tangent at the vertex. Suggest how, with the aid of a set square, this property may be used to draw a parabola.

11. Prove that $ST = SP$ and deduce that $SPMT$ is a rhombus. (Notice that this gives the parabolic reflector property.)

12. Prove that $ST = GS$ and deduce that NG is of fixed length.

13. Prove that NT is bisected at O.

14. Prove that $SZ^2 = SO.SP$.

15. Prove that $TZ = ZN$ and that $GP = 2SZ$.

16. PQ is a focal chord of a parabola. If the tangents at P and Q meet at T and the normals at P and Q meet at U prove that TU is parallel to the axis of the parabola.

17. If R is the mid-point of the focal chord PQ of a parabola and V is the foot of the perpendicular from R on to the directrix, prove that $RV = \frac{1}{2}PQ$.

18. P is a variable point on a parabola, focus S, and R is the mid-point of PS. Prove that the locus of R is a parabola and find its axis, focus and directrix. What checks can you apply to your result?

19. The chord PQ of a parabola subtends a right angle at the vertex. Find the locus of the mid-point of PQ.

20. In Figure 22.6 the line UOV is perpendicular to the tangent PT, meeting PT at U and the parabola at O and V. Prove that $UO.OV$ is constant for all positions of P on the parabola.

21. P is a variable point of the parabola $y^2 = 4ax$. Find the locus of the mid-point of OP and sketch in the same diagram the given parabola and the locus.

22. With the notation of Figure 22.6, K is the mid-point of PG. Prove that SK is parallel to TP and find the locus of K.

23. Prove that the equation of the tangent at the point $(at^2, 2at)$ on the parabola $y^2 = 4ax$ is

$$x - ty + at^2 = 0.$$

A and B are any two points on the parabola $y^2 = 4ax$, and the tangents at A and B meet at P. The line PM is parallel to the axis of the parabola and meets the line AB at M. Prove that M is the mid-point of AB.

If the parameters of the points A, B are t and $2t$ respectively and the tangents meet at P, find the coordinates of P and show that it always lies on the parabola $y^2 = 9ax/2$. (O & C)

24. Show that any line through the fixed point (ha, ka) may be written in the form

$$y - ka = m(x - ha),$$

where m is a parameter.

Deduce that, if the chord PQ of the parabola $y^2 = 4ax$ passes through a fixed point, then the tangents at P and Q meet on a fixed line.

25. P_1P_2 and Q_1Q_2 are parallel chords of a parabola. If Q_1Q_2 produced cuts the tangents at P_1 and P_2 at R_1 and R_2 then $Q_1R_1 = Q_2R_2$.

26. Prove that the equation of the chord joining the points $(at^2, 2at)$ and $(aT^2, 2aT)$ on the parabola $y^2 = 4ax$ is

$$2x - (t+T)\,y + 2atT = 0,$$

and deduce the equation of the tangent at the point whose parameter is t.

The parameters of the three points A, B and C on the parabola are t_1, t_2 and t_3. The tangents at A and B meet the tangent at C at Q and P respectively; the tangents at A and B meet at R. Prove that if P is the mid-point of QC, then

$$2t_2 = t_1 + t_3.$$

Prove also that RP is parallel to AC. (O & C)

27. Prove that the equation of the normal at the point $P(at^2, 2at)$ to the parabola $y^2 = 4ax$ is

$$tx+y = at^3+2at,$$

and that it meets the parabola again at $Q(aT^2, 2aT)$ where $T = -t-2/t$.

The tangents at P and Q meet again at R; prove that, if P is a variable point on the parabola, the locus of R is

$$y^2(x+2a)+4a^2 = 0. \qquad \text{(O \& C)}$$

28. The tangents at two points P and Q on the parabola $y^2 = 4ax$ meet at R. Find the equation of the locus of R in each of the following two distinct cases:

(i) when the sum of the ordinates of P and Q is $4a$;

(ii) when the sum of the squares of the ordinates of P and Q is $4a^2$. (O & C)

(The *ordinate* of a point is the y coordinate.)

29. The chord PQ of the parabola $y^2 = 4ax$ subtends a right angle at the vertex, P and Q being the points $(at^2, 2at)$ and $(aT^2, 2aT)$. Prove that $tT+4 = 0$ and that the locus of the point of intersection of the normals at P and Q is

$$y^2 = 16a(x-6a). \qquad \text{(O \& C)}$$

30. Show that the differential equation

$$\mathrm{d}^2y/\mathrm{d}x^2 = a,$$

where a is a constant, defines a *family of parabolas* (that is, any solution of the given equation represents a parabola). If $a = 6$, find the equation of the parabola of the family which has its vertex at the point $(\frac{2}{3}, -\frac{4}{3})$.

31. Prove that, in general, a unique parabola with its axis parallel to a given direction may be drawn through three given points. What exceptional cases may arise?

Find the equation of the parabola with axis parallel to the y axis and passing through the points $(-1, 5)$, $(0, 2)$, $(1, 3)$.

32. A man rows a boat across a river flowing at a constant speed v. If his speed in still water is also v and he always aims at a point on the further bank exactly opposite his starting point, show that his path is a parabola.

4. THE RECTANGULAR HYPERBOLA

The parabola proved particularly amenable to analytical treatment because of its simple parametric form when referred to suitable axes. Another curve with a simple parametric representation is the rectangular hyperbola. We shall see later that the rectangular hyperbola is a special case of a more general type of curve, namely, the hyperbola, and, for this reason, we shall postpone a geometrical definition (see Chapter 27) and content ourselves with a purely analytical treatment by defining a *rectangular hyperbola* as the set of points

$$\{(x, y): xy = c^2\},$$

where the coordinates are referred to perpendicular axes Ox, Oy. (c is a constant which appears here as c^2 since it will later be interpreted as a length. It is not necessary for the constant to be positive and the set of points

$$\{(x, y): xy = -c^2\}$$

also represents a rectangular hyperbola.)

**Ex.* 23. Show that $x = ct$, $y = c/t$ is a suitable parametric representation of the rectangular hyperbola.

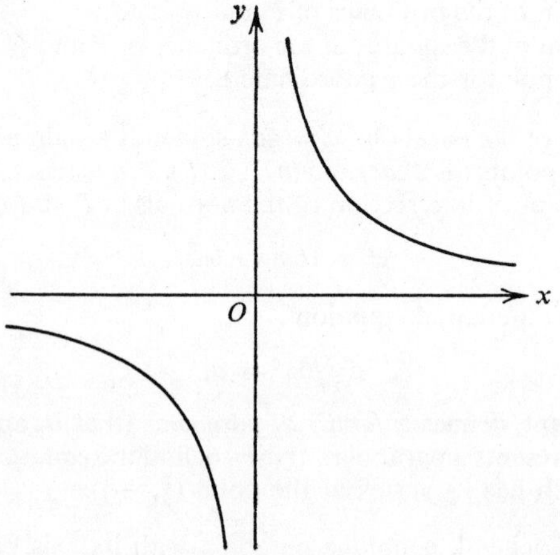

Fig. 22.7

The shape of the curve is shown in Figure 22.7. It is seen to consist of two *branches* (notice that, if (h, k) lies on the curve, so also does $(-h, -k)$).

The point O is called the *centre* of the rectangular hyperbola (see Ex. 24) and the curve is symmetrical about the lines $x-y = 0$ and $x+y = 0$.

When $x = 0$, there is no corresponding value of y and vice-versa; in fact, as $x \to 0$ through positive values, $y \to \infty$ and as $x \to 0$ through negative values, $y \to -\infty$. Thus, the line $x = 0$ is an asymptote and similarly so is $y = 0$.

Ex. 24. Show that, if any line through O cuts the rectangular hyperbola at P and Q, then O is the mid-point of PQ. (This corresponds to a property of the centre of a circle.)

Ex. 25. Trace the position of the point $P(ct, c/t)$ as t varies from $-\infty$ to $+\infty$.

The chord joining the points $P_1(ct_1, c/t_1)$ and $P_2(ct_2, c/t_2)$ on the rectangular hyperbola has equation

$$y-\frac{c}{t_1} = \frac{c(t_2-t_1)}{ct_1t_2(t_1-t_2)}(x-ct_1)$$

which reduces to

$$x+t_1t_2y = c(t_1+t_2).$$

The tangent at P_1 may be obtained immediately by letting $t_2 \to t_1$:

$$x+t_1^2y = 2ct_1.$$

**Ex.* 26. Obtain the gradient of the tangent at P_1 by differentiating the equation $xy = c^2$.

**Ex.* 27. Obtain the equation of the chord P_1P_2 by the method of Ex. 18.

Ex. 28. Show that the equation of the normal to the curve at P_1 is

$$t_1^2x-y = c(t_1^4-1)/t_1.$$

Problems concerning the rectangular hyperbola are solved by methods reminiscent of those employed for the parabola. We conclude this chapter with a worked example.

Example 3. *A rectangular hyperbola has centre O. Through a fixed point A lines are drawn to cut the hyperbola at P and Q. Show that the mid-point, M, of PQ lies on another rectangular hyperbola and find its centre and asymptotes.*

Set up coordinate axes with O as origin and the asymptotes of the given rectangular hyperbola as the axes of coordinates, labelled in such a way that one branch of the hyperbola lies in the first quadrant (positive x, positive y). Let A be the point (α, β) and let PQ be any chord through A, where P is the point $(ct_1, c/t_1)$ and Q is the point $(ct_2, c/t_2)$. Then PQ has equation

$$x+t_1t_2y = c(t_1+t_2)$$

(as shown above) and thus, since PQ contains the point A,

$$\alpha+t_1t_2\beta = c(t_1+t_2). \tag{1}$$

Let (h, k) be the mid-point of the chord PQ; then

$$2h = c(t_1+t_2), \quad 2k = c\left(\frac{1}{t_1}+\frac{1}{t_2}\right) = \frac{c(t_1+t_2)}{t_1t_2},$$

giving
$$t_1+t_2 = \frac{2h}{c}, \quad t_1t_2 = \frac{c(t_1+t_2)}{2k} = \frac{h}{k}.$$

Hence, using (1) we see that h, k are connected by the equation

$$\alpha+\frac{h}{k}\beta = 2h,$$

which may be rewritten as

$$(h-\tfrac{1}{2}\alpha)(k-\tfrac{1}{2}\beta) = \tfrac{1}{4}\alpha\beta.$$

Thus (h, k) lies on the curve

$$(x-\tfrac{1}{2}\alpha)(y-\tfrac{1}{2}\beta) = \tfrac{1}{4}\alpha\beta \tag{2}$$

which represents the required locus (or, more accurately, *contains* the locus: see Ex. 29).

By translating the coordinate axes to pass through the new origin $(\frac{1}{2}\alpha, \frac{1}{2}\beta)$, this is seen to represent a rectangular hyperbola, centre $(\frac{1}{2}\alpha, \frac{1}{2}\beta)$, with axes parallel to the axes of coordinates. Thus, M lies on a rectangular hyperbola, centre the mid-point of OA and with asymptotes parallel to those of the given hyperbola.

The reader should compare the choice of parameters in Example 3 with that in the similar problem of Example 2.

Ex. 29. The locus sought in Example 3 may be definitely only part of the curve $(x-\frac{1}{2}\alpha)(y-\frac{1}{2}\beta) = \frac{1}{4}\alpha\beta$. Show this by taking A as the origin.

Exercise 22(c)

1. The tangent to a rectangular hyperbola at the point P meets the asymptotes at Q and R. Prove that P is the mid-point of QR.

2. An axis of symmetry cuts a rectangular hyperbola at A, A'. P is any point of the hyperbola and N is the foot of the perpendicular from P to AA'. Prove that $PN^2 = AN.NA'$.

3. With the notation of Question 2, if O is the centre of the hyperbola and G is the point of intersection of the normal at P with the line AA', prove that $ON = NG$.

4. With the notation of Questions 2 and 3, if Z is the foot of perpendicular from O to the tangent at P, prove that $OZ.OP = OA^2$.

5. With the notation of Questions 2 and 3, if the tangent at P meets AA' at T, prove that $ON.OT = OA^2$.

6. Show that the equation

$$(x-a)(y-b) = c^2$$

represents a rectangular hyperbola and find its centre and asymptotes.

Find the centre and asymptotes of the rectangular hyperbola

$$xy-x+3y-7 = 0.$$

7. A variable line passes through the point (1, 1) and meets the x and y axes at P, Q respectively. The rectangle $OPRQ$ is completed. Prove that R lies on a rectangular hyperbola, and find its centre and asymptotes.

8. The tangent to a rectangular hyperbola at the point P meets the asymptotes at Q and R. If O is the centre of the hyperbola, prove that $QR = 2OP$ and that the area of the triangle OQR does not depend upon the position of P.

9. P is a variable point on a rectangular hyperbola and Q is the foot of the perpendicular from P on to one of the asymptotes. If R divides PQ in a fixed ratio show that R lies on a rectangular hyperbola with the same asymptotes as the original hyperbola.

10. Find the equations of the normals to the hyperbola $xy = c^2$ which are parallel to the line $x-4y = 0$.

11. The tangents at P, Q to a rectangular hyperbola meet at R. If O is the centre of the hyperbola, show that OR bisects PQ.

12. P_1, P_2, P_3, P_4 are four points on a rectangular hyperbola. Prove that, if P_1P_2 is perpendicular to P_3P_4, then each of the four points is the orthocentre of the triangle formed by the other three.

13. The normal to the rectangular hyperbola $xy = c^2$ at the point $P_1(ct_1, c/t_1)$ meets the curve again at $P_2(ct_2, c/t_2)$. Prove that $t_1^3t_2 = -1$ and deduce that the locus of the mid-point M of P_1P_2 has equation

$$4x^3y^3+c^2(x^2-y^2)^2 = 0.$$

Miscellaneous Exercise 22

1. Prove that the line $px+qy = 1$ is a tangent to the parabola

$$y^2 = 4a(a-x) \quad \text{if} \quad a(p^2+q^2) = p.$$

A line l and a point O not on l are given; P is a variable point on l and Q divides OP in a fixed ratio. Prove that the line through Q perpendicular to OP touches a fixed parabola with focus at O and directrix parallel to l. (O & C)

2. The coordinates of the mid-point of the line joining the points $(au^2, 2au)$, $(av^2, 2av)$ are (X, Y). Express $u+v$ and uv in terms of X and Y.

A variable chord of the parabola $y^2 = 4ax$ passes through the fixed point $(b, 0)$. Prove that the locus of the mid-point of the chord is a parabola, and find the coordinates of its vertex and focus. (O & C)

3. The normal at a point P meets the parabola again at Q. Find the length of PQ in terms of the acute angle θ between the normal and the axis of the parabola.

4. Two parabolas, $y^2 = 4a(x+b)$ and $x^2 = 4a(y+b)$, where $a > 0$, are given. Prove that each point of intersection lies either on the line $y = x$ or on the line

$$y+x+4a = 0.$$

Hence, or otherwise, prove that, if the parabolas have four real points of intersection, then $b > 3a$. (O & C)

5. Lines are drawn through a variable point P of the parabola $y^2 = 4ax$, making angles α and $\pi-\alpha$ with the axis of the parabola. The lines meet the parabola again at Q and R. Prove that:

(i) the point of intersection of the tangents at Q and R lies on the parabola $y^2 = 4a(x+4a\cot^2\alpha)$;

(ii) the line QR touches the parabola $y^2 = 4a(x-4a\cot^2\alpha)$. (O & C)

6. Prove that the line $y = mx+a/m$ touches the parabola $y^2 = 4ax$ for all values of m. Deduce that, if two tangents to a parabola are perpendicular, then their intersection lies on the directrix.

7. Two directed lines meet at A, and points P and P' are taken on the two lines so that $AP+AP'$ is constant, sense being taken into account. It is required to prove that PP' touches a fixed parabola. Prove this by first showing that the equations of the lines may be taken in the form

$$y = (x+c)\tan\alpha \quad \text{and} \quad y = -(x+c)\tan\alpha,$$

where 2α is the angle between the lines and

$$AP+AP' = 2c\sec\alpha,$$

and that the coordinates of P and P' may be taken in the form

$$(t\cos\alpha,\ t\sin\alpha+c\tan\alpha) \quad \text{and} \quad (-t\cos\alpha,\ t\sin\alpha-c\tan\alpha).$$

Deduce that PP' touches the parabola $y^2 = 4cx\tan^2\alpha$. (O & C)

8. Prove that the mid-points of parallel chords of a parabola lie on a straight line (called a *diameter* of the parabola) parallel to the axis.

Given a parabola traced on paper, find a construction for its focus.

9. If a point is marked on a rectangular sheet of paper, one of whose sides is AB and the paper is then folded in such a way that AB passes through the point, show that the crease will always touch a fixed parabola.

10. The tangents l, m from a point P to a parabola meet the directrix at L, M respectively. The other tangents to the parabola from L, M meet m, l at Q, R respectively. Prove that QR passes through the focus, S, and that the angle PSQ is a right angle.

11. A straight line passes through the point $P_1(x_1, y_1)$ and has gradient $m = \tan\alpha$. Prove that a general point of the line has coordinates $(x_1+r\cos\alpha, y_1+r\sin\alpha)$, where r is a parameter, and that the values of r giving the points of intersection of the line with the parabola $y^2 = 4ax$ are the roots of the quadratic equation

$$r^2\sin^2\theta+2r(y_1\sin\theta-2a\cos\theta)+y_1^2-4ax_1 = 0.$$

Deduce that, if P_1 is the mid-point of the chord QR, then the gradient of QR is $2a/y_1$ and its equation is

$$y_1y-2ax = y_1^2-2ax_1.$$

12. Two chords P_1Q_1, P_2Q_2 of a parabola intersect at K. Prove that the value of the fraction

$$(KP_1.KQ_1)/(KP_2.KQ_2)$$

depends on the directions of the two chords, but not upon the position of the point K.

[This result, known as *Newton's Theorem*, holds for all conics (see Chapter 27). Use the analysis of Questions 11 to prove the result for the parabola.]

13. Follow through the work of Questions 11 and 12 to prove Newton's Theorem for the rectangular hyperbola $xy = c^2$.

14. Three points $P(ap^2, 2ap)$, $Q(aq^2, 2aq)$, $R(ar^2, 2ar)$ taken on the parabola $y^2 = 4ax$ are such that PQ subtends a right angle at R. Show that

$$(p+r)(q+r)+4 = 0.$$

Show that every chord of the parabola which subtends a right angle at R intersects the normal at R at the same point F. Show also that, as R varies, F describes another parabola. (London)

15. The chord AB joining points $A(ct_1, c/t_1)$ and $B(ct_2, c/t_2)$ on the rectangular hyperbola $xy = c^2$ is of constant length l. Show that, as the position of the chord varies, the centroid G of the triangle AOB, where O is the origin, moves on the curve

$$(9xy-4c^2)(x^2+y^2) = l^2xy.$$

Find the area of the triangle AOB when the coordinates of G are $(c, 2c)$. (London)

16. Prove that the equation of the tangent PT at the point $P(ct, c/t)$ on the rectangular hyperbola $xy = c^2$ is

$$x+t^2y = 2ct.$$

The perpendicular to PT from the origin O meets PT at Q and the normal at P meets the hyperbola again at R.

Prove that:

(i) as P varies, the locus of Q is

$$(x^2+y^2)^2 = 4c^2xy;$$

and

(ii) $$c^2.PR = OP^3.$$ (O & C)

17. Prove that the equation of the chord joining the points $P(ct, c/t)$ and $Q(cT, c/T)$ on the rectangular hyperbola $xy = c^2$ is

$$x+tTy = c(t+T).$$

M is the mid-point of PQ and PQ meets the x axis at N. Prove that $OM = MN$, where O is the origin.

The line through N parallel to OM meets the hyperbola at R and S whose parameters are t_1 and t_2; C is the mid-point of RS. Prove that:

(i) CM is parallel to the y axis;

(ii) $t_1t_2 = -tT$. (O & C)

18. Prove that the equation of the tangent at the point (h, k) on the parabola $y^2 = 4ax$ and the equation of the tangent at the point (H, K) on the rectangular hyperbola $xy = c^2$ are, respectively,

$$ky = 2a(x+h) \quad \text{and} \quad Kx+Hy = 2c^2.$$

Find the coordinates of the point of intersection, P, of the parabola $y^2 = 4ax$ and the rectangular hyperbola $xy = 4a^2\sqrt{2}$ and prove that the tangent to the parabola at P is the normal to the rectangular hyperbola at P.

Prove that, if this normal meets the rectangular hyperbola again at Q, then the abscissa of Q is $-4a$. (O & C)

19. Prove that the normal to the rectangular hyperbola $xy = c^2$ at the point $P(ct, c/t)$ meets the curve again at $Q(-c/t^3, -ct^3)$.

The circle on PQ as diameter meets the hyperbola at R; prove that PR passes through the origin. (O & C)

20. Given a point P on a rectangular hyperbola, prove that one and only one real chord PQ can be drawn which is normal to the hyperbola at Q. Deduce that there is only one (real) chord AB of the hyperbola which is normal at both A and B and locate the two points A, B.

21. Points P, Q, R, S are taken on the rectangular hyperbola $xy = c^2$. Prove that $PQRS$ is a rectangle if and only if the parameters of the four points are of the form $t, t^{-1}, -t, -t^{-1}$. Deduce that it is impossible to inscribe a square in a rectangular hyperbola.

22. The tangent to a rectangular hyperbola Γ at the point P meets the asymptotes at Q and R. O is the centre of the hyperbola and $OQSR$ is a rectangle. If SQ, SR cut Γ at Q', R' prove that $Q'R'$ touches a second rectangular hyperbola whose asymptotes coincide with those of Γ.

23. Prove that all chords of a parabola which subtend a right angle at a point P on the parabola pass through a fixed point Q.

Q is called the *Frégier point* of P. If a given chord subtends right angles at two points, P and P' of the parabola, prove that the join of the two Frégier points, Q and Q', is parallel to the given chord.

24. Find the equation of the tangent to the parabola S_1: $y^2 = 4ax$ at the point $P(at^2, 2at)$.

If this tangent cuts the rectangular hyperbola S_2: $xy = k^2$ at *real* points Q, R and M is the mid-point of QR show that M lies on part of the parabola Σ:

$$2y^2 + ax = 0.$$

Sketch in the same diagram the curves S_1 and S_2 and the set of all such points M in the case $a > 0$. Mark in your diagram the line other than $x = 0$ which is a tangent to both S_1 and S_2.

23. *Polynomial equations*

1. SOME PRELIMINARY OBSERVATIONS

The results we have obtained in Chapter 18 generalize very readily to general polynomial equations. In particular, it is possible to find expressions for symmetric functions of the roots of polynomial equations in terms of the coefficients. We shall pursue the question of symmetric functions in Section 2; in this section we shall discuss a number of other results connected with the solution of general polynomial equations of degree n.

If we are given a specific equation, say

$$z^4+(3+2\mathrm{j})\,z^2-(1-\mathrm{j})\,z+4 = 0$$

it is not at all obvious whether or not the equation has a solution. Indeed, two questions arise:

(i) Does every equation possess a solution? (That is, is there a complex number which satisfies the equation?)

(ii) If a solution exists, can it be found by processes similar to those employed for solving a quadratic equation?

The answer to the first question is 'Yes' but the proof of this is rather difficult (but not beyond the understanding of a really enthusiastic pupil: see, for example, Hardy's *Pure Mathematics*, Appendix II or Courant and Robbins *What is Mathematics?*, Chapter 5). The result was first proved, like so many other central theorems of mathematics, by C. F. Gauss and is generally known as 'The Fundamental Theorem of Algebra': explicitly, it states that any polynomial equation with complex numbers as coefficients (and in particular, any equation with real coefficients) has a root which is a complex number (R is regarded as a subset of C in this result). We shall content ourselves with assuming the truth of this theorem.

The answer to the second question is 'Yes, if the degree of the given equation is less than five' but 'No, in general, if the degree is five or more'. The first part of this result is comparatively easy to prove and is due to a number of Italian mathematicians of the Sixteenth Century; the second part is very much harder and is due to two mathematicians, E. Galois (1811–32) and N. H. Abel (1802–29). (The truly dedicated reader may care to follow the matter up in, say, Birkhoff and Maclane, '*Survey of Modern Algebra*. Brief readable accounts of the life and work of Galois and Abel may be found in Bell, *Men of Mathematics*, Volume II.)

Ex. 1. Although it is not possible, in general, to solve a given quintic equation except by approximate numerical methods, this is not to say that no quintic equation is soluble. Solve the equation $z^5 - 32\mathrm{j} = 0$.

If $P(z)$ is a polynomial of degree n, and if $P(z)$ has a factor $(z-\alpha)^r$ (but not $(z-\alpha)^{r+1}$) where r is an integer $\leqslant n$, the equation $P(z) = 0$ is said to have a *root of multiplicity r*. For example, the quartic equation $(z-1)(z+2\mathrm{j})^3 = 0$ has a root $z = 1$ and a root $z = -2\mathrm{j}$, of multiplicity 3.

Ex. 2. Solve the cubic equation $z^3+3z^2-4 = 0$, given that it has an integral root of multiplicity 2.

**Ex.* 3. Assuming that every polynomial equation has a root, prove, by induction, that a polynomial equation of degree n has n roots, where a root of multiplicity r is counted as r roots.

**Ex.* 4. By writing $P(z) \equiv (z-\alpha)^r Q(z)$ prove that, if $P(z) = 0$ has a root of multiplicity r, then $P'(z) = 0$ has a root of multiplicity $(r-1)$.

Prove that the converse is false.

**Ex.* 5. If $P(z) = 0$ has a root of multiplicity $r > 2$, what can you say about the equation $P''(z) = 0$?

As with quadratic equations, additional restrictions upon the coefficients of a polynomial equation enable us to make further assertions about the roots. In particular, if all the coefficients are real we have the following three important results which enable us to decide whether we have an odd or even number of roots and to locate such roots.

Theorem 23.1. *If all the coefficients of*

$$P(z) \equiv a_0 z^n + a_1 z^{n-1} + \ldots + a_n$$

are real and if the complex number z_1 is a root of $P(z) = 0$ so also is z_1^*.

Proof. Since

$$a_0 z_1^n + a_1 z_1^{n-1} + \ldots + a_n = 0,$$

we have

$$a_0(z_1^*)^n + a_1(z_1^*)^{n-1} + \ldots + a_n = 0,$$

on taking the complex conjugate of each side, and noting that $a_i^* = a_i$, since the a_i are real. The result follows immediately.

Ex. 6. Given that $-1+\mathrm{j}$ is a root of the equation

$$z^4+3z^3+5z^2+4z+2 = 0,$$

solve the equation completely.

**Ex.* 7. Prove that a polynomial equation of odd degree has at least one real root.

Theorem 23.2. *If all the coefficients of*

$$P(z) \equiv a_0 z^n + a_1 z^{n-1} + \dots + a_n \quad (a_0 \neq 0)$$

are real and if the equation $P(z) = 0$ *has no real root, then* $P(z)$ *is always positive or always negative for all real* z.

Proof. (The theorem is intuitively obvious from graphical considerations.)

By Theorem 23.1 the roots of $P(z) = 0$ occurs in conjugate pairs:

$$\alpha_1 \pm j\beta_1, \quad \alpha_2 \pm j\beta_2, \quad \dots.$$

But
$$[z-(\alpha_r+j\beta_r)]\,[z-(\alpha_r-j\beta_r)] = (z-\alpha_r)^2+\beta_\sigma^2$$

and thus
$$P(z) = a_0[(z-\alpha_1)^2+\beta_1^2]\,[(z-\alpha_2)^2+\beta_2^2] \dots$$

and every square bracket is positive for all real z. The sign of $P(z)$ is thus determined by the sign of a_0.

**Ex.* 8. Prove that, if a_0 and a_n have opposite signs, the equation $P(z) = 0$ has a real root.

Theorem 23.3. *If all the coefficients of*

$$P(z) \equiv a_0 z^n + a_1 z^{n-1} + \dots + a_n \quad (a_0 \neq 0)$$

are real and if x_1, x_2 *are real numbers such that* $P(x_1), P(x_2)$ *have opposite signs, then* $P(z) = 0$ *has an odd number of real roots between* x_1 *and* x_2 *(and, in particular, of course, at least one real root). For the purposes of this theorem a root of multiplicity k is counted as k roots.*

Proof. (Again, the result is intuitively obvious from graphical considerations.)

Let $\alpha_1, \alpha_2, \dots, \alpha_r$ be all the real roots of $P(z) = 0$. Then

$$P(z) = a_0(z-\alpha_1)\,(z-\alpha_2) \dots (z-\alpha_r)\,Q(z),$$

where $Q(z)$ is either positive or negative for all real z (by Theorem 23.2). Thus $(x_1-\alpha_1)\,(x_1-\alpha_2) \dots (x_1-\alpha_r)$ and $(x_2-\alpha_1)\,(x_2-\alpha_2) \dots (x_2-\alpha_r)$ have opposite signs and

$$[(x_1-\alpha_1)\,(x_2-\alpha_1)]\,[(x_1-\alpha_2)\,(x_2-\alpha_2)] \dots [(x_1-\alpha_r)\,(x_2-\alpha_r)] < 0.$$

Thus $(x_1-\alpha_i)\,(x_2-\alpha_i)$ is negative in an odd number of cases and the result follows.

Ex. 9. Prove that $x^3-x+3 = 0$ has a root between $x = -2$ and $x = -1$.

We conclude this Section with two worked examples. The first shows how one equation may be transformed into another in such a way that the roots of the two equations bear a given relationship to one another. The second example indicates how calculus methods may be used to solve problems on polynomial equations.

Example 1. *If α, β, γ are the roots of the equation*

$$3z^3-5z^2+2z+2 = 0$$

form the equation: (i) *with roots $\alpha-2, \beta-2, \gamma-2$;*

(ii) *with roots $2\alpha, 2\beta, 2\gamma$;*

(iii) *with roots $\alpha^2, \beta^2, \gamma^2$.*

(i) As in Example 2, Chapter 21, $\alpha-2, \beta-2, \gamma-2$ are the roots of the equation

$$3(w+2)^3-5(w+2)^2+2(w+2)+2 = 0.$$

This may be expanded directly. Alternatively, since we are essentially making the transformation $z \to w+2$ we may rewrite the polynomial

$$3z^3-5z^2+2z+2$$

in the form

$$a(z-2)^3+b(z-2)^2+c(z-2)+d$$

and the required equation will be

$$aw^3+bw^2+cw+d = 0.$$

By Horner's Method (Chapter 18, Section 1) we have

3	−5	2	2 (2
	6	2	8
3	1	4	10 (2
	6	14	
3	7	18	(2
	6		
3	13		

giving the required equation

$$3w^3+13w^2+18w+10 = 0.$$

(ii) In a similar way, to form the equation with roots 2α, 2β, 2γ we make the transformation $z \to w/2$, giving the required equation:

$$3(\tfrac{1}{2}w)^3-5(\tfrac{1}{2}w)^2+2(\tfrac{1}{2}w)+2 = 0.$$

or

$$3w^3-10w^2+8w+16 = 0.$$

(iii) If $w = z^2$, then either $z = w^{\frac{1}{2}}$ or $z = -w^{\frac{1}{2}}$. Thus

$$
\begin{aligned}
& & 3z^3-5z^2+2z+2 &= 0 \\
\Leftrightarrow & & z(3z^2+2) &= 5z^2-2 \\
\Leftrightarrow & \text{ either} & w^{\frac{1}{2}}(3w+2) &= 5w-2 \\
& \text{or} & -w^{\frac{1}{2}}(3w+2) &= 5w-2 \\
\Leftrightarrow & & \{5w-2-w^{\frac{1}{2}}(3w+2)\}\{5w-2+w^{\frac{1}{2}}(3w+2)\} &= 0 \\
\Leftrightarrow & & 9w^3-13w^2+24w-4 &= 0,
\end{aligned}
$$

on multiplying out and rearranging terms, and this equation has roots α^2, β^2, γ^2.

Example 2. Discuss the nature of the roots of the equation

$$3x^4-4x^3-12x^2+k = 0$$

for different real values of k.

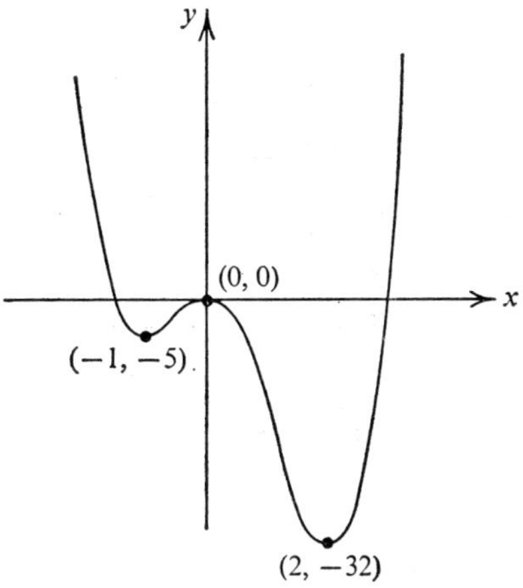

Fig. 23.1

$$
\begin{aligned}
&\text{Write} & P(x) &\equiv 3x^4-4x^3-12x^2; \\
&\text{then} & P'(x) &\equiv 12x^3-12x^2-24x \\
&\text{and thus} & P'(x) &= 0 \quad \text{when} \quad x = -1, 0, \text{or } 2.
\end{aligned}
$$

Since $P(x) > 0$ for sufficiently large x, the values -1, 0, 2 for x yield respectively a minimum, maximum and minimum of $P(x)$. The graph of $y = P(x)$ thus assumes roughly the shape shown in Figure 23.1.

Now the given equation may be written as

$$P(x) = -k$$

and the number of real roots of the equation may be deduced immediately by considering the intersections of the curve $y = P(x)$ and the straight line $y = -k$:

$k < 0$: the equation has two unequal real roots and two complex conjugate roots.

$k = 0$: the equation has three unequal real roots, one of multiplicity 2.

$0 < k < 5$: the equation has four unequal real roots.

$k = 5$: the equation has three unequal real roots, one of multiplicity 2.

$5 < k < 32$: the equation has two unequal real roots and two conjugate complex roots.

$k = 32$: the equation has a real root of multiplicity 2 and two conjugate (complex) roots.

$k > 32$: the equation has two pairs of conjugate complex roots.

Ex. 10. What are the roots of the equation in Example 2 in the cases
(i) $k = 0$; (ii) $k = 5$; (iii) $k = 32$?

Exercise 23(a)

1. Form the equation with roots $1, -2, 3-\sqrt{5}, 3+\sqrt{5}$.

2. Form the equation with roots $1+2\text{j}, 1-2\text{j}, -2, 3$.

3. Form the equation with a root -1 of multiplicity 3 together with a pair of roots $+\text{j}, -\text{j}$.

4. Solve the equation
$$z^4-2z^3+3z^2-4z+2 = 0,$$
given that it has a root of multiplicity 2.

5. Use the Remainder Theorem to solve the equation
$$z^3-3z^2-8z+30 = 0,$$
given that it has an integral root.

6. The equation
$$24z^4+100z^3+126z^2+27z-27 = 0$$
has a root of multiplicity three. Solve the equation completely.

7. If α, β, γ are the roots of the equation
$$z^3-z-4 = 0,$$
form the equation with roots:

(i) $\frac{1}{\alpha}, \frac{1}{\beta}, \frac{1}{\gamma}$; (ii) $\frac{\alpha+1}{\alpha}, \frac{\beta+1}{\beta}, \frac{\gamma+1}{\gamma}$.

8. If α, β, γ are the roots of the equation

$$z^3 - 3z^2 - 3z - 4 = 0,$$

form the equation with roots:
(i) $2\alpha, 2\beta, 2\gamma$; (ii) $\alpha^2, \beta^2, \gamma^2$.

9. If α, β, γ are the roots of the equation

$$z^3 + jz + 1 + j = 0,$$

form the equation with roots:
(i) $\frac{1}{3}\alpha, \frac{1}{3}\beta, \frac{1}{3}\gamma$; (ii) $(1+j)\alpha, (1+j)\beta, (1+j)\gamma$.

10. If α, β, γ are the roots of the equation

$$2z^3 - z^2 + 4z + 7 = 0,$$

form the equation with roots
(i) $\alpha+2, \beta+2, \gamma+2$; (ii) $\alpha-3, \beta-3, \gamma-3$.

11. If $\alpha, \beta, \gamma, \delta$ are the roots of the equation

$$z^4 - z - 5 = 0,$$

form the equation:
(i) with roots $-\alpha, -\beta, -\gamma, -\delta$; (ii) with roots $\alpha+2, \beta+2, \gamma+2, \delta+2$;
(iii) with roots $\alpha^2, \beta^2, \gamma^2, \delta^2$.

12. If α, β, γ are the roots of the equation

$$3z^3 - 20z^2 + 39z - 18 = 0,$$

form the equation with roots $\alpha-3, \beta-3, \gamma-3$. Hence find α, β, γ.

13. Solve the equation

$$z^4 + 4z^3 + 5z^2 + 2z - 2 = 0.$$

by first increasing the roots by 1.

14. Solve the equation

$$z^3 + 3jz^2 - 5z - 3j = 0$$

by first increasing the roots by j.

15. Show that the cubic equation

$$z^3 + 4z^2 - 3z + 25 = 0$$

has three real roots, one of which is positive and the other two negative.

16. Show that the cubic equation

$$z^3 - 3z + k = 0$$

has three distinct real roots if and only if k is real and $-2 < k \leqslant 2$.

17. Discuss the nature of the roots of the equation

$$2z^3 - 15z^2 + 24z + k = 0$$

for different real values of k.

18. Discuss the nature of the roots of the equation

$$z^4 - 4z^3 - 2z^2 + 12z + k = 0$$

for different real values of k.

19. Discuss the nature of the roots of the equation

$$3z^4-4z^3-24z^2+48z+k = 0$$

for different real values of k.

20. Find a necessary and sufficient condition that the three roots of the cubic equation

$$z^3+3Hz+G = 0$$

should be real and unequal.

2. RELATIONS BETWEEN ROOTS AND COEFFICIENTS FOR POLYNOMIAL EQUATIONS

Consider first the cubic equation

$$a_0z^3+a_1z^2+a_2z+a_3 = 0$$

with roots α, β, γ. We have

$$\begin{aligned} a_0z^3+a_1z^2+a_2z+a_3 & \\ &\equiv a_0(z-\alpha)(z-\beta)(z-\gamma) \\ &\equiv a_0z^3-a_0(\alpha+\beta+\gamma)z^2+a_0(\beta\gamma+\gamma\alpha+\alpha\beta)z-a_0\alpha\beta\gamma \end{aligned}$$

and, equating coefficients of z^2, z^1 and z^0,

$$\begin{aligned} \alpha+\beta+\gamma &= -a_1/a_0, \\ \beta\gamma+\alpha\alpha+\alpha\beta &= +a_2/a_0, \\ \alpha\beta\gamma &= -a_3/a_0. \end{aligned}$$

Similarly, for the quartic equation

$$a_0z^4+a_1z^3+a_2z^2+a_3z+a_4 = 0,$$

with roots α, β, γ, δ we have

$$\begin{aligned} a_0z^4+a_1z^3+a_2z^2+a_3z+a_4 & \\ &\equiv a_0(z-\alpha)(z-\beta)(z-\gamma)(z-\delta) \\ &\equiv a_0z^4-a_0(\alpha+\beta+\gamma+\delta)z^3+a_0(\alpha\beta+\alpha\gamma+\alpha\delta+\beta\gamma+\beta\delta+\gamma\delta)z^2 \\ &\quad -a_0(\beta\gamma\delta+\gamma\delta\alpha+\delta\alpha\beta+\alpha\beta\gamma)z+a_0\alpha\beta\gamma\delta \end{aligned}$$

from which, by equating coefficients,

$$\begin{aligned} \alpha+\beta+\gamma+\delta &= -a_1/a_0, \\ \alpha\beta+\alpha\gamma+\alpha\delta+\beta\gamma+\beta\delta+\gamma\delta &= +a_2/a_0, \\ \beta\gamma\delta+\gamma\delta\alpha+\delta\alpha\beta+\alpha\beta\gamma &= -a_3/a_0, \\ \alpha\beta\gamma\delta &= +a_4/a_0. \end{aligned}$$

In words, the sum of the roots one at a time is $-a_1/a_0$, two at a time is $+a_2/a_0$, three at a time is $-a_3/a_0$, and four at a time is a_4/a_0. For brevity, we denote $\alpha+\beta+\gamma+\delta$ by $\Sigma\alpha$, $\alpha\beta+\alpha\gamma+\alpha\delta+\beta\gamma+\beta\delta+\gamma\delta$ by $\Sigma\alpha\beta$, etc.; no ambiguity arises provided we know the degree of the equation under consideration (in this case *four*).

Ex. 11. The roots of the equation

$$2z^3-3z^2-5z-13 = 0$$

are α, β, γ. Write down the numerical values of $\Sigma\alpha$, $\Sigma\alpha\beta$, $\alpha\beta\gamma$.

Ex. 12. The roots of the equation

$$z^4-4jz+1 = 0$$

are $\alpha, \beta, \gamma, \delta$. Write down the numerical values of $\Sigma\alpha$, $\Sigma\alpha\beta$, $\Sigma\alpha\beta\gamma$, $\alpha\beta\gamma\delta$.

**Ex.* 13. If the coefficients of a cubic equation are all real, it follows that $\Sigma\alpha$, $\Sigma\alpha\beta$ and $\alpha\beta\gamma$ are all real. Explain how this can be so, even if the equation has complex roots. Illustrate your answer by considering the equation $z^3-1 = 0$.

Now suppose that the general polynomial equation

$$a_0z^n+a_1z^{n-1}+a_2z^{n-2}+\ldots+a_n = 0$$

has roots $\alpha_1, \alpha_2, \ldots, \alpha_n$. Then it may be shown by induction that the results proved previously in the particular cases $n = 2, 3, 4$ hold generally.

$$\begin{aligned}\Sigma\alpha_i &= -a_1/a_0\\ \Sigma\alpha_i\alpha_j &= +a_2/a_0\\ \Sigma\alpha_i\alpha_j\alpha_k &= -a_3/a_0\\ &\vdots\\ \alpha_1\alpha_2\alpha_3\ldots\alpha_n &= (-1)^n\, a_n/a_0.\end{aligned}$$

(Let the equation with roots $\alpha_1, \alpha_2, \ldots, \alpha_{n-1}$ be

$$a_0\, z^{n-1}+b_1z^{n-2}+b_2z^{n-3}+\ldots+b_{n-1} = 0$$

and consider the identity

$$(z-\alpha_n)\,(a_0z^{n-1}+b_1z^{n-2}+b_2z^{n-3}+\ldots+b_{n-1}) \equiv a_0z^n+a_1z^{n-1}+a_2z^{n-2}+\ldots+a_n.)$$

Ex. 14. By considering the equation $z^n-1 = 0$, prove that the sum of the nth roots of unity is zero ($n \geqslant 2$).

Example 3. *Find the quartic equation with roots* 1, 3, −2, −4.

We have

$$\begin{aligned}\Sigma\alpha &= -2,\\ \Sigma\alpha\beta &= 3-2-4-6-12+8 = -13,\\ \Sigma\alpha\beta\gamma &= 24+8-12-6 = 14,\\ \alpha\beta\gamma\delta &= 24,\end{aligned}$$

and the required equation is

$$z^4+2z^3-13z^2-14z+24 = 0.$$

Example 4. If the roots of the equation $z^3-4z-6 = 0$ *are* α, β, γ *find* (i) $\Sigma\alpha^2$, (ii) $\Sigma\alpha^2\beta$, (iii) $\Sigma\alpha^3$, (iv) $\Sigma\alpha^4$.

We have $\Sigma\alpha = 0$, $\Sigma\alpha\beta = -4$, $\alpha\beta\gamma = 6$:

(i) $\Sigma\alpha^2 = (\Sigma\alpha)^2-2\Sigma\alpha\beta$

$= 0+8.$

$= 8.$

(ii) Notice first that $\Sigma\alpha^2\beta$ contains *six* terms:

$$\Sigma\alpha^2\beta = \alpha^2\beta+\alpha^2\gamma+\beta^2\gamma+\beta^2\alpha+\gamma^2\alpha+\gamma^2\beta.$$

Consider the product $(\Sigma\alpha)(\Sigma\alpha\beta)$: each term such as $\alpha^2\beta$ occurs just *once*, but the terms $\alpha\beta\gamma$ appears *three* times, as $\alpha.\beta\gamma$, $\beta.\gamma\alpha$ and $\gamma.\alpha\beta$. Thus

$$\Sigma\alpha^2\beta = \Sigma\alpha\Sigma\alpha\beta-3\alpha\beta\gamma$$
$$= -18.$$

(iii) Since α is a root of the given equation, we have

$$\alpha^3-4\alpha-6 = 0$$

and similarly for β and γ. Adding these three results

$$\Sigma\alpha^3-4\Sigma\alpha-18 = 0.$$

Thus $$\Sigma\alpha^3 = 18.$$

(iv) By an argument similar to that adopted in (iii),

$$\Sigma\alpha^4-4\Sigma\alpha^2-6\Sigma\alpha = 0.$$

Thus $$\Sigma\alpha^4 = 32,$$

on using the result of part (i).

Example 5. Solve the equation

$$8z^3-12z^2-66z+35 = 0$$

given that the roots are consecutive terms of an arithmetic sequence.

From the equation we have

$$\Sigma\alpha = \tfrac{3}{2}, \quad \Sigma\alpha\beta = -\tfrac{33}{4}, \quad \alpha\beta\gamma = -\tfrac{35}{8}.$$

Since the roots are consecutive terms of an arithmetic sequence, we take

$$\alpha = a-d, \quad \beta = a, \quad \gamma = a+d.$$

Then $\Sigma\alpha = 3a$, $\Sigma\alpha\beta = 3a^2-d^2$, giving

$$3a = \tfrac{3}{2},\quad 3a^2-d^2 = -\tfrac{33}{4}.$$

Thus $a = \frac{1}{2}$, $d = \pm 3$ and the roots are $-2\frac{1}{2}, \frac{1}{2}, 3\frac{1}{2}$.

Example 6. *Solve the simultaneous equations*

$$\begin{aligned} x+y+z &= 2,\\ x^2+y^2+z^2 &= 30,\\ x^3+y^3+z^3 &= 116. \end{aligned}$$

Let x, y, z be the roots of the cubic equation

$$t^3-at^2+bt-c = 0.$$

Then

$$\begin{aligned} a &= \Sigma x = 2,\\ b &= \Sigma xy = \tfrac{1}{2}\{(\Sigma x)^2-\Sigma x^2\} = \tfrac{1}{2}\{4-30\} = -13. \end{aligned}$$

To find c, we make use of the identity

$$x^3+y^3+z^3-3xyz \equiv (x+y+z)(x^2+y^2+z^2-yz-zx-xy);$$

$$\begin{aligned} c = xyz &= \tfrac{1}{3}\{\Sigma x^3-\Sigma x[\Sigma x^2-\Sigma yz]\}\\ &= \tfrac{1}{3}\{116-2[30+13]\}\\ &= 10. \end{aligned}$$

Thus x, y, z are the roots of the cubic equation

$$t^3-2t^2-13t-10 = 0.$$

By observation, this has a root $t = -1$:

$$(t+1)(t^2-3t-10) = 0,$$

i.e.

$$(t+1)(t+2)(t-5) = 0,$$

and $x = -1$, $y = -2$, $z = 5$ (or any equivalent permutation).

Example 7. *If the roots of the cubic equation* $z^3+3Hz+G = 0$ *are* α, β, γ, *form the cubic equation with roots* $(\beta-\gamma)^2$, $(\gamma-\alpha)^2$, $(\alpha-\beta)^2$.

$$w = (\beta-\gamma)^2$$

$$\Rightarrow\quad w = \Sigma\alpha^2-\alpha^2-2\beta\gamma$$

$$\Rightarrow \alpha w = \alpha(\Sigma\alpha)^2-2\alpha\Sigma\alpha\beta-\alpha^3-2\alpha\beta\gamma \quad (\text{since } (\Sigma\alpha)^2 = \Sigma\alpha^2+2\Sigma\alpha\beta),$$

$$\Rightarrow \alpha w = -6H\alpha-\alpha^3+2G \quad (\text{since } \Sigma\alpha = 0,\ \Sigma\alpha\beta = 3H),$$

$$\Rightarrow\quad \alpha = 3G/(w+3H) \quad (\text{since } \alpha^3 = -3H\alpha-G)$$

$$\Rightarrow 27G^3+9HG(w+3H)^2+G(w+3H)^3 = 0 \quad (\text{since } \alpha^3+3H\alpha+G = 0)$$

$$\Rightarrow w^3+18Hw^2+81H^2w+27(G^2+4H^3) = 0.$$

Thus, $(\beta-\gamma)^2$ is a root of the equation

$$w^3+18Hw^2+81H^2w+27(G^2+4H^3) = 0$$

and, by symmetry, so also are $(\gamma-\alpha)^2$, $(\alpha-\beta)^2$.

Note: the equation derived in Example 7 is of importance in studying the nature of the roots of cubic equations—see Section 4.

Exercise 23(b)

1. If α, β, γ are the roots of the equation

$$z^3-4z^2+z+2 = 0,$$

find:

(i) $\Sigma\alpha^2$; (ii) $\Sigma\alpha^3$; (iii) $\Sigma\alpha^4$; (iv) $\Sigma\alpha^2\beta$;
(v) $\Sigma\alpha^2\beta^2$; (vi) $(\alpha+2)(\beta+2)(\gamma+2)$; (vii) $\Sigma 1/\alpha$;
(viii) $\Sigma\alpha\beta(\alpha^2+\beta^2)$; (ix) $\Sigma 1/\alpha\beta$; (x) $\Sigma(\alpha+\beta)^2$.

2. If $\alpha, \beta, \gamma, \delta$ are the roots of the equation

$$z^4-4z^3-z+9 = 0,$$

find:

(i) $\Sigma\alpha^2$; (ii) $\Sigma\alpha^2\beta$; (iii) $\Sigma 1/\alpha$; (iv) $\Sigma\alpha^3$;
(v) $\Sigma\alpha^3\beta$; (vi) $\Sigma(\alpha+\beta+\gamma)^2$; (vii) $\Sigma(\alpha+1)(\beta+1)$.

3. If α, β, γ are the roots of the equation

$$z^3+pz+q = 0,$$

find:

(i) $\Sigma\alpha^4$; (ii) $\Sigma 1/\alpha^4$; (iii) $\Sigma(\alpha+\beta-2\gamma)$.

4. Solve the equation

$$2z^3-3z^2-23z+12 = 0,$$

given that its roots are successive terms of an arithmetic sequence.

5. Solve the equation

$$32z^3+48z^2+6z-5 = 0,$$

given that its roots are successive terms of an arithmetic sequence.

6. Solve the equation

$$24z^3-38z^2+19z-3 = 0,$$

given that its roots are successive terms of a geometric sequence.

7. Solve the equation

$$2z^3-11z^2-5z+50 = 0,$$

given that one of the roots is twice a second root.

8. Solve the equations:

$$x+y+z = -4,$$
$$yz+zx+xy = 1,$$
$$xyz = 6.$$

9. Solve the equations:

$$x+y+z = 1,$$
$$x^2+y^2+z^2 = 29,$$
$$x^3+y^3+z^3 = -29.$$

10. Solve the equations:

$$x+y+z = 4,$$
$$x^2+y^2+z^2 = 14,$$
$$x^3+y^3+z^3 = 34.$$

11. The roots of the equation:

$$ax^3+3bx^2+3cx+d = 0$$

are in arithmetic progression. Express c in terms of a, b and d. (O & C)

12. Prove that, if the cubic equation

$$ax^3+bx^2+cx+d = 0$$

has a pair of reciprocal roots (that is, α and $1/\alpha$), then

$$a^2-d^2 = ac-bd.$$

Verify that this condition is satisfied for the equation

$$6x^3+11x^2-24x-9 = 0$$

and hence, or otherwise, solve the equation. (O & C)

13. If α, β, γ are the roots of the equation

$$z^3+pz+q = 0,$$

form the equations with roots:
(i) $\alpha^2, \beta^2, \gamma^2$;
(ii) $\beta\gamma, \gamma\alpha, \alpha\beta$;
(iii) $\beta+\gamma, \gamma+\alpha, \alpha+\beta$;
(iv) $\beta+\gamma-\alpha, \gamma+\alpha-\beta, \alpha+\beta-\gamma$.

14. If α, β, γ are the roots of the equation

$$z^3+pz+q = 0,$$

form the equations with roots:
(i) $\alpha^3, \beta^3, \gamma^3$;
(ii) $\alpha(\beta+\gamma), \beta(\gamma+\alpha), \gamma(\alpha+\beta)$.

15. If α, β, γ are three non-zero complex numbers such that $\alpha+\beta+\gamma = 0$, prove that $\alpha^2-\beta\gamma = \beta^2-\gamma\alpha = \gamma^2-\alpha\beta$.

16. Given that α, β, γ are the roots of the equation

$$x^3+px^2+qx+r = 0,$$

and that $a = \beta\gamma-\alpha^2$, $b = \gamma\alpha-\beta^2$, $c = \alpha\beta-\gamma^2$, prove that

$$a = p\alpha+q, \quad b = p\beta+q, \quad c = p\gamma+q.$$

Hence, or otherwise, prove that

$$a\alpha+b\beta+c\gamma = (a+b+c)(\alpha+\beta+\gamma).$$ (O & C)

17. If α, β, γ are the roots of the equation

$$x^3+px^2+qx+r = 0,$$

find the equations whose roots are:

(i) $\alpha^2-\beta\gamma, \beta^2-\gamma\alpha, \gamma^2-\alpha\beta$; (ii) $\beta+\gamma-2\alpha, \gamma+\alpha-2\beta, \alpha+\beta-2\gamma$.

What can be said about α, β, γ if

(i) $q^3 = p^3r$; (ii) $2p^3-9pq+27r = 0$? (O & C)

18. By considering the product $(\beta+\gamma)(\gamma+\alpha)(\alpha+\beta)$, find a necessary and sufficient condition that two of the roots of the equation

$$z^3+pz^2+qz+r = 0 \quad (r \neq 0)$$

should be equal in magnitude but of opposite signs.

3. THE CUBIC EQUATION

For practical purposes an algebraic solution of the general cubic is of limited value since, in practice, only approximate numerical solutions are required and these are best found by the iterative methods to be described in Chapter 26. Nevertheless, a brief survey of the method of attack seems worthwhile on at least three counts: having seen a solution to the general quadratic it is natural to wonder whether a similar method holds for the cubic; such a question did occur to mathematicians of the past and the study of the cubic is of considerable historical interest; finally, the question whether general polynomial equations are soluble leads one to some of the central ideas of modern mathematics.

The general cubic equation was first solved in the sixteenth century by Tartaglia and published by Cardan in one of the most famous of all acts of mathematical plagiarism. (See *Cardan: the Gambling Scholar* by O. Ore for a biography of this unsavoury yet fascinating character.) The general quartic soon followed (Ferrari) but, as has been pointed out already in Section 1, the general quintic evaded all attempts at solution, for the reason discovered independently by Abel and Galois in the nineteenth century. An account of the Italian mathematicians of the sixteenth century whose work we have referred to will be found in *History of Mathematics* by D. E. Smith, Volume 1, Chapter 8.

We shall take the general cubic in the form

$$a_0z^3+3a_1z^2+3a_2z+a_3 = 0. \tag{1}$$

If we apply the transformation $z \to (w-a_1)/a_0$ we have

$$a_0z^3+3a_1z^2+3a_2z+a_3 = 0$$

$$\Rightarrow \quad a_0^3z^3+3a_1a_0^2z^2+3a_2a_0a_0z+a_3a_0^2 = 0$$

$$\Rightarrow \quad (w-a_1)^3+3a_1(w-a_1)^2+3a_2a_0(w-a_1)+a_3a_0^2 = 0$$

$$\Rightarrow \quad w^3+3w(a_2a_0-a_1^2)+(2a_1^3-3a_0a_1a_2+a_3a_0^2) = 0$$

$$\Rightarrow w^3+3Hw+G = 0, \quad \text{where} \quad H = a_2a_0-a_1^2,\ G = 2a_1^3-3a_0a_1a_2+a_3a_0^2.$$

Thus any cubic equation may be reduced to

$$w^3+3Hw+G = 0 \tag{2}$$

and we shall, from now on, regard this as the general cubic equation which it is our intention to solve.

Ex. 15. Show how to reduce the equations:

(i) $x^3-9x^2+22x-3 = 0$, (ii) $2x^3+6x^2+10x-1 = 0$,

to the standard form $x^3+3Hx+G = 0$.

If the roots of (1) are z_1, z_2, z_3 and the roots of (2) are w_1, w_2, w_3, then the geometrical interpretation of the transformation $z \to (w-a_1)/a_0$ is that it transforms the triangle $z_1z_2z_3$ into the similar triangle $w_1w_2w_3$ *which has its centroid at the origin* ($w_1+w_2+w_3 = 0$). (See Figure 23.2.)

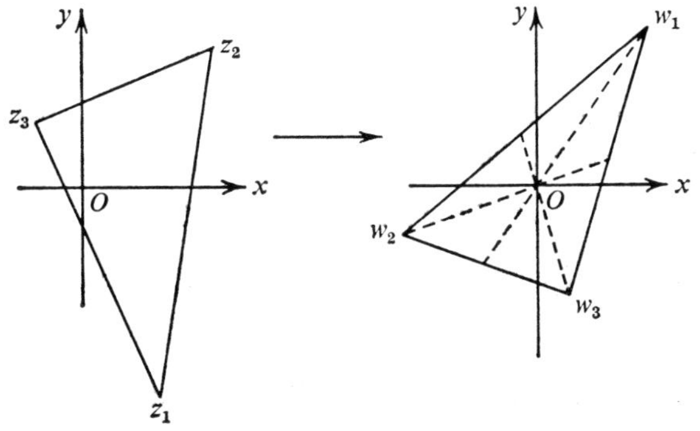

Fig. 23.2

Ex. 16. Why are the triangles $z_1z_2z_3$ and $w_1w_2w_3$ similar?

Cardan's solution of (2) consists in expressing the roots as the sums of pairs of complex numbers lying at the vertices of two *equilateral* triangles with centroid at the origin. (There is no 'completing the cube' process in general and thus it is not possible to find a further transformation that maps the triangle $w_1w_2w_3$ into an equilateral triangle.)

Since

$$w^3-3wpq-(p^3+q^3) = (w-p-q)(w-\omega p-\omega^2 q)(w-\omega^2 p-\omega q)$$

(see Exercise 19(*b*), Question 12) we have

$$w^3-3wpq-(p^3+q^3) = 0$$

$$\Leftrightarrow w = p+q \text{ or } w = \omega p+\omega^2 q \text{ or } w = \omega^2 p+\omega q. \tag{3}$$

Now write $$-pq = H, \quad -(p^3+q^3) = G; \tag{4}$$

if we can find values p, q satisfying (4) then we shall have a complete solution of the cubic equation.

From (4), p^3 and q^3 are the roots of the quadratic equation

$$Z^2+GZ-H^3 = 0. \tag{5}$$

Let α, β be the roots of the quadratic equation (5). Take p_1, q_1 as any cube roots of α and β; then

$$p_1^3+q_1^3 = \alpha+\beta = -G,$$

$$p_1^3q_1^3 = \alpha\beta = -H^3.$$

If $H = 0$, the original cubic is immediately soluble; if $H \neq 0$, then

$$\left(\frac{p_1q_1}{-H}\right)^3 = 1,$$

giving

$$\frac{p_1q_1}{-H} = \epsilon,$$

where $\epsilon = 1$, ω or ω^2, and p_1 and q_1/ϵ are suitable values to take for p and q.

In summary, to solve the equation

$$w^3+3Hw+G = 0 \quad (H \neq 0),$$

let α, β be the roots (repeated if necessary) of the quadratic equation

$$z^2+Gz-H^3 = 0,$$

and let $p = \alpha^{\frac{1}{3}}$, $q = \beta^{\frac{1}{3}}\epsilon$ where $\epsilon = \alpha^{\frac{1}{3}}\beta^{\frac{1}{3}}/(-H)$. Then

$$p+q, \quad \omega p+\omega^2 q, \quad \omega^2 p+\omega q \tag{6}$$

are the required roots for the cubic equation.

For

$$(w-(p+q))\,(w-(\omega p+\omega^2 q))\,(w-(\omega^2 p+\omega q)) \equiv w^3-3wpq-(p^3+q^3),$$

and p, q have been so chosen that

$$pq = -H, \quad p^3+q^3 = -G.$$

Example 8. *Solve the equation* $z^3-6z+6 = 0$.

We seek p, q such that

$$p^3+q^3 = -6,$$

$$pq = 2.$$

Thus, p^3 and q^3 are the roots of the quadratic

$$Z^2+6Z+8 = 0,$$

from which it follows that $p^3 = 2$, $q^3 = 4 = 2^2$.

Thus
$$z^3-6z+6 = 0$$
$$\Rightarrow z^3+(2^{\frac{1}{3}})^3+(2^{\frac{2}{3}})^3-3z(2^{\frac{1}{3}})(2^{\frac{2}{3}}) = 0$$
$$\Rightarrow (z+2^{\frac{1}{3}}+2^{\frac{2}{3}})(z+\omega 2^{\frac{1}{3}}+\omega 2^{\frac{2}{3}})(z+\omega^2 2^{\frac{1}{3}}+\omega 2^{\frac{2}{3}}) = 0$$
$$\Rightarrow z = -2^{\frac{1}{3}}-2^{\frac{2}{3}} \quad \text{or} \quad -\omega 2^{\frac{1}{3}}-\omega^2 2^{\frac{2}{3}} \quad \text{or} \quad -\omega^2 2^{\frac{1}{3}}-\omega 2^{\frac{2}{3}}$$
$$\Rightarrow z = -(2^{\frac{1}{3}}+2^{\frac{2}{3}}) \quad \text{or} \quad \{(2^{\frac{1}{3}}+2^{\frac{2}{3}})\pm j(2^{\frac{2}{3}}-2^{\frac{1}{3}})\sqrt{3}\}/2.$$

**Ex.* 17. Show that $w = Z-(H/Z)$ transforms equation (2) into
$$Z^6+GZ^3-H^3 = 0$$
the roots of which are p, ωp, $\omega^2 p$, q, ωq, $\omega^2 q$.

**Ex.* 18. With the notation of Ex. 17, show that, to each w correspond two values of Z and that to the triangle $w_1 w_1 w_3$ formed by the roots of (2) correspond *two equilateral* triangles.

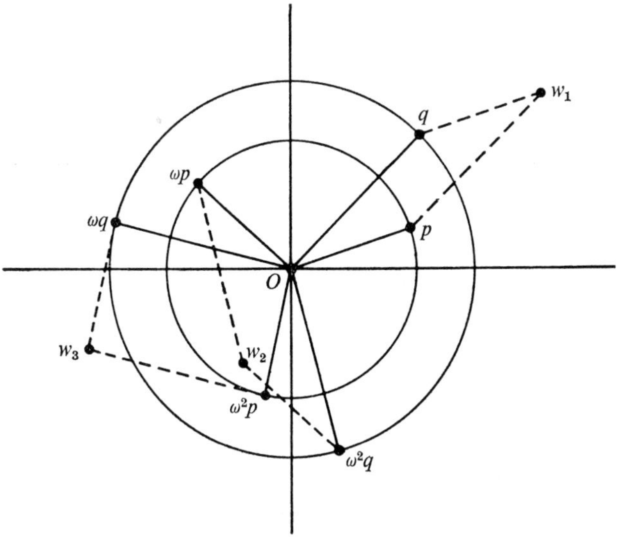

Fig. 23.3

Suppose now we have determined p and q as two complex numbers. Then, by (6), we have
$$w_1 = p+q, \quad w_2 = \omega p+\omega^2 q, \quad w_3 = \omega^2 p+\omega q.$$
(See Figure 23.3.)

Ex. 19. Show that w_2, w_3 may be found by constructing equilateral triangles on $-p$, $-q$ as base.

From now on we shall suppose that the coefficients of (2) *are real:* then either one root is real and the other two are conjugate (complex) numbers, or all three roots are real. From (5), since G, H are real, either p^3 or q^3 are both real or else p^3 and q^3 are complex conjugates, in which case we must have $|p| = |q|$.

Case (i). p^3 and q^3 both real (and unequal)—in which case we may take p and q as both real and the geometrical construction (Figure 23.4) shows that, of the solutions of (2), one is real and two are complex conjugates.

Case (ii). p^3 and q^3 are complex conjugates (and thus p, q may be taken as complex conjugates). In this case, we must have $|p| = |q|$ and the geometrical construction (Figure 23.5) shows that all three roots of (2) are real.

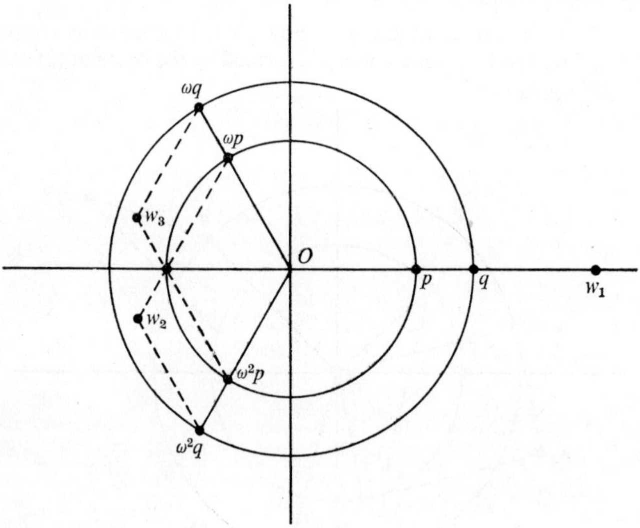

Fig. 23.4

Since this exhausts all the possibilities, we have arrived at the famous paradox that, if only one root of (2) is real, then p and q are both real, but, if all three roots of (2) are real, then they arise from complex values of p and q (*the irreducible case*). Put another way, if all three roots of a cubic equation are real then the working of Cardan's solution necessarily involves complex numbers p and q at the intermediate stages.

Ex. 20. Explain geometrically how (i) a root of multiplicity two, (ii) a root of multiplicity three, arise.

The expression
$$\Delta = [(w_2 - w_3)(w_3 - w_1)(w_1 - w_2)]^2$$

is called the *discriminant* of the equation

$$w^3+3Hw+G = 0.$$

From Example 7,
$$\Delta = -27(G^2+4H^3).$$

Consider now the various cases that may arise (remember that we are considering cubic equations with *real* coefficients).

(i) One real root α, two conjugate complex roots $\beta \pm j\gamma$.

$$\begin{aligned}\Delta &= (2j\gamma)^2(\alpha-\beta-j\gamma)^2(\beta-j\gamma-\alpha)^2\\ &= -4\gamma^2(-\alpha^2-\beta^2-\gamma^2+2\alpha\beta)^2\\ &< 0.\end{aligned}$$

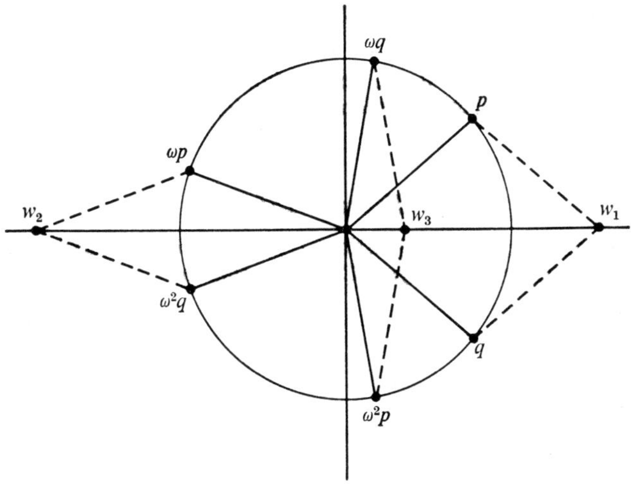

Fig. 23.5

(ii) A multiple (real) root. $\Delta = 0.$

(iii) Three unequal real roots $\Delta > 0.$

Thus, the nature of the roots determines the sign of Δ.

**Ex.* 21. Show, conversely, that the sign of Δ determines the nature of the roots of the real cubic equation $w^3+3Hw+G = 0$.

The case $\Delta > 0$ leads to three real solutions via complex values of p and q: the consequent working in Cardan's solution is frequently rather tedious, but can be avoided in this case by employing the trigonometric solution illustrated in Example 10, which employs the identity

$$\cos 3\theta \equiv 4\cos^3\theta - 3\cos\theta.$$

(See p. 244.)

Example 9. *Solve the equation* $z^3 - 6z + 4 = 0$.

Here $H = -2$, $G = 4$ and $\Delta > 0$: there are three real roots. Put

$$z = 2(2)^{\frac{1}{2}}\cos\theta.$$

$$16\sqrt{2}\cos^3\theta - 12\sqrt{2}\cos\theta = -4$$

$$\therefore \quad 4\cos^3\theta - 3\cos\theta = -\sqrt{\tfrac{1}{2}}$$

$$\therefore \quad \cos 3\theta = -\sqrt{\tfrac{1}{2}},$$

giving $\theta = \frac{1}{4}\pi$ or $\frac{5}{12}\pi$ or $\frac{11}{12}\pi$ and $z = 2$ or $2^{\frac{3}{2}}\cos\frac{5}{12}\pi$ or $2^{\frac{3}{2}}\cos\frac{11}{12}\pi$.

**Ex*. 22. Show that if $\Delta > 0$, the substitution $z = 2(-H)^{\frac{1}{2}}\cos\theta$ will always yield the solution of the cubic equation $z^3 + 3Hz + G = 0$.

Exercise 23(c)

Solve the cubic equations in Questions 1–5, using a trigonometric substitution if possible and otherwise employing Cardan's solution.

1. $z^3 - 2z + 4 = 0$.

2. $4z^3 - 5z + 6 = 0$.

3. $z^3 - 3z - 1 = 0$.

4. $12z^3 - 9z - 2 = 0$.

5. $z^3 - 9z + 12 = 0$.

6. Reduce the equation

$$z^3 - 6z^2 + 18z - 22 = 0$$

to the form $z^3 + 3Hz + G = 0$, and hence solve it completely.

7. Show that the equation

$$az^3 + 3bz^2 + 3cz + d = 0$$

is reduced to the equation

$$p(w-q)^3 = q(w-p)^3$$

by the transformation $w = az + b$ where p, q are the roots (supposed unequal) of the equation

$$(ac - b^2)\,\lambda^2 + (a^2d - 3abc + 2b^3)\,\lambda - (ac - b^2)^2 = 0.$$

Solve the equation

$$z^3 - 15z^2 + 57z - 5 = 0.$$ (O & C)

8. Solve the equation $z^3 + 5z^2 + 5z + 4 = 0$.

9. Solve the equation $z^3+5z+2j = 0$.

10. If p, q are real, find the turning points (if any) of the graph of $y = x^3+px+q$. Deduce a necessary and sufficient condition for all the roots of the equation

$$x^3+px+q = 0$$

to be real.

Prove that the roots of the cubic equation $\det(\mathbf{A}-\lambda\mathbf{I}) = 0$ where

$$\mathbf{A} = \begin{pmatrix} a & h & g \\ h & a & f \\ g & f & a \end{pmatrix}$$

are all real. Can two of the roots be equal?

11. (*The Quartic Equation.*) Show that the general quartic equation

$$z^4+4pz^3+6qz^2+4rz+s = 0$$

can be reduced to the form

$$w^4+aw^2+bw+c = 0.$$

By identifying this equation with the equation

$$(w^2+\alpha)^2 = \beta(w+\gamma)^2$$

show that α is a root of the *cubic* equation

$$8\alpha^3-4a\alpha^2-8c\alpha+4ac-b^2 = 0.$$

If one of the roots of this equation is α_0 and the corresponding values of β, γ are β_0, γ_0, show that

$$w^4+aw^2+bw+c = (w^2-\beta_0^{\frac{1}{2}}w+\alpha_0-\gamma_0\beta_0^{\frac{1}{2}})\,(w^2+\beta_0^{\frac{1}{2}}w+\alpha_0+\gamma_0\beta_0^{\frac{1}{2}}).$$

Solve the quartic equation

$$x^4+4x-1 = 0.$$

Miscellaneous Exercise 23

1. Prove, by graphical methods or otherwise, that a cubic equation (with real coefficients) of the form

$$ax^3+bx^2+cx+d = 0$$

has at least one real root.

Prove that, if λ is real, the equation

$$x^3-3x^2+3x+\lambda = 0$$

is satisfied for one and only one real value of x. (O & C)

2. By considering two alternative expansions of the expression

$$\ln\{(1+\alpha x)\,(1+\beta x)\,(1+\gamma x)\}$$

(α, β, γ all real), prove that

$$\alpha^4+\beta^4+\gamma^4 = p^4-4p^2q+4pr+2q^2,$$

where α, β, γ are the roots of the equation

$$x^3+px^2+qx+r = 0.$$

3. In the equation

$$x^4+ax^3-20x^2+bx-576 = 0,$$

a and b are to be chosen so that two of the roots are equal and the sum of the other two roots is zero.

Find all possible values of a and b, supposing them chosen so that the roots of the given equation are all real. (O & C)

4. Find the value or values of a for which the roots of the equation

$$2x^3+6x^2+5x+a = 0$$

are in arithmetical progression. (O & C)

5. Given that $-2+jk$ is a root of the equation

$$2x^4+8x^3+11x^2+4x+5 = 0,$$

find k and the other three roots of the equation.

6. Given that α, β, γ are the roots of the equation

$$x^3-px^2+qx-r = 0,$$

and that $s_n = \alpha^n+\beta^n+\gamma^n$, prove that

$$s_1 = p, \quad s_2 = p^2-2q, \quad s_3 = ps_2-qs_1+3r.$$

Calculate s_2 and s_3 for the equation

$$x^3-12x^2+30x-20 = 0.$$

Show that $(s_2)^{\frac{1}{2}}$ and $(s_3)^{\frac{1}{3}}$ are both near to 9 and verify that 9 approximates to a root of the equation. (O & C)

7. If ω is a root other than 1 of the equation $\omega^7 = 1$, prove that the other roots are $\omega^2, \omega^3, \omega^4, \omega^5, \omega^6$.

Prove that $1+\omega+\omega^2+\omega^3+\omega^4+\omega^5+\omega^6 = 0$.

If $\alpha = \omega+\omega^6, \beta = \omega^2+\omega^5, \gamma = \omega^3+\omega^4$, prove that the equation with roots α, β, γ is

$$z^3+z^2-2z-1 = 0.$$

Hence, or otherwise, find the values of

(i) $\cos\frac{2}{7}\pi+\cos\frac{4}{7}\pi+\cos\frac{6}{7}\pi$. (ii) $\cos\frac{2}{7}\pi\cos\frac{4}{7}\pi\cos\frac{6}{7}\pi$. (O & C)

8. When $f(x)$ and $g(x)$ are given polynomials having no common factor, prove that the values of the constant k for which the equation

$$f(x)-kg(x) = 0$$

has a repeated root are given by $f(\alpha)/g(\alpha)$ where α is a root of the equation

$$f(x)\,g'(x)-f'(x)\,g(x) = 0.$$

Hence, or otherwise, find the values of k for which the equation

$$x^3-3x^2+3kx-1 = 0$$

has a repeated root, and find also all the roots of the equation for each case. (O & C)

9. Discuss the reality of the solution of the equation

$$2x^3+3x^2-36x+k = 0$$

for different real values of k.

10. If $f(x)$ is a polynomial in x which has a factor $(x-\alpha)^2$, prove that $(x-\alpha)$ is a factor of $f'(x)$. Prove conversely that, if $f(x)$ and $f'(x)$ have a common factor $x-\alpha$, then $(x-\alpha)^2$ is a factor of $f(x)$.

Solve the equation
$$4x^4+4x^3-31x^2-66x-36 = 0$$
given that it has a repeated root. (O & C)

11. If α, β, γ are the roots of the equation
$$x^3+px+q = 0,$$
find the equation with roots $\beta/\gamma+\gamma/\beta$, $\gamma/\alpha+\alpha/\gamma$, $\alpha/\beta+\beta/\alpha$. (O & C)

12. The sum of the kth powers of the roots of the equation
$$x^4+px^2+qx+r = 0$$
is denoted by s_k. Prove that
$$s_2 = -2p, \quad s_3 = -3q.$$
$$s_{n+4}+ps_{n+2}+qs_{n+1}+rs_n = 0 \quad (n \geqslant 0),$$
and deduce that
$$s_4 = 2p^2-4r, \quad s_7 = 7q(r-p^2).$$

Given that the roots a, b, c, d of the equation are all real and that $s_7 = 0$, prove that $q = 0$ and deduce that
$$(a+b)(a+c)(a+d) = 0.$$
(O & C)

13. Show that, if $z = 2(-H)^{\frac{1}{2}} \cos\theta$, where z is a root of the cubic equation $z^3+3Hz+G = 0$, then
$$\cos 3\theta = \sqrt{(-G^2/4H^3)}.$$

Deduce that, if $H < 0$ and $G^2+4H^3 < 0$ the cubic equation has three real roots.

By using the expression
$$\cos 3\theta = \tfrac{1}{2}(e^{3j\theta}+e^{-3j\theta}),$$
solve the equation when
$$G^2+4H^3 > 0,$$
treating separately the cases
$$H < 0 \quad \text{and} \quad H > 0.$$

24. *Vector products and their applications*

1. THE VECTOR PRODUCT

We have already seen, in Chapter 11, that it is frequently useful to find a vector which is perpendicular to two given vectors; for example, to find the equation of the plane through three given points A, B, C it is sufficient to know the unit normal vector; that is, a unit vector perpendicular to both AB and AC. We therefore introduce a new method of combining two vectors $\mathbf{a}$ and $\mathbf{b}$ to form a third vector perpendicular to both of them and called their *vector product*, written $\mathbf{a} \wedge \mathbf{b}$ (or $\mathbf{a} \times \mathbf{b}$). It is defined by

$$\mathbf{a} \wedge \mathbf{b} = |\mathbf{a}||\mathbf{b}| \sin \theta \hat{\mathbf{c}},$$

where θ is the angle between $\mathbf{a}$ and $\mathbf{b}$ (see Chapter 11) and $\hat{\mathbf{c}}$ is a unit vector perpendicular to both $\mathbf{a}$ and $\mathbf{b}$ and with sense that makes $\mathbf{a}$, $\mathbf{b}$, $\hat{\mathbf{c}}$ (*in that order!*) a right-handed triple (see Chapter 3). The relative positions of $\mathbf{a}$, $\mathbf{b}$, and $\mathbf{a} \wedge \mathbf{b}$ are shown in Figure 24.1 where $\mathbf{a}$, $\mathbf{a} \wedge \mathbf{b}$ lie on the page and $\mathbf{b}$ points into the page.

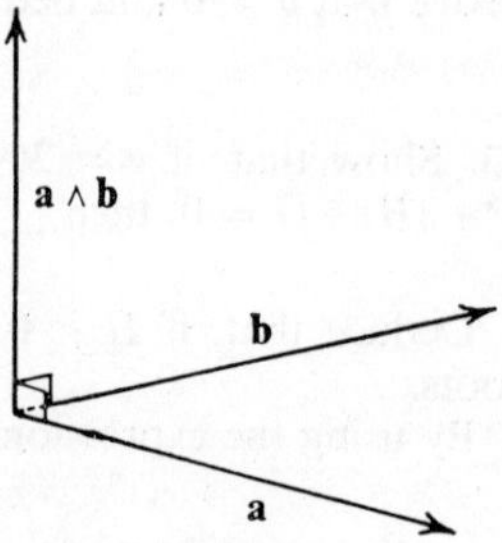

Fig. 24.1

**Ex*. 1. Show that $\mathbf{a} \wedge \mathbf{a} = \mathbf{0}$.

**Ex*. 2. Show that $\mathbf{a} \wedge \lambda\mathbf{b} = \lambda(\mathbf{a} \wedge \mathbf{b})$.

**Ex*. 3. Show that, if $\mathbf{a}$ and $\mathbf{b}$ are parallel, then $\mathbf{a} \wedge \mathbf{b} = \mathbf{0}$.

**Ex*. 4. Show that, if $a \wedge \mathbf{b} = 0$, then *either* $\mathbf{a}$ and $\mathbf{b}$ are parallel *or* one or other or both of $\mathbf{a}$, $\mathbf{b}$ is zero.

**Ex*. 5. Show that the vector product is *anti-commutative*; that is, that

$$\mathbf{a} \wedge \mathbf{b} = -\mathbf{b} \wedge \mathbf{a}.$$

**Ex*. 6. Verify that $\mathbf{i} \wedge \mathbf{i} = \mathbf{j} \wedge \mathbf{j} = \mathbf{k} \wedge \mathbf{k} = \mathbf{0}$ and that

$$\mathbf{j} \wedge \mathbf{k} = -\mathbf{k} \wedge \mathbf{j} = \mathbf{i}, \quad \mathbf{k} \wedge \mathbf{i} = -\mathbf{i} \wedge \mathbf{k} = \mathbf{j}, \quad \mathbf{i} \wedge \mathbf{j} = -\mathbf{j} \wedge \mathbf{i} = \mathbf{k}.$$

Ex. 7. $\mathbf{a}$ is a vector of magnitude 2 units pointing due east; $\mathbf{b}$ has magnitude 3 units and points N 60° E. $\hat{\mathbf{c}}$ is a vertical unit vector. Find $\mathbf{a} \wedge \mathbf{b}$ and $\mathbf{b} \wedge \mathbf{a}$.

So far, apart from the result of Ex. 2, the vector product appears remarkably dissimilar to the product of ordinary numbers. However, there is one rule which vector products obey and which, in a sense, justifies the use of the word 'product': it is *distributive over vector addition*; that is

$$\mathbf{a} \wedge (\mathbf{b}+\mathbf{c}) = (\mathbf{a} \wedge \mathbf{b})+(\mathbf{a} \wedge \mathbf{c}).$$

We shall deduce this very important result in a manner reminiscent of that used to prove the analogous result for scalar products:

$$\mathbf{a}.(\mathbf{b}+\mathbf{c}) = (\mathbf{a}.\mathbf{b})+(\mathbf{a}.\mathbf{c}).$$

In Chapter 11, we saw that the scalar product of $\mathbf{a} = \mathbf{OA}$ with a unit vector $\hat{\mathbf{n}}$ could be interpreted geometrically as the (scalar) projection of $\mathbf{a}$

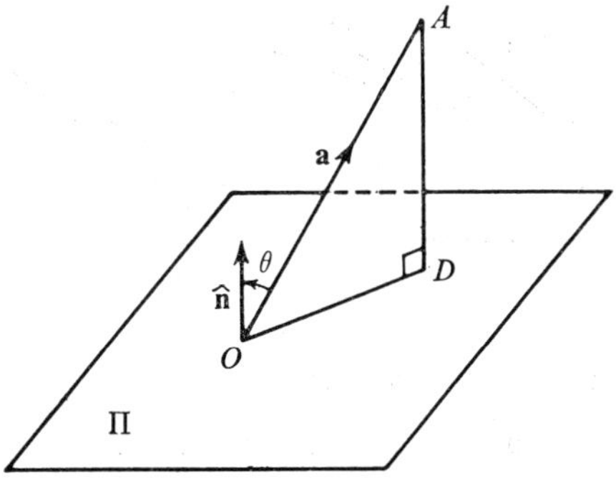

Fig. 24.2

in the direction $\hat{\mathbf{n}}$ and it was this fact that enabled us to deduce the distributive law for scalar products. Consider the plane, Π, perpendicular to $\hat{\mathbf{n}}$ and through O; let the foot of the perpendicular from A to Π be D; then the vector $\mathbf{OD}$ is called the (*vector*) *projection of* $\mathbf{a}$ *on the plane* Π (see Figure 24.2).

It follows from the definition of the vector product that

$$OD = |\mathbf{a} \wedge \hat{\mathbf{n}}|,$$

but $$\mathbf{OD} \neq \mathbf{a} \wedge \hat{\mathbf{n}},$$

since $\mathbf{OD}$ and $\mathbf{a} \wedge \hat{\mathbf{n}}$ are perpendicular to one another.

As in Chapter 11, Section 1, the vector projection of $\mathbf{a}$ on Π is unique and, if the projection of $\mathbf{a}_1$ and $\mathbf{a}_2$ on Π are $\mathbf{OD}_1$ and $\mathbf{OD}_2$, then the projection of $\mathbf{OB} = \mathbf{a}_1+\mathbf{a}_2$ is $\mathbf{OC} = \mathbf{OD}_1+\mathbf{OD}_2$ (see Figure 24.3 and consider the mid-point of A_1A_2).

In words, the (vector) projection, in a plane, of the sum of two vectors is the sum of the separate projections of the two vectors.

Now consider $\hat{\mathbf{n}} \wedge (\mathbf{a}_1+\mathbf{a}_2)$: from what has just been demonstrated,

$$|\hat{\mathbf{n}} \wedge (\mathbf{a}_1+\mathbf{a}_2)| = |(\hat{\mathbf{n}} \wedge \mathbf{a}_1)+(\hat{\mathbf{n}} \wedge \mathbf{a}_2)|.$$

But $\hat{\mathbf{n}} \wedge (\mathbf{a}_1+\mathbf{a}_2)$ is perpendicular to the plane OBC and thus to $\mathbf{OC}$, $\hat{\mathbf{n}} \wedge \mathbf{a}_1$ is perpendicular to $\mathbf{OD}_1$, $\hat{\mathbf{n}} \wedge \mathbf{a}_2$ is perpendicular to $\mathbf{OD}_2$ and thus, by rotating the parallelogram OD_1CD_2 through a right angle in the plane Π, it follows that

$$\hat{\mathbf{n}} \wedge (\mathbf{a}_1+\mathbf{a}_2) = \hat{\mathbf{n}} \wedge \mathbf{a}_1+\hat{\mathbf{n}} \wedge \mathbf{a}_2.$$

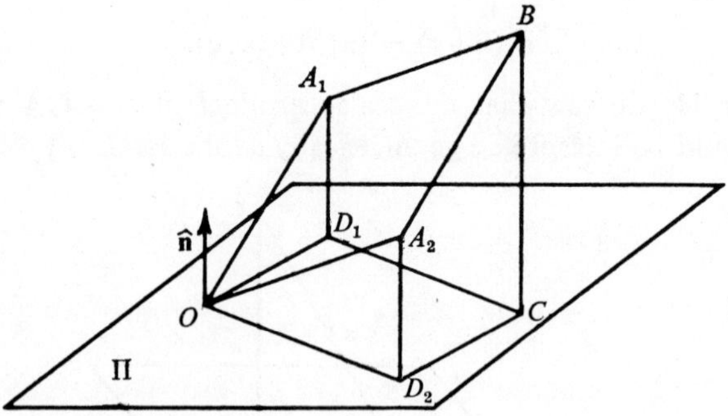

Fig. 24.3

**Ex.* 8. By writing $\mathbf{a} = \lambda\hat{\mathbf{a}}$, deduce that

$$\mathbf{a} \wedge (\mathbf{b}+\mathbf{c}) = (\mathbf{a} \wedge \mathbf{b})+(\mathbf{a} \wedge \mathbf{c}).$$

Show also that $$(\mathbf{b}+\mathbf{c}) \wedge \mathbf{a} = (\mathbf{b} \wedge \mathbf{a})+(\mathbf{c} \wedge \mathbf{a}).$$

With the distributive rules behind us, it is an easy matter to express the vector product $\mathbf{a} \wedge \mathbf{b}$ in terms of components in three mutually perpendicular directions $\mathbf{i}, \mathbf{j}, \mathbf{k}$. If $\mathbf{a} = a_1\mathbf{i}+a_2\mathbf{j}+a_3\mathbf{k}$ and $\mathbf{b} = b_1\mathbf{i}+b_2\mathbf{j}+b_3\mathbf{k}$, then

$$\begin{aligned}\mathbf{a} \wedge \mathbf{b} &= (a_1\mathbf{i}+a_2\mathbf{j}+a_3\mathbf{k}) \wedge (b_1\mathbf{i}+b_2\mathbf{j}+b_3\mathbf{k})\\ &= (a_2b_3-a_3b_2)\,\mathbf{i}+(a_3b_1-a_1b_3)\,\mathbf{j}+(a_1b_2-a_2b_1)\,\mathbf{k},\end{aligned}$$

on repeated application of the distributive rule and using the results proved in Ex. 6. The result is best remembered written formally as a determinant:

$$\mathbf{a} \wedge \mathbf{b} = \begin{vmatrix} \mathbf{i} & \mathbf{j} & \mathbf{k} \\ a_1 & a_2 & a_3 \\ b_1 & b_2 & b_3 \end{vmatrix}.$$

Ex. 9. If $\mathbf{a} = \mathbf{i}+2\mathbf{j}+3\mathbf{k}$ and $\mathbf{b} = \mathbf{i}-\mathbf{j}-\mathbf{k}$, find $\mathbf{a} \wedge \mathbf{b}$.

Since $|\mathbf{a} \wedge \mathbf{b}| = |\mathbf{a}|\,|\mathbf{b}| \sin\theta$, we see that the magnitude of the vector product gives us twice the area of ΔOAB (see Figure 24.4). The vector

$\frac{1}{2}(\mathbf{a} \wedge \mathbf{b})$ is often referred to as the *vector area* of ΔOAB. (Notice that the vector area of ΔOAB is minus the vector area of ΔOBA.) More generally, the vector area of the triangle ABC is $\frac{1}{2}(\mathbf{AB} \wedge \mathbf{AC})$.

Ex. 10. $OABC$ is a tetrahedron; prove that the sum of the vector areas of the triangles OBC, OCA, OAB is the vector area of the triangle ABC.

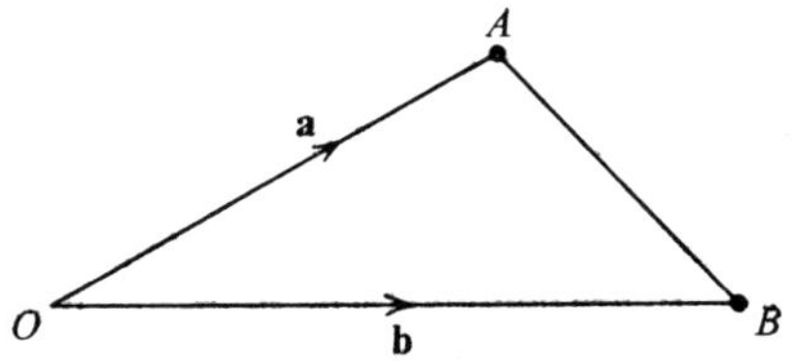

Fig. 24.4

Example 1. *ABCD is a plane quadrilateral whose sides CD, BA intersect at O; if P, Q are the mid-points of the diagonals AC, BD, prove that the area of ΔOPQ is one-quarter of the area of the quadrilateral ABCD.*

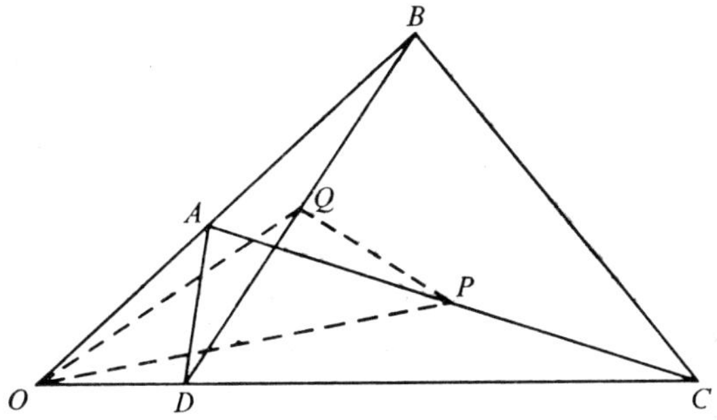

Fig. 24.5

Take origin O and $\mathbf{a} = \mathbf{OA}$, $\mathbf{b} = \lambda\mathbf{a}$, $\mathbf{c} = \mathbf{OC}$, $\mathbf{d} = \mu\mathbf{c}$. We have

$$\mathbf{p} = \tfrac{1}{2}(\mathbf{a}+\mathbf{c}),$$

$$\mathbf{q} = \tfrac{1}{2}(\lambda\mathbf{a}+\mu\mathbf{c}),$$

and the vector area of OPQ

$$= \tfrac{1}{2}\mathbf{p} \wedge \mathbf{q}$$

$$= \tfrac{1}{8}(\mathbf{a}+\mathbf{c}) \wedge (\lambda\mathbf{a}+\mu\mathbf{c})$$

$$= \tfrac{1}{8}(\mu-\lambda)(\mathbf{a} \wedge \mathbf{c})$$

on using the distributive and anti-commutative rules. But since $ABCD$ is a plane quadrilateral, its area is the magnitude of the sum of the vector areas of ΔADB and ΔBDC (notice that we letter the triangles in the same sense). Thus

$$\begin{aligned} ABCD &= \tfrac{1}{2}(\mathbf{d}\wedge\mathbf{b}+\mathbf{b}\wedge\mathbf{a}+\mathbf{a}\wedge\mathbf{d})+\tfrac{1}{2}(\mathbf{d}\wedge\mathbf{c}+\mathbf{c}\wedge\mathbf{b}+\mathbf{b}\wedge\mathbf{d}) \\ &= \tfrac{1}{2}(\mathbf{b}\wedge\mathbf{a}+\mathbf{a}\wedge\mathbf{d}+\mathbf{d}\wedge\mathbf{c}+\mathbf{c}\wedge\mathbf{b}) \\ &= \tfrac{1}{2}[\mathbf{a}\wedge(\mathbf{d}-\mathbf{b})-\mathbf{c}\wedge(\mathbf{d}-\mathbf{b})] \\ &= \tfrac{1}{2}(\mathbf{a}-\mathbf{c})\wedge(\mathbf{d}-\mathbf{b}) \\ &= \tfrac{1}{2}(\mathbf{a}-\mathbf{c})\wedge(\mu\mathbf{c}-\lambda\mathbf{a}) \\ &= \tfrac{1}{2}(\mu-\lambda)(\mathbf{a}\wedge\mathbf{c}), \quad \text{and the result follows.} \end{aligned}$$

Ex. 11. The position vectors, relative to some origin O, of the vertices of a convex polygon $P_1P_2P_3\ldots P_n$ are denoted by $\mathbf{p}_1, \mathbf{p}_2, \mathbf{p}_3, \ldots, \mathbf{p}_n$. Extend the definition of the vector area of a triangle to the vector area of a convex polygon and show that the vector area of $P_1P_2P_3\ldots P_n$ is

$$\tfrac{1}{2}(\mathbf{p}_1\wedge\mathbf{p}_2+\mathbf{p}_2\wedge\mathbf{p}_3+\ldots+\mathbf{p}_{n-1}\wedge\mathbf{p}_n+\mathbf{p}_n\wedge\mathbf{p}_1).$$

Prove the result of Example 1 by considering the vector areas of $ABCD$ and OPQ.

2. PRODUCTS OF THREE VECTORS

If from two vectors, **b** and **c**, we form the vector product $\mathbf{b}\wedge\mathbf{c}$, we may multiply **a** by this new vector in two quite different ways, to give either a scalar or a vector:

(i) *the scalar triple product*, $\mathbf{a}.(\mathbf{b}\wedge\mathbf{c})$;

(ii) *the vector triple product*, $\mathbf{a}\wedge(\mathbf{b}\wedge\mathbf{c})$.

The component form for $\mathbf{a}.(\mathbf{b}\wedge\mathbf{c})$ is easily found. Recall that

$$\mathbf{x}.\mathbf{y} = x_1y_1+x_2y_2+x_3y_3;$$

thus, since $\mathbf{a} = a_1\mathbf{i}+a_2\mathbf{j}+a_3\mathbf{k}$ and

$$\mathbf{b}\wedge\mathbf{c} = \begin{vmatrix} \mathbf{i} & \mathbf{j} & \mathbf{k} \\ b_1 & b_2 & b_3 \\ c_1 & c_2 & c_3 \end{vmatrix},$$

we have

$$\mathbf{a}.(\mathbf{b}\wedge\mathbf{c}) = \begin{vmatrix} a_1 & a_2 & a_3 \\ b_1 & b_2 & b_3 \\ c_1 & c_2 & c_3 \end{vmatrix}.$$

**Ex.* 12. Prove that $\mathbf{a}.(\mathbf{b}\wedge\mathbf{c}) = \mathbf{b}.(\mathbf{c}\wedge\mathbf{a}) = \mathbf{c}.(\mathbf{a}\wedge\mathbf{b})$ (that is, the scalar triple product is unaffected by cyclic interchange of the letters) and that

$$\mathbf{a}.(\mathbf{b}\wedge\mathbf{c}) = (\mathbf{a}\wedge\mathbf{b}).\mathbf{c}$$

(the . and $\wedge$ may be interchanged).

The scalar triple product $\mathbf{a}.(\mathbf{b} \wedge \mathbf{c})$ has an interesting and important geometrical interpretation. Suppose $\hat{\mathbf{n}}$ is perpendicular to the plane OBC (see Figure 24.6). Then $\mathbf{b} \wedge \mathbf{c} = S\hat{\mathbf{n}}$, where S is the area of the parallelogram $OBDC$. Thus

$$\begin{aligned}\mathbf{a}.(\mathbf{b} \wedge \mathbf{c}) &= (\mathbf{a}.\hat{\mathbf{n}})\, S \\ &= \text{volume of the parallelepiped formed by } \mathbf{a}, \mathbf{b}, \mathbf{c}.\end{aligned}$$

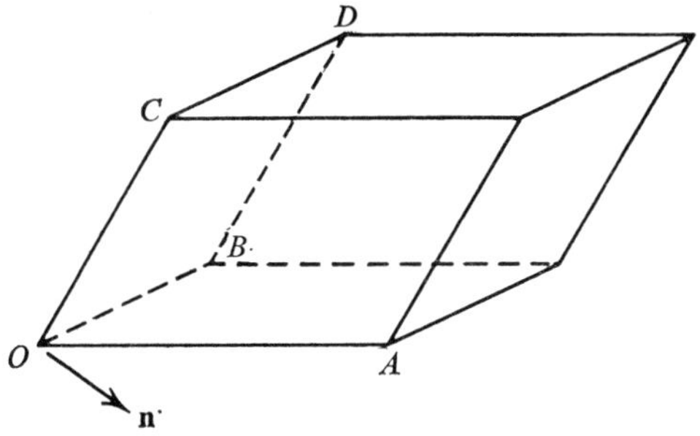

Fig. 24.6

Ex. 13. Show that $\mathbf{a}.(\mathbf{b} \wedge \mathbf{c}) = 0 \Leftrightarrow O, A, B, C$, lie in a plane.

Ex. 14. Prove that $\mathbf{x}.(\mathbf{y} \wedge \mathbf{z}) = (\mathbf{x} \wedge \mathbf{y}).\mathbf{z}$ without using components (and thus without assuming the distributive law $\mathbf{a} \wedge (\mathbf{b}+\mathbf{c}) = \mathbf{a} \wedge \mathbf{b}+\mathbf{a} \wedge \mathbf{c}$).

Ex. 15. The fact that $\mathbf{x}.(\mathbf{y} \wedge \mathbf{z}) = (\mathbf{x} \wedge \mathbf{y}).\mathbf{z}$ may be used to verify the distributive law: define $\mathbf{d} = \mathbf{a} \wedge (\mathbf{b}+\mathbf{c})-\mathbf{a} \wedge \mathbf{b}-\mathbf{a} \wedge \mathbf{c}$ and form the product $\hat{\mathbf{u}}.\mathbf{d}$, where $\hat{\mathbf{u}}$ is an arbitrary unit vector. Show that $\hat{\mathbf{u}}.\mathbf{d} = 0$. Why does this imply that $\mathbf{d} = \mathbf{0}$?

**Ex.* 16. Show that the volume of the tetrahedron $OABC$ is $\frac{1}{6}\mathbf{a}.(\mathbf{b} \wedge \mathbf{c})$.

**Ex.* 17. Prove that, if any one of the vectors $\mathbf{a}, \mathbf{b}, \mathbf{c}$ is a scalar multiple of one of the other two, then $\mathbf{a}.(\mathbf{b} \wedge \mathbf{c}) = 0$.

The vector triple product $\mathbf{a} \wedge (\mathbf{b} \wedge \mathbf{c})$ is a little more difficult to express in component form. First observe that, if $\mathbf{a}$ is perpendicular to $\mathbf{b}$ and also perpendicular to $\mathbf{c}$, then it is parallel to $\mathbf{b} \wedge \mathbf{c}$ and thus $\mathbf{a} \wedge (\mathbf{b} \wedge \mathbf{c}) = \mathbf{0}$; let us assume that none of $\mathbf{a}, \mathbf{b}, \mathbf{c}$ is zero and that $\mathbf{a}$ is not perpendicular to $\mathbf{b}$ and thus that $\mathbf{a}.\mathbf{b} \neq 0$. Again, if $\mathbf{b} = \lambda\mathbf{c}$, $\mathbf{a} \wedge (\mathbf{b} \wedge \mathbf{c}) = \mathbf{0}$; let us assume that $\mathbf{b} \neq \lambda\mathbf{c}$. Since $\mathbf{b} \wedge \mathbf{c}$ is perpendicular to both $\mathbf{b}$ and $\mathbf{c}$ and since $\mathbf{a} \wedge (\mathbf{b} \wedge \mathbf{c})$ is perpendicular to $\mathbf{b} \wedge \mathbf{c}$ it follows that $\mathbf{a} \wedge (\mathbf{b} \wedge \mathbf{c})$ lies in the plane of $\mathbf{b}$ and $\mathbf{c}$. Thus, $\mathbf{a} \wedge (\mathbf{b} \wedge \mathbf{c})$ can be expressed uniquely as $\lambda\mathbf{b}+\mu\mathbf{c}$; we have to determine λ and μ. By Ex. 17,

$$\mathbf{a}.[\mathbf{a} \wedge (\mathbf{b} \wedge \mathbf{c})] = 0,$$

$$\therefore\ \lambda(\mathbf{a}.\mathbf{b})+\mu(\mathbf{a}.\mathbf{c}) = 0,$$

$$\therefore\ \frac{\lambda}{\mu} = -\frac{(\mathbf{a}.\mathbf{c})}{(\mathbf{a}.\mathbf{b})} \quad (\text{since } \mathbf{a}.\mathbf{b} \neq 0),$$

and thus $\mathbf{a} \wedge (\mathbf{b} \wedge \mathbf{c}) = \nu[(\mathbf{a}.\mathbf{c})\,\mathbf{b} - (\mathbf{a}.\mathbf{b})\,\mathbf{c}]$.

To find ν, consider the **i** component of both sides:

$$a_2(b_1 c_2 - b_2 c_1) - a_3(b_3 c_1 - b_1 c_3)$$
$$= \nu[(a_1 c_1 + a_2 c_2 + a_3 c_3)\, b_1 - (a_1 b_1 + a_2 b_2 + a_3 b_3)\, c_1],$$

$$\therefore\; b_1(a_2 c_2 + a_3 c_3) - c_1(a_2 b_2 + a_3 b_3) = \nu[b_1(a_2 c_2 + a_3 c_3) - c_1(a_2 b_2 - a_3 b_3)].$$

$$\therefore\; \nu = 1.$$

Thus we have finally

$$\mathbf{a} \wedge (\mathbf{b} \wedge \mathbf{c}) = (\mathbf{a}.\mathbf{c})\,\mathbf{b} - (\mathbf{a}.\mathbf{b})\,\mathbf{c}.$$

**Ex.* 18. Prove that the vector triple product is not associative; that is

$$\mathbf{a} \wedge (\mathbf{b} \wedge \mathbf{c}) \neq (\mathbf{a} \wedge \mathbf{b}) \wedge \mathbf{c}.$$

Ex. 19. Prove that $\mathbf{a} \wedge (\mathbf{b} \wedge \mathbf{c}) + \mathbf{b} \wedge (\mathbf{c} \wedge \mathbf{a}) + \mathbf{c} \wedge (\mathbf{a} \wedge \mathbf{b}) = \mathbf{0}$.

Exercise 24(a)

1. Prove that $(\mathbf{a}-\mathbf{b}) \wedge (\mathbf{a}+\mathbf{b}) = 2\mathbf{a} \wedge \mathbf{b}$ and interpret the result geometrically in terms of areas.

2. What can you deduce from the equation $\mathbf{r} \wedge \mathbf{a} = \mathbf{r} \wedge \mathbf{b}$?

3. Prove that $\mathbf{a}+\mathbf{b}+\mathbf{c} = \mathbf{0} \Rightarrow \mathbf{b} \wedge \mathbf{c} = \mathbf{c} \wedge \mathbf{a} = \mathbf{a} \wedge \mathbf{b}$ and interpret the result geometrically.

Is it true that $\mathbf{b} \wedge \mathbf{c} = \mathbf{c} \wedge \mathbf{a} = \mathbf{a} \wedge \mathbf{b} \Rightarrow \mathbf{a}+\mathbf{b}+\mathbf{c} = \mathbf{0}$?

4. Find $\mathbf{a} \wedge \mathbf{b}$ in the following cases:

(i) $\mathbf{a} = 3\mathbf{i}-\mathbf{j}+\mathbf{k}$, $\mathbf{b} = 2\mathbf{i}+\mathbf{j}+\mathbf{k}$; (ii) $\mathbf{a} = 5\mathbf{i}+4\mathbf{j}+3\mathbf{k}$, $\mathbf{b} = 3\mathbf{i}+4\mathbf{j}+5\mathbf{k}$;
(iii) $\mathbf{a} = \mathbf{i}-\mathbf{k}$, $\mathbf{b} = \mathbf{j}+\mathbf{k}$; (iv) $\mathbf{a} = 2\mathbf{i}-\mathbf{j}-\mathbf{k}$, $\mathbf{b} = 2\mathbf{i}+\mathbf{j}+\mathbf{k}$;
(v) $\mathbf{a} = a\mathbf{i}+c\mathbf{k}$, $\mathbf{b} = b\mathbf{j}$.

5. If $\mathbf{a} = 3\mathbf{i}+\mathbf{j}+2\mathbf{k}$, $\mathbf{b} = \mathbf{i}+2\mathbf{j}+3\mathbf{k}$, $\mathbf{c} = \mathbf{i}-3\mathbf{j}-4\mathbf{k}$, find $\mathbf{a}.(\mathbf{b} \wedge \mathbf{c})$. What can you deduce about the points A, B, C?

6. If $\mathbf{a} = \mathbf{i}+\mathbf{j}$, $\mathbf{b} = 2\mathbf{i}+\mathbf{j}$, $\mathbf{c} = \mathbf{i}+\mathbf{k}$, form the vector products $\mathbf{a} \wedge \mathbf{b}$ and $(\mathbf{a} \wedge \mathbf{b}) \wedge \mathbf{c}$ and mark the position of the five vectors $\mathbf{a}$, $\mathbf{b}$, $\mathbf{c}$, $\mathbf{a} \wedge \mathbf{b}$, $(\mathbf{a} \wedge \mathbf{b}) \wedge \mathbf{c}$ in a rough sketch.

7. The points A, B, C have position vectors $\mathbf{a} = \mathbf{i}-4\mathbf{j}+3\mathbf{k}$, $\mathbf{b} = 2\mathbf{j}-3\mathbf{k}$, $\mathbf{c} = 3\mathbf{i}-6\mathbf{k}$. Find the unit vector perpendicular to the plane OAB and deduce the perpendicular distance from C to this plane.

8. With the notation of Question 7, find the area of ΔOAB and hence the volume of the tetrahedron $OABC$.

9. If A, B, C are non-collinear points with position vectors $\mathbf{a}$, $\mathbf{b}$, $\mathbf{c}$, prove that the vector $\mathbf{b} \wedge \mathbf{c}+\mathbf{c} \wedge \mathbf{a}+\mathbf{a} \wedge \mathbf{b}$ is normal to the plane ABC.

If $\mathbf{d}$ is the position vector of the point D and $\mathbf{a} = 5\mathbf{i}+\mathbf{j}+\mathbf{k}$, $\mathbf{b} = \mathbf{i}+\mathbf{j}+3\mathbf{k}$, $\mathbf{c} = 3\mathbf{i}+4\mathbf{j}-\mathbf{k}$, $\mathbf{d} = 8\mathbf{i}+3\mathbf{j}+2\mathbf{k}$, find:

(i) the unit normal to the plane ABC;
(ii) the area of the triangle ABC;
(iii) the volume of the tetrahedron $ABCD$.

10. If $\mathbf{a} = a_1\mathbf{i}+a_2\mathbf{j}+a_3\mathbf{k}$, $\mathbf{b} = b_1\mathbf{i}+b_2\mathbf{j}+b_3\mathbf{k}$, $\mathbf{c} = c_1\mathbf{i}+c_2\mathbf{j}+c_3\mathbf{k}$, prove that the volume of the tetrahedron $OABC$ is

$$\frac{1}{6}\begin{vmatrix} a_1 & a_2 & a_3 \\ b_1 & b_2 & b_3 \\ c_1 & c_2 & c_3 \end{vmatrix} \text{ cubic units.}$$

If $\mathbf{d} = d_1\mathbf{i}+d_2\mathbf{j}+d_3\mathbf{k}$, write down the volume of the tetrahedron $ABCD$.

What is the condition that A, B, C, D should be coplanar. Is this a necessary and sufficient condition?

11. If $\mathbf{a}$, $\mathbf{b}$ are perpendicular vectors and if $\mathbf{r} \wedge \mathbf{a} = \mathbf{b}$, prove that

$$\mathbf{r} = \lambda\mathbf{a}+\mu(\mathbf{a} \wedge \mathbf{b}),$$

where λ may be chosen arbitrarily. Find μ in terms of $|\mathbf{a}|$.

12. If $l_1 \ldots m_3$ are real numbers such that

$$l_1^2+l_2^2+l_3^2 = m_1^2+m_2^2+m_3^2 = 1 \quad \text{and} \quad l_1m_1+l_2m_2+l_3m_3 = 0$$

prove that $\quad (l_2m_3-l_3m_2)^2+(l_3m_1-l_1m_3)^2+(l_1m_2-l_2m_1)^2 = 1.$

13. Prove that $(\mathbf{a} \wedge \mathbf{b}).(\mathbf{c} \wedge \mathbf{d}) = (\mathbf{b}.\mathbf{d})(\mathbf{a}.\mathbf{c})-(\mathbf{b}.\mathbf{c})(\mathbf{a}.\mathbf{d})$. What trigonometric identity follows if O, A, B, C, D are coplanar and the angles AOB, COD are both right angles?

14. The vectors $\mathbf{a} = \mathbf{OA}$, $\mathbf{b} = \mathbf{OB}$, $\mathbf{c} = \mathbf{OC}$ are non-coplanar; prove that a given vector $\mathbf{r}$ may be expressed uniquely in the form

$$\mathbf{r} = \alpha\mathbf{b} \wedge \mathbf{c}+\beta\mathbf{c} \wedge \mathbf{a}+\gamma\mathbf{a} \wedge \mathbf{b}$$

and find α, β, γ in terms of $\mathbf{a}$, $\mathbf{b}$, $\mathbf{c}$ and $\mathbf{r}$.

If $\mathbf{r} = \mathbf{OP}$, locate P in terms of O, A, B, C: (i) if $\alpha = \beta = \gamma = 1$; (ii) if $\alpha = 0$, $\beta = \gamma = 1$.

15. Points Q, R are taken on the sides CA, AB (*not* produced) respectively of a triangle ABC in such a manner that the triangles ABQ, ARC are equal in area. Prove that RQ is parallel to BC.

16. G is the centroid of the triangle ABC, and E, F are the mid-points of CA, AB respectively. Prove that the quadrilateral $AFGE$ and the triangle GBC have the same area.

17. $ABCD$ is a skew quadrilateral; the mid-points of the opposite sides BC, DA are P, Q, while the mid-points of the diagonals AC, BD are R, S. Prove that $PRQS$ is a parallelogram whose area is $\frac{1}{2}|\Delta ABC-\Delta ABD|$.

18. ABC is a triangle and points L, M, N are taken on BC, CA, AB respectively such that

$$\frac{BL}{LC} = \lambda, \quad \frac{CM}{MA} = \mu, \quad \frac{AN}{NB} = \nu.$$

Prove that $\Delta LMN = (1+\lambda\mu\nu)\,\Delta ABC$ and deduce the theorem of Menelaus, that, if LMN is a straight line, $\lambda\mu\nu = -1$.

19. The sides (and sides produced) of the parallelogram *ABCD* separate the plane into nine regions (see Figure 24.7). Prove that, if a point *O* is taken within region I, II, VI or IX the area of *OAC* is the sum of the areas of *OAB* and *OAD* but that, if *O* is taken within any other region, the area of *OAC* is the difference of the areas of *OAB* and *OAD*.

(The reader who has studied mechanics may be able to interpret this result in terms of moments.)

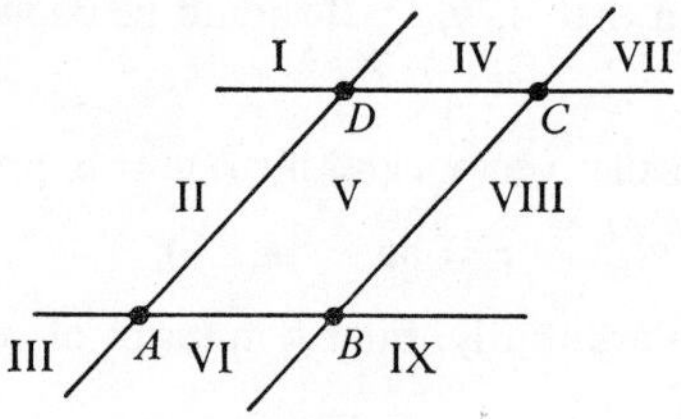

Fig. 24.7

20. *ABCD* is a tetrahedron. The lengths of *AB* and *CD* are *a* and *b* and the angle between *AB* and *CD* is θ. A plane is drawn parallel to *AB* and *CD* and meets the edges *CA*, *CB*, *BD*, *DA* at *P*, *Q*, *R*, *S*. Prove that *PQRS* is a parallelogram and that its area is

$$\frac{ab}{h^2}x(h-x)\sin\theta,$$

where *x* is the distance of the plane from *AB* and *h* is the distance between *AB* and *CD*.

Hence, or otherwise, prove that the volume of the tetrahedron is

$$\tfrac{1}{6}abh\sin\theta.$$

3. APPLICATIONS TO COORDINATE GEOMETRY

Throughout this section we shall assume that an origin *O* and a right-handed set of coordinate axes *Ox*, *Oy*, *Oz* specified by the unit vectors **i**, **j**, **k** are given. The position vector of the point *A*... with respect to *O* will, as usual, be denoted by **a** ....

Consider first the line through *A* in the direction of the unit vector $\hat{\mathbf{u}}$. If *R* is any point on this line, then $\mathbf{AR} = \lambda\hat{\mathbf{u}}$ and thus

$$\mathbf{AR} \wedge \hat{\mathbf{u}} = \mathbf{0},$$

giving the equation of the line in the form

$$(\mathbf{r}-\mathbf{a}) \wedge \hat{\mathbf{u}} = \mathbf{0}.$$

If $\mathbf{r} = x\mathbf{i}+y\mathbf{j}+z\mathbf{k}$, $\mathbf{a} = a_1\mathbf{i}+a_2\mathbf{j}+a_3\mathbf{k}$, $\hat{\mathbf{u}} = l\mathbf{i}+m\mathbf{j}+n\mathbf{k}$, we can transform this to (or from) the familiar coordinate form as follows:

$$(\mathbf{r}-\mathbf{a}) \wedge \hat{\mathbf{u}} = \mathbf{0}$$

$$\Rightarrow [(x-a_1)\,\mathbf{i}+(y-a_2)\,\mathbf{j}+(z-a_3)\,\mathbf{k}] \wedge (l\mathbf{i}+m\mathbf{j}+n\mathbf{k}) = \mathbf{0}$$

$$\Leftrightarrow (y-a_2)\,n-(z-a_3)\,m = (z-a_3)\,l-(x-a_1)\,n = (x-a_1)\,m-(y-a_2)\,l = 0$$

$$\Leftrightarrow \frac{x-a_1}{l} = \frac{y-a_2}{m} = \frac{z-a_3}{n}.$$

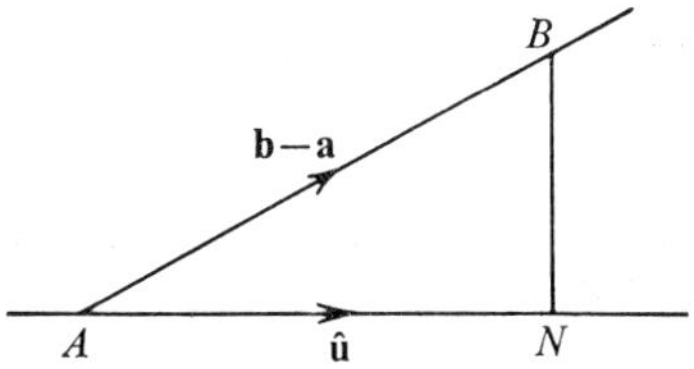

Fig. 24.8

The distance from the point B to the line, the length BN (see Figure 24.8), is

$$|(\mathbf{b}-\mathbf{a}) \wedge \hat{\mathbf{u}}|.$$

If B is the point (b_1, b_2, b_3) this gives

$$BN = |[(b_2-a_2)\,n-(b_3-a_3)\,m]\,\mathbf{i}+[(b_3-a_3)\,l-(b_1-a_1)\,n]\,\mathbf{j} \\ +[(b_1-a_1)\,m-(b_2-a_2)\,l]\,\mathbf{k}|.$$

Ex. 20. Show that the perpendicular distance from the point (3, 3, 2) to the line

$$\frac{x-1}{1} = \frac{y-2}{2} = \frac{z-3}{3} \quad \text{is} \quad \sqrt{\frac{83}{14}}.$$

The vector product is particularly useful in giving us the vector perpendicular to two given vectors. For example, consider the plane through the (non-collinear) points A, B, C. The vector

$$\mathbf{d} = \mathbf{AB} \wedge \mathbf{AC}$$

is perpendicular to this plane. Thus

$$\begin{aligned}\mathbf{d} &= (\mathbf{b}-\mathbf{a}) \wedge (\mathbf{c}-\mathbf{a})\\ &= \mathbf{b} \wedge \mathbf{c}+\mathbf{c} \wedge \mathbf{a}+\mathbf{a} \wedge \mathbf{b}\end{aligned}$$

and the equation of the plane is now found easily.

Example 2. *Find the equation of the plane through the points* $A(1, 1, -3)$, $B(2, -1, 10)$, $C(-1, -1, 1)$.

$$\mathbf{AB} = \mathbf{i}-2\mathbf{j}+13\mathbf{k},$$

$$\mathbf{AC} = -2\mathbf{i}-2\mathbf{j}+4\mathbf{k},$$

$$\mathbf{AB} \wedge \mathbf{AC} = 18\mathbf{i}-30\mathbf{j}-6\mathbf{k},$$

and thus the vector $3\mathbf{i}-5\mathbf{j}-\mathbf{k}$ is perpendicular to ABC. Since A lies in the plane the equation of ABC is therefore

$$3(x-1)-5(y-1)-(z+3) = 0$$

which gives, on simplification,

$$3x-5y-z = 1.$$

**Ex*. 21. If $\mathbf{r} = x\mathbf{i}+y\mathbf{j}+z\mathbf{k}$ is the position vector of the point P show that the volume of the tetrahedron $PABC$ is

$$\tfrac{1}{6}(\mathbf{r}-\mathbf{a}).[(\mathbf{b}-\mathbf{a}) \wedge (\mathbf{c}-\mathbf{a})].$$

(See Ex. 15.)

Hence obtain the equation of the plane ABC of Example 2 in determinant form, and verify that the expansion of this determinant gives the same equation as that obtained above.

Example 3. *Find the equation of the plane through the point* $(2, -5, -1)$ *perpendicular to the line of intersection of the planes*

$$x+7y+z = 0, \quad x-5y-2z = 0.$$

The line of intersection is perpendicular to both of the vectors

$$\mathbf{i}+7\mathbf{j}+\mathbf{k} \quad \text{and} \quad \mathbf{i}-5\mathbf{j}-2\mathbf{k}$$

and thus its direction is given by

$$(\mathbf{i}+7\mathbf{j}+\mathbf{k}) \wedge (\mathbf{i}-5\mathbf{j}-2\mathbf{k}) = -9\mathbf{i}+3\mathbf{j}-12\mathbf{k}.$$

Thus $3\mathbf{i}-\mathbf{j}+4\mathbf{k}$ is normal to the required plane, whose equation must be

$$3(x-2)-(y+5)+4(z+1) = 0$$

or

$$3x-y+4z = 7.$$

Example 4. *Find the length of the common perpendicular to the two skew lines*

$$\frac{x}{1} = \frac{y+1}{2} = \frac{z+2}{4}; \quad \frac{x-2}{-1} = \frac{y-6}{4} = \frac{z-3}{2}.$$

The direction of the common perpendicular is given by

$$(\mathbf{i}+2\mathbf{j}+4\mathbf{k}) \wedge (-\mathbf{i}+4\mathbf{j}+2\mathbf{k}) = -12\mathbf{i}-6\mathbf{j}+6\mathbf{k}.$$

A unit vector in the direction of the common perpendicular is thus

$$\hat{\mathbf{u}} = (2\mathbf{i}+\mathbf{j}-\mathbf{k})/\sqrt{6}.$$

The lines pass through the points $A(0, -1, -2)$, $B(2, 6, 3)$ respectively and we have the length of the common perpendicular equal to the projection of **AB** in the direction $\hat{\mathbf{u}}$

$$\begin{aligned}
&= (\mathbf{b}-\mathbf{a}).\hat{\mathbf{u}} \\
&= (2\mathbf{i}+7\mathbf{j}+5\mathbf{k}).(2\mathbf{i}+\mathbf{j}-\mathbf{k})/\sqrt{6} \\
&= (4+7-5)/\sqrt{6} \\
&= \sqrt{6}.
\end{aligned}$$

The vector product may be used to obtain an alternative method of solving three linear simultaneous equations. Consider the three equations

$$\begin{aligned}
3x+5y+8z &= -1 \qquad P_1, \\
2x+3y+4z &= 3 \qquad P_2, \\
4x+2y+5z &= -2 \qquad P_3.
\end{aligned}$$

These three equations represent three planes, P_1, P_2, P_3; their solution consists in finding the point Q common to each plane (assuming such a point exists).

From equations P_1 and P_2 we obtain $(3P_1+P_2)$, the equation of the plane through the line of intersection of P_1, P_2 containing the origin. Thus

$$11x+18y+28z = 0 \qquad P_4$$

is a plane containing O and Q. Similarly, $(2P_1-P_3)$,

$$2x+8y+11z = 0 \qquad P_5$$

is a plane containing O and Q.

Thus OQ is perpendicular to both $11\mathbf{i}+18\mathbf{j}+28\mathbf{k}$ and $2\mathbf{i}+8\mathbf{j}+11\mathbf{k}$; that is

$$\begin{aligned}
\mathbf{OQ} &= \lambda(11\mathbf{i}+18\mathbf{j}+28\mathbf{k}) \wedge (2\mathbf{i}+8\mathbf{j}+11\mathbf{k}) \\
&= -13\lambda(2\mathbf{i}+5\mathbf{j}-4\mathbf{k}),
\end{aligned}$$

and Q has coordinates $(-26\lambda, -65\lambda, 52\lambda)$ where λ is some scalar. Writing $-13\lambda = \mu$, Q has coordinates $(2\mu, 5\mu, -4\mu)$. Substituting in P_1, we find that $\mu = 1$ giving Q as the point $(2, 5, -4)$. The solution of the given set of equations is thus

$$x = 2, \quad y = 5, \quad z = -4.$$

Ex. 22. Use the method outlined above to find a necessary and sufficient condition for the existence of a non-trivial solution to the set of three homogeneous linear equations:

$$a_1x+b_1y+c_1z = 0,$$
$$a_2x+b_2y+c_2z = 0,$$
$$a_3x+b_3y+c_3z = 0.$$

Exercise 24(b)

1. Find the equations of the planes through the following sets of points:
(i) $(1, -1, 1), (2, 1, 2), (3, 1, -1)$;
(ii) $(1, 1, 1), (0, -3, -2), (-1, 3, -1)$;
(iii) $(1, 1, 5), (1, -1, -3), (2, 1, 1)$;
(iv) $(4, 0, 1), (7, 3, 5), (-10, -4, -1)$;
(v) $(a, b, c), (3a, b, 0), (4a, 2b, c)$.

2. Find, in each part, the equation of the plane through the given point and perpendicular to the line of intersection of the given planes:
(i) $(1, -1, 3)$; $4x-y+4z = 0$, $12x-5y+8z = 0$;
(ii) $(-5, -4, -1)$, $3x+4y+3z = 15$, $9x-2y-12z = 3$;
(iii) $(2, 3, 4)$, $5x+y+2z = 7$, $x+y+z = -4$;
(iv) $(9, 3, 1)$, $3x+2y+2z = 0$, $5x+2y+z = 4$;
(v) $(1, 1, -1)$, $3x+2y-4z = 0$, $4x+y+3z = 0$.

3. Find the length of the common perpendicular to the following pairs of skew lines:

(i) $\dfrac{x}{-3} = \dfrac{y-7}{2} = \dfrac{z-6}{2}$, $\quad \dfrac{x+3}{2} = \dfrac{y-6}{-5} = \dfrac{z+4}{6}$;

(ii) $x-6 = y-3 = z-1$, $\quad \dfrac{x-4}{3} = \dfrac{y-7}{5} = \dfrac{z-3}{2}$;

(iii) $x-8 = y-7 = \dfrac{z}{2}$, $\quad \dfrac{x+1}{2} = \dfrac{y+1}{-1} = \dfrac{z-2}{2}$.

4. Find the length of the common perpendicular to the two lines

$$\frac{x-4}{1} = \frac{y}{-2} = \frac{z+3}{1} \quad \text{and} \quad \frac{x-6}{3} = \frac{y+1}{-3} = \frac{z+5}{-1}.$$

What deductions can you make from your answer?

5. Find the equation of the plane containing the points $A(6, 3, 0)$, $B(-2, 0, -5)$ which is parallel to the vector $\mathbf{i}+\mathbf{j}+\mathbf{k}$.

6. Find the equation of the plane containing the points $A(3, 5, 10)$, $B(4, 3, -3)$ which is parallel to the vector $3\mathbf{i}+5\mathbf{j}-6\mathbf{k}$.

7. Find the equation of the plane parallel to the two lines

$$\tfrac{1}{3}(x-6) = \tfrac{1}{5}(y-6) = \tfrac{1}{4}(z-1), \qquad \tfrac{1}{3}x = -(y-2) = z+4$$

which passes through the point $(2, 0, -1)$.

8. Given the plane P: $2x+3y-z=-1$ and the line L: $\frac{1}{2}x=-y=-(z-5)$, find:

(i) the coordinates of the point of intersection of P and L;
(ii) a vector parallel to P and perpendicular to L;
(iii) a vector in the direction of the projection of L in P;
(iv) the equation of the projection of L in P.

9. Show that the equation of the plane containing the points A, B, C has equation

$$(\mathbf{r}-\mathbf{a}).(\mathbf{b}-\mathbf{a}) \wedge (\mathbf{c}-\mathbf{a}) = 0.$$

Write down, in determinant form, the equation of the plane through the points $A(a_1, a_2, a_3)$, $B(b_1, b_2, b_3)$, $C(c_1, c_2, c_3)$.

10. Find a vector perpendicular to both $a\mathbf{i}+b\mathbf{j}+c\mathbf{k}$ and $l\mathbf{i}+m\mathbf{j}+n\mathbf{k}$. Hence write down the equation of the plane through the origin containing the line

$$\frac{x-a}{l} = \frac{y-b}{m} = \frac{z-c}{n}.$$

Use this result to find the line through the origin which intersects each of the lines

$$\tfrac{1}{2}(x+1) = y-2 = \tfrac{1}{4}(z+2),$$
$$\tfrac{1}{3}(x+1) = \tfrac{1}{2}(y-4) = (z-3),$$

giving your answer in standard form.

11. Find the vector equation of the line through the point C (position vector $\mathbf{c}$) which intersects both of the lines

$$\mathbf{r} = \mathbf{a}_1+\lambda\mathbf{b}_1, \quad \mathbf{r} = \mathbf{a}_2+\mu\mathbf{b}_2.$$

Discuss any special cases that might arise.

12. Find the volume of the tetrahedron formed by the planes

$$\beta y+\gamma z = 0, \quad \gamma z+\alpha x = 0, \quad \alpha x+\beta y = 0, \quad \alpha x+\beta y+\gamma z = \delta.$$

13. A square piece of paper $ABCD$ of side a is folded along the diagonal BD so that the planes of the triangles ABD, CBD are perpendicular. Find the shortest distance between the edges AB and CD.

25. *Continuous probability distributions*

1. DENSITY FUNCTIONS

In Chapters 7, 10 and 16 we discussed probability distributions in discrete outcome spaces, that is, in spaces whose elementary events we could count by putting them into one-to-one correspondence with the set N of natural numbers. Associated with each elementary event s_i of the outcome space S was a number x_i, usually an integer—the value of our associated random variable X.

Random experiments occur, however, in which it is mathematically more convenient to take as outcome space a set which cannot be counted and as random variable a subset X of the set R of real numbers. For example, suppose we take a point P at random on a line AB of length l m. Since we can only measure lengths to within some given limits of accuracy, say 0·5 mm, we could take as values of a (discrete) random variable the number of intervals of length 0·5 mm P is from A, giving 201 possible values in all (0–200). Mathematically the situation may be simpler to analyse if we take as values of our (continuous) random variable the supposed 'exact' length of AP, measured as a real number of mm; that is

$$X = \{x \in R: 0 \leqslant x \leqslant 100\}.$$

If we are dealing with a subset of R, calculus methods are available and it is, for example, easier in general to integrate a function than it is to sum a series.

Given a continuous random variable X, it will generally be impossible to assign a probability to a specific value of X; rather we assign a probability to each *interval*

$$E = \{x \in X; \quad x_1 < x < x_2\}$$

as follows. Let f be an integrable function such that

(i) $f(x) \geqslant 0$, for all $x \in X$;

(ii) $\int_X f(x)\,\mathrm{d}x = 1$;

where the notation $\int_X$ denotes that the integration is to be performed over the set X. Then we define

$$\Pr(E) = \int_{x_1}^{x_2} f(x)\,\mathrm{d}x.$$

The function f so defined is called a *probability density function.* As in the discrete case the choice of f constitutes an assumption about our random experiment; that is, to choose f is to choose a particular mathematical model. Geometrically, *probabilities are measured as areas under the curve* $y = f(x)$.

Consider again the random experiment of marking a point on the line AB. To analyse the situation we must first set up a mathematical model; in other words, we seek to make the vague statement 'we mark a point P at random on AB' mathematically precise by stating explicitly what we feel 'at random' could mean.

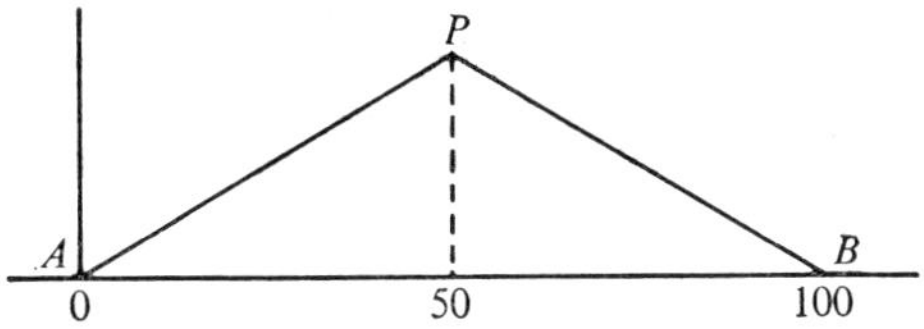

Fig. 25.1

We may regard AB as an interval on the x axis, with A as origin and B 100 units to the right. Suppose we conjecture that a point is more likely to be chosen towards the middle of AB than near the ends of the interval. Taking, as is customary, the whole set R as the range of our random variable X,† a possible choice of $f(x)$ would be

$$f(x) = \begin{cases} 0 & \text{if } x < 0 \\ kx & \text{if } 0 \leqslant x < 50 \\ k(100-x) & \text{if } 50 \leqslant x < 100 \\ 0 & \text{if } x \geqslant 100, \end{cases}$$

where k is some number. Figure 25.1 shows the graph of $y = f(x)$.

$f(x) = 0$ if $x < 0$ or $x > 100$ simply means that we are restricted to a choice of P within AB. $f(x) = kx$ if $0 < x < 50$ and $f(x) = k(100-x)$ if $50 < x < 100$ ensures that our probability distribution is *symmetrical* (reflecting a feeling that we are as likely to take x to be between, say, 35 and 45 as we are to take x to be between 55 and 65) and that the probability of selecting a point P within an interval of some specified length becomes progressively greater the nearer the interval is to the centre of AB (see Figure 25.1 again).

† For consistency we shall take R to be the range of X throughout this chapter.

To find k in this case, we must have

$$\int_0^{100} f(x)\,dx = 1$$

$$\Rightarrow \int_0^{50} kx\,dx + \int_{50}^{100} k(100-x)\,dx = 1$$

$$\Rightarrow \tfrac{1}{2}k[x^2]_0^{50} + k[100x - \tfrac{1}{2}x^2]_{50}^{100} = 1$$

$$\Rightarrow \tfrac{1}{2}.k.50^2 + k.100.50 - \tfrac{1}{2}k.150.50 = 1$$

$$\Rightarrow k = \frac{1}{2500}.$$

(Figure 25.1 is not drawn to scale!) Since $\int_0^{100} f(x)\,dx$ represents the area of ΔAPB, integration is a heavy tool to use here—we employ it to illustrate the general method.

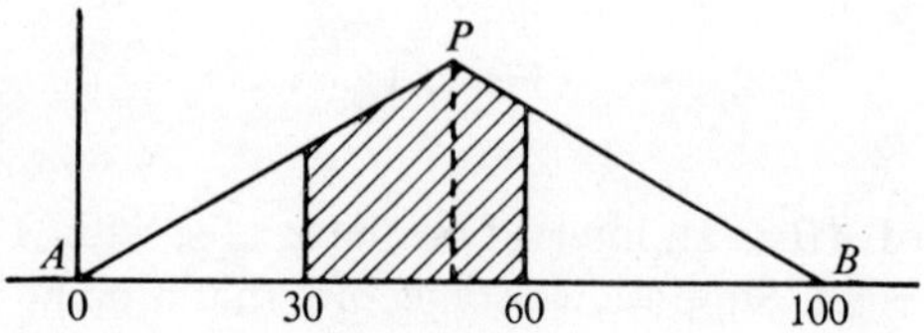

Fig. 25.2

Now suppose we wish to find the probability that P lies between 30 and 60 mm from A. We define the event E by

$$E = \{x \in R: 30 \leqslant x \leqslant 60\}$$

and obtain

$$\Pr(E) = \int_{30}^{50} \tfrac{1}{2500}\,x\,dx + \int_{50}^{60} \tfrac{1}{2500}\,(100-x)\,dx$$

$$= \tfrac{27}{50} = 0{\cdot}54 \quad \text{(see Figure 25.2).}$$

(Again, integration is heavy in the circumstances.)

A different mathematical model, that is, a different choice for f, leads to a different estimate of the probability of the event E as we shall now show. We shall suppose a function f defined thus:

$$f(x) = \begin{cases} 0 & \text{if } x < 0 \\ kx\,(100-x) & \text{if } 0 \leqslant x \leqslant 100 \\ 0 & \text{if } x > 100 \end{cases}$$

(see Figure 25.3). This distribution is again symmetrical about $x = 50$ and gives a greater probability to points near the centre. To find Pr (E) we must first calculate k.

We have

$$\int_0^{100} kx(100-x)\,\mathrm{d}x = 1$$

$$\Rightarrow k[50x^2 - \tfrac{1}{3}x^3]_0^{100} = 1$$

$$\Rightarrow k = \frac{3}{500000}.$$

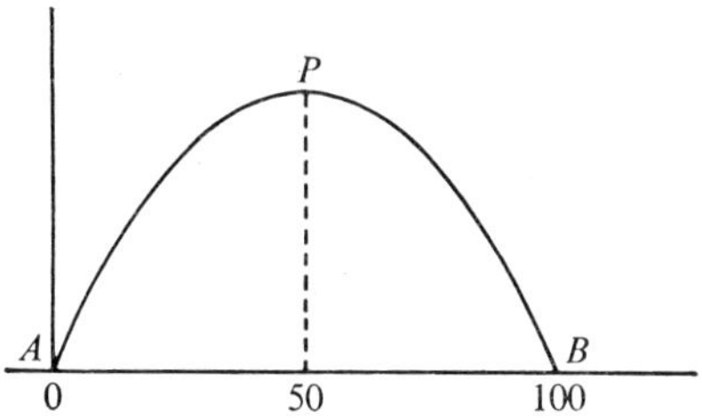

Fig. 25.3

(Again, Figure 25.3 is not drawn to scale!) Defining E as before we have:

$$\Pr(E) = \int_{30}^{60} \frac{3}{500000}\,x(100-x)\,\mathrm{d}x$$

$$= 6 \times 10^{-6}[50x^2 - \tfrac{1}{3}x^3]_{30}^{60}$$

$$= \tfrac{54}{125} = 0{\cdot}432.$$

Ex. 1. Why would you expect the second answer for Pr (E) to be less than the first?

Ex. 2. Write down the value of Pr (E) on the assumption that all points of the interval are equally likely. What form does $f(x)$ take in this case?

Ex. 3. If $f(x) = kx$, $0 \leqslant x \leqslant 100$ and $f(x) = 0$ otherwise, find k and Pr (E).

A *mode* of a random variable with given density function f is a value of the random variable for which f has a local maximum. The *median*, m, is the value of the random variable defined by the equation

$$\int_{-\infty}^{m} f(x)\,\mathrm{d}x = \tfrac{1}{2}.$$

Similarly, the *lower and upper quartile scores L, Q*, are defined by

$$\int_{-\infty}^{L} f(x)\,dx = \tfrac{1}{4}, \quad \int_{-\infty}^{Q} f(x)\,dx = \tfrac{3}{4}.$$

Ex. 4. Interpret the median and quartiles in terms of probabilities.

2. DISTRIBUTION FUNCTIONS

The density function, f, of a continuous probability distribution bears the same relation to probability as mass density does to mass: mass density is the mass per unit volume, probability density is the probability per unit interval. It must be remembered that $f(x)$ itself is not a probability, but $\delta p = f(x)\,\delta x$ is.

We now introduce a function, closely related to the density function, whose values do represent probabilities. Suppose the domain of our random variable is R and that we are given a density function $f: R \to R$. Then we know

$$\text{(i)}\quad f(x) > 0,$$

$$\text{(ii)}\quad \int_{-\infty}^{+\infty} f(x)\,dx = 1.$$

Let E be the set of real numbers less than, or equal to x, that is, the set of real numbers to the left of x on the real axis:

$$E = \{y \in R : y \leqslant x\}.$$

Then
$$\Pr(E) = \int_{-\infty}^{x} f(x)\,dx$$

depends upon x and we may define a function $F: R \to R$ where

$$F(x) = \int_{-\infty}^{x} f(x)\,dx = \Pr(E).$$

F is called the *distribution function* for the probability distribution; $F(x)$ is the probability that, given a probability distribution with density $f(x)$, a point chosen at random will be less than x.

From the original definition we have (provided f is continuous)

$$\frac{dF}{dx} = f.$$

Since $f(x) \geqslant 0$ for all x, $F(x)$ increases as x increases. Also we have

$$\lim_{x\to-\infty} F(x) = 0, \quad \lim_{x\to+\infty} F(x) = 1,$$

which gives the range of F as the set $\{x \in R : 0 \leqslant x \leqslant 1\}$.

Again, given an event $E = \{x \in R: x_1 < x < x_2\}$ we have

$$\Pr(\mathrm{E}) = \int_{x_1}^{x_2} f(x)\,\mathrm{d}x$$
$$= F(x_2) - F(x_1).$$

Since the sets E and $\{x \in R: x < x_1\}$ are mutually exclusive, this illustrates the result that the probability of the union of mutually exclusive events is the sum of the probabilities of the separate events.

**Ex.* 5. If $E_1, E_2, \ldots, E_n$ are mutually exclusive events in a continuous outcome space, prove that

$$\Pr(E_1) + \Pr(E_2) + \ldots + \Pr(E_n) = \Pr(E_1 \cup E_2 \cup \ldots \cup E_n).$$

Example 1. *The density function, f, is defined by*

$$f(x) = \begin{cases} 0 & \text{if } x < 0 \\ kx(1-x) & \text{if } 0 \leqslant x \leqslant 1 \\ 0 & \text{if } x > 1. \end{cases}$$

Find

(i) *the value of k;*
(ii) *the distribution function F;*
(iii) *the density and distribution functions, g and G, of the random variable W where*

$$w = x^2.$$

Sketch the graphs of f and F.

(i) Since $\int_{-\infty}^{+\infty} f(x)\,\mathrm{d}x = 1$,

we have
$$\int_0^1 kx(1-x)\,\mathrm{d}x = 1$$
$$\Rightarrow \qquad k = 6.$$

(ii) If $x < 0$, $F(x) = 0$; if $x > 1$, $F(x) = 1$.
If $0 \leqslant x \leqslant 1$,

$$F(x) = \int_0^x 6x(1-x)\,\mathrm{d}x$$
$$= [3x^2 - 2x^3]_0^x$$
$$= x^2(3-2x).$$

Thus F is the function $F: R \rightarrow \{y \in R: 0 \leqslant y \leqslant 1\}$ defined by

$$F(x) = \begin{cases} 0 & \text{if } x < 0 \\ x^2(3-2x) & \text{if } 0 \leqslant x \leqslant 1 \\ 1 & \text{if } x > 1. \end{cases}$$

(iii) If $w < 0$, $G(w) = 0$; if $w > 1$, $G(w) = 1$.
If $0 \leqslant w \leqslant 1$,

$$G(w) = \Pr\{z \in R: 0 \leqslant z \leqslant w\}$$

$$= w(3-2\sqrt{w}) \quad \text{(since } x \text{ is positive, } x = \sqrt{w}\text{)}.$$

Thus

$$G(w) = \begin{cases} 0 & \text{if } w < 0 \\ 3w-2w^{\frac{3}{2}} & \text{if } 0 \leqslant w \leqslant 1 \\ 1 & \text{if } w > 1. \end{cases}$$

Also,

$$g = \frac{dG}{dw},$$

giving

$$g(w) = \begin{cases} 0 & \text{if } w < 0 \\ 3(1-\sqrt{w}) & \text{if } 0 \leqslant w \leqslant 1 \\ 0 & \text{if } w > 1. \end{cases}$$

The graphs of f, F are shown in Figures 25.4(i) and (ii).

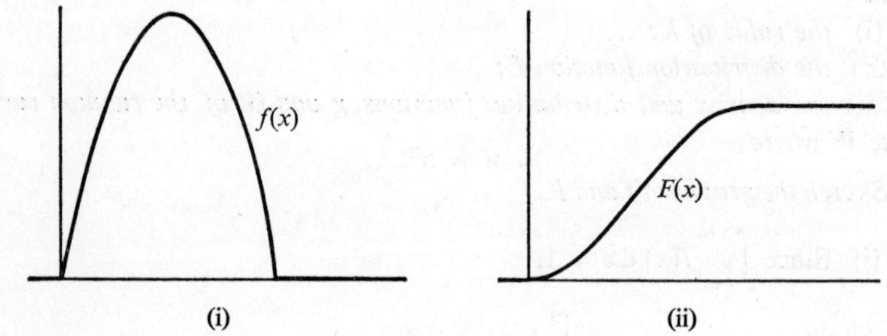

Fig. 25.4

Notice in this question that, since $\sqrt{x}$ increases as x increases, as we move from left to right along the x axis so we move from left to right along the w axis. More generally, if a transformation $y = h(x)$ of a random variable is made, the method of Example 1 may be used directly if y either increases or decreases steadily with x but, if this is not the case, the domain of f should be split into those parts for which it is an increasing function and those for which it is decreasing. (See Example 4.)

The value of transforming from one random variable to another is that a complicated form for the density function may be changed into a simpler (or more familiar) density function.

F is sometimes referred to as the *cumulative distribution* function, to emphasize the property it has of increasing from left to right. The graph of F, which usually has a shape very similar to that shown in Figure 25.4, is called a *cumulative probability curve* (or, occasionally, an *ogive* curve).

Ex. 6. Explain how the cumulative probability curve may be used to find the median and upper and lower quartiles.

In the case of discrete distributions, if we had a random variable X which could take all integral values r in the range $1 \leqslant r \leqslant n$, where

$$\Pr(X = r) = p_r,$$

we defined the expectation of X by

$$\mathscr{E}(X) = \sum_{r=1}^{n} r p_r.$$

For continuous distributions, integration takes the place of summation and we define the expectation of the random variable X, whose density function is f, by

$$\mathscr{E}(X) = \int_{-\infty}^{+\infty} x f(x)\, dx.$$

$\mathscr{E}(X)$ is usually referred to as the *mean* of X, and is denoted by μ.

Example 2. *Find the mean of the random variable X whose density function, f, is defined by*

$$f(x) = \begin{cases} 0 & \textit{if}\ \ x < 0 \\ 4x(1-x^2) & \textit{if}\ \ 0 \leqslant x \leqslant 1 \\ 0 & \textit{if}\ \ x > 1. \end{cases}$$

We have

$$\begin{aligned} \mu = \mathscr{E}(X) &= \int_{-\infty}^{+\infty} x f(x)\, dx \\ &= \int_0^1 4x^2(1-x^2)\, dx \quad (\text{since } f(x) = 0 \text{ outside the interval } 0 \leqslant x \leqslant 1), \\ &= \tfrac{8}{15}. \end{aligned}$$

More generally, if we have a function $g: R \to R$, the expectation of the function g of the random variable X is defined by

$$\mathscr{E}[g(x)] = \int_{-\infty}^{+\infty} g(x) f(x)\, dx.$$

Particularly important is the case where

$$g(x) = (x-\mu)^2.$$

The *variance* of X is the expectation of the function g so defined:

$$\sigma^2 = \mathscr{E}[(x-\mu)^2] = \int_{-\infty}^{+\infty} (x-\mu)^2 f(x)\,dx.\dagger$$

Since

$$\int_{-\infty}^{+\infty} (x-\mu)^2 f(x)\,dx = \int_{-\infty}^{+\infty} (x^2-2\mu x+\mu^2) f(x)\,dx$$

$$= \int_{-\infty}^{+\infty} x^2 f(x)\,dx - 2\mu \int_{-\infty}^{+\infty} x f(x)\,dx + \mu^2 \int_{-\infty}^{+\infty} f(x)\,dx$$

$$= \int_{-\infty}^{+\infty} x^2 f(x)\,dx - 2\mu.\mu+\mu^2.1,$$

we have

$$\sigma^2+\mu^2 = \int_{-\infty}^{+\infty} x^2 f(x)\,dx,$$

a result which may be compared with that on page 178.

**Ex.* 7. Show that the variance of the random variable in Example 2 is given by

$$\sigma^2+\tfrac{64}{225} = \int_0^1 4x^3(1-x^2)\,dx.$$

Deduce that $\sigma^2 = \frac{11}{225}$.

**Ex.* 8. The *mean deviation* about the mean, η, is defined as the expectation of the function g where

$$g(x) = |x-\mu|.$$

Find η if

$$f(x) = \begin{cases} 0, & x < -1 \\ \tfrac{1}{2}, & -1 \leqslant x \leqslant 1 \\ 0, & x > 1. \end{cases}$$

Exercise 25(a)

1. $f(x) = kx^{-2}$ if $x \geqslant 1$, $f(x) = 0$ if $x < 1$; find the value of k and hence find $\Pr(x < 2)$.

2. $f(x) = k\mathrm{e}^{-2x}$ if $x \geqslant 0$, $f(x) = 0$ if $x < 0$; find k and hence find

$$\Pr(1 < x < 2).$$

3. If $f(x) = k \sin \pi x$ if $0 \leqslant x \leqslant 1$ and is zero otherwise, find k and hence find $\Pr(x > \frac{1}{3})$.

4. Find the distribution function for each of the density functions in Questions 1–3.

† The use of σ^2 to denote variance is sometimes insufficiently explicit. If it is necessary to refer to the random variable, X, under consideration, the associated variance may be written σ_x^2 or, more commonly, $\mathscr{V}(X)$.

Find the mean, median, mode and variance for the distribution with the density function defined as in Questions 5–8.

5.
$$f(x) = \begin{cases} 0 & \text{if } x < 0 \\ \tfrac{3}{4}x(2-x) & \text{if } 0 \leqslant x \leqslant 2 \\ 0 & \text{if } x > 2. \end{cases}$$

6.
$$f(x) = \begin{cases} 0 & \text{if } x < 0 \\ (6x+1)/4 & \text{if } 0 \leqslant x \leqslant 1 \\ 0 & \text{if } x > 1. \end{cases}$$

7.
$$f(x) = \begin{cases} 0 & \text{if } x < -\tfrac{1}{2} \\ \tfrac{1}{2}\pi \cos \pi x & \text{if } -\tfrac{1}{2} \leqslant x \leqslant \tfrac{1}{2} \\ 0 & \text{if } x > \tfrac{1}{2}. \end{cases}$$

8.
$$f(x) = \begin{cases} 0 & \text{if } x < 0 \\ \lambda e^{-\lambda x} & \text{if } x \geqslant 0. \end{cases}$$

9. If the density function, f, of a distribution is given by

$$f(x) = \begin{cases} 0 & \text{if } x < 0 \\ \tfrac{1}{2}x & \text{if } 0 \leqslant x \leqslant 1 \\ \tfrac{1}{2} & \text{if } 1 < x \leqslant 2 \\ \tfrac{1}{2}(3-x) & \text{if } 2 < x \leqslant 3 \\ 0 & \text{if } x > 3 \end{cases}$$

find the distribution function, F, and sketch the graphs of f and F.

Find the density and distribution functions, g and G, of the new random variable Y, where $y = x^2$, and sketch the graphs of g and G.

10. The density function, f, of a probability distribution is given by

$$f(x) = \begin{cases} 0 & \text{if } x < 0 \\ x & \text{if } 0 \leqslant x \leqslant 1 \\ 2-x & \text{if } 1 < x \leqslant 2 \\ 0 & \text{if } x > 2. \end{cases}$$

Find the density function, g, of the new random variable, Y, where $y = \sqrt{x}$ if $x \geqslant 0$, $y = 0$ if $x < 0$ and sketch the graphs of f and g.

Find the mean and variance of X and the mean of Y.

11. A probability density function of a random variable X is defined as follows:

$$f(x) = \begin{cases} x(x-1)(x-2) & \text{for } 0 \leqslant x < 1 \\ \lambda & \text{for } 1 \leqslant x \leqslant 3 \\ 0 & \text{otherwise,} \end{cases}$$

where λ is a suitable constant. Calculate the expectation μ of x. What is the probability that x is less than or equal to μ? (M.E.I.)

12. A continuous probability distribution has density function f, where

$$f(x) = a+bx+cx^2 \quad \text{for} \quad 0 \leqslant x \leqslant 1, \quad \text{and} \quad f(x) = 0 \quad \text{outside this range.}$$

The mean is 2/3 and the variance is 4/45. Find the values of a, b, and c.

13. The probability density function $p(t)$ of the length of life, t hours, of a certain component is given by

$$p(t) = k\mathrm{e}^{-kt} \quad (0 \leqslant t < \infty),$$

where k is a positive constant. Show that the mean and standard deviation of this distribution are each equal to $1/k$.

Find the probability that the life of a component will be at least t_0 hours. Given that a particular component is already t_1 hours old and has not failed, show that the probability that it will last at least a further t_0 hours is e^{-kt_0}.

An apparatus contains three components of this type and the failure of one may be assumed independent of the failure of the others. Find the probability that:

(i) none will have failed at t_0 hours;

(ii) exactly one will fail in the first t_0 hours, another in the next t_0 hours and a third after more than $2t_0$ hours. (M.E.I.)

14. A random variable X has the distribution function

$$F(x) = 0 \quad (x \leqslant 0),$$

$$F(x) = kx^3 \quad (0 \leqslant x \leqslant 2),$$

$$F(x) = 1 \quad (x \geqslant 2),$$

where k is a constant. Find the mean, median and variance of X. (M.E.I.)

15. A random variable X has probability density given by

$$f(x) = A/(1+x^2) \quad \text{if} \quad 0 \leqslant x \leqslant 1,$$

and

$$f(x) = 0 \quad \text{otherwise.}$$

Find A the mean and the variance of the distribution.

Find also the median and the 90th percentile.

(The 90th percentile is the value of X for which $F(x) = \frac{9}{10}$.) (O & C adapted)

3. SOME IMPORTANT CONTINUOUS DISTRIBUTIONS

(i) *The uniform (or rectangular) distribution*

If a continuous random variable can assume all real values between $x = a$ and $x = b$ and if the density function, f, is independent of position within this interval, then the random variable is said to possess a *uniform* distribution. f and F are thus given, for the *uniform* distribution over the interval $a \leqslant x \leqslant b$, by

$$f(x) = \begin{cases} 0 & \text{if } x < a \\ \dfrac{1}{b-a} & \text{if } a \leqslant x \leqslant b \\ 0 & \text{if } x > b; \end{cases}$$

$$F(x) = \begin{cases} 0 & \text{if } x < a \\ \dfrac{x-a}{b-a} & \text{if } a \leqslant x \leqslant b \\ 1 & \text{if } x > b. \end{cases}$$

**Ex.* 9. Sketch the graphs of f, F as defined above.

**Ex.* 10. Prove that the mean and variance of the uniform distribution defined over the interval $a \leqslant x \leqslant b$ are given by

$$\mu = \tfrac{1}{2}(a+b), \quad \sigma^2 = \tfrac{1}{12}(a-b)^2.$$

The continuous uniform distribution arises as a direct extension of the discrete uniform distribution. As in the discrete case, the assumption of a uniform distribution is often not made explicit in the formulation of problems. This is a weakness in the statement of the problem and the reader must be on his guard. For example, such a statement as 'a point is chosen at random on the line AB' is likely to imply, in the eyes of an examiner, that a uniform distribution is the required mathematical model, although there may be little physical justification for this. The dangers of a loose enunciation of a problem are illustrated in the following very well-known example, due to J. Bertrand.

Example 3. *A chord of a circle is drawn at random. Find the probability that its length will exceed that of the side of the inscribed equilateral triangle:*

(i) *if one end, A, of the chord is regarded as fixed and the other end, P, obeys a uniform distribution on the circumference of the circle;*

(ii) *if the direction of the chord is regarded as fixed and the perpendicular distance, x, of the chord from the centre obeys a uniform distribution over the interval* $0 \leqslant x \leqslant a$, *where a is the radius of the circle.*

(i) Let ABC be an equilateral triangle inscribed in the circle (Figure 25.5). The chord AP will exceed the length AB if P lies on the arc BC and the probability of this occurring, on the assumption that P is distributed uniformly on the circumference of the circle, is $\frac{1}{3}$.

(ii) Let AB, CD be the two chords in the given direction which are distant $\frac{1}{2}a$ from the centre of the circle (and thus are of length $a\sqrt{3}$, the side of the inscribed equilateral triangle). If x is the distance of a random chord from O in the given direction, and if x is uniformly distributed in the interval $-a \leqslant x \leqslant a$, the chord will exceed $a\sqrt{3}$ in length (e.g. PQ, in Figure 25.6) if $-\frac{1}{2}a \leqslant x \leqslant \frac{1}{2}a$ giving a probability of $\frac{1}{2}$.

Ex. 11. If, in Example 3, the mid-point of the chord is taken to have a uniform distribution over the interior of the circle, prove that the probability that the chord will exceed the length of the side of the inscribed equilateral triangle is $\frac{1}{4}$.

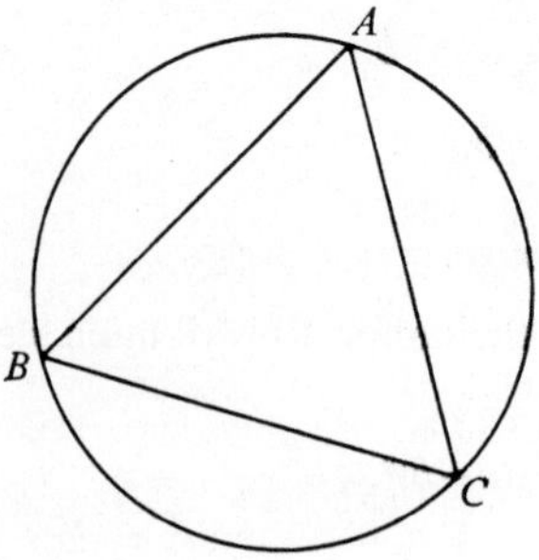

Fig. 25.5

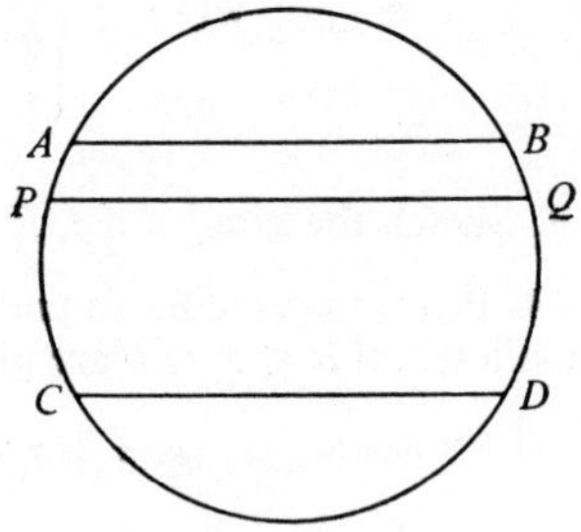

Fig. 25.6

Example 4. *A point P is taken on a line AB, of length 2a, the distribution being uniform. A rectangle is then drawn with adjacent sides of lengths AP, PB. Find:*

(i) *the expected value of the area of this rectangle;* (ii) *the distribution function, G, for the random variable, Y, whose value represents this area.*

Let $AP = x$ and the area of the corresponding rectangle $= y$. Then

$$y = x(2a-x).$$

Now, if f is the density function of X, the random variable whose value for given P is the length AP,

$$f(x) = \begin{cases} 0 & \text{if } x < 0 \\ 1/2a & \text{if } 0 \leqslant x \leqslant 2a \\ 0 & \text{if } x > 2a, \end{cases}$$

and we have

$$\mathscr{E}(Y) = \int_0^{2a} \frac{1}{2a} x(2a-x)\,dx, \quad \text{by the definition of expectation,}$$

$$= \frac{1}{2a}\left[ax^2 - \tfrac{1}{3}x^3\right]_0^{2a}$$

$$= \tfrac{2}{3}a^2.$$

As x varies from 0 to a, y varies from 0 to its maximum value, a^2, and, as x varies from a to $2a$, y varies from a^2 to 0. If F is the distribution function of X,

$$F(x) = \begin{cases} 0 & \text{if } x < 0 \\ x/2a & \text{if } 0 \leqslant x \leqslant 2a \\ 1 & \text{if } x > 2a \end{cases}$$

and, by the symmetry of the curve $y = x(2a-x)$ about the line $x = a$ (see Figure 25.7), we have, for any point (x, y) on the curve,

$$\begin{aligned} \Pr(z \in R: z < y) &= \Pr(z \in R: z < x) + \Pr(z \in R: z > 2a-x) \\ &= F(x) + [1 - F(2a-x)] \\ &= \frac{x}{2a} + 1 - \frac{2a-x}{2a} \\ &= \frac{x}{a}. \end{aligned}$$

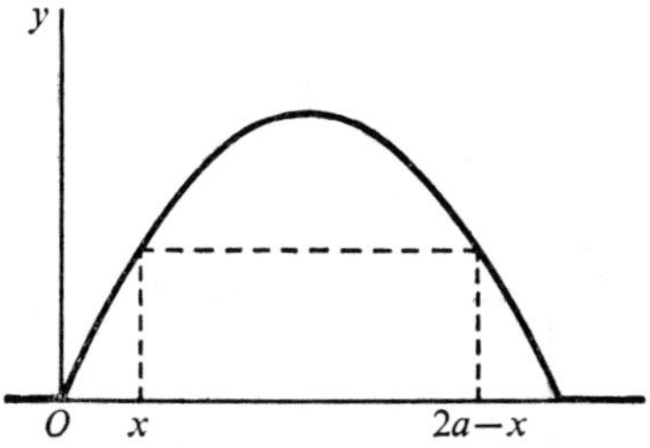

Fig. 25.7

But, since $y = 2ax - x^2$,

we have $x = a - \sqrt{(a^2-y)} \qquad (x < a)$,

giving

$$\begin{aligned} G(y) &= \Pr(z \in R: z < y) \\ &= \frac{a - \sqrt{(a^2-y)}}{a}. \end{aligned}$$

Thus we have

$$G(y) = \begin{cases} 0 & \text{if } y < 0 \\ 1 - \sqrt{\left(1 - \dfrac{y}{a^2}\right)} & \text{if } 0 \leqslant y \leqslant a^2 \\ 1 & \text{if } y > a^2. \end{cases}$$

The graphs of F and G are shown in Figure 25.8.

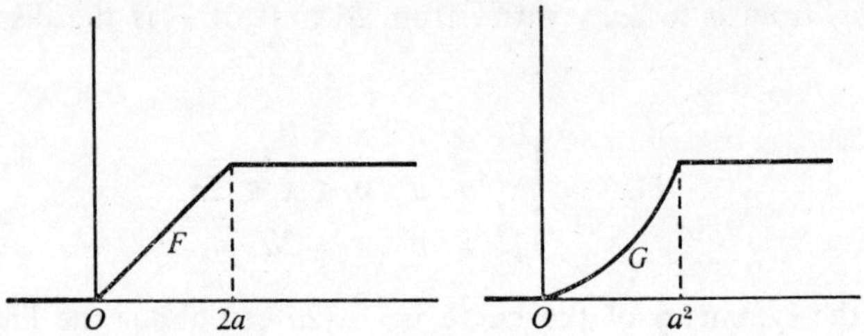

Fig. 25.8

Example 5. *X, Y are both uniformly distributed between* 0 *and* 1. *Values x, y are chosen independently at random and a new random variable, Z, is formed where*

$$z = x+y.$$

Show that Z has a triangular distribution (that is, a distribution with density function whose graph is triangular).

Like most problems concerned with independent uniformly distributed random variables, it is best to consider this question graphically. Since X, Y are distributed independently, the choice of two numbers, x and y, may be represented by plotting the point $R(x, y)$, all points of the square $OABC$ (see Figure 25.9) being equally likely. If $z = x+y$,

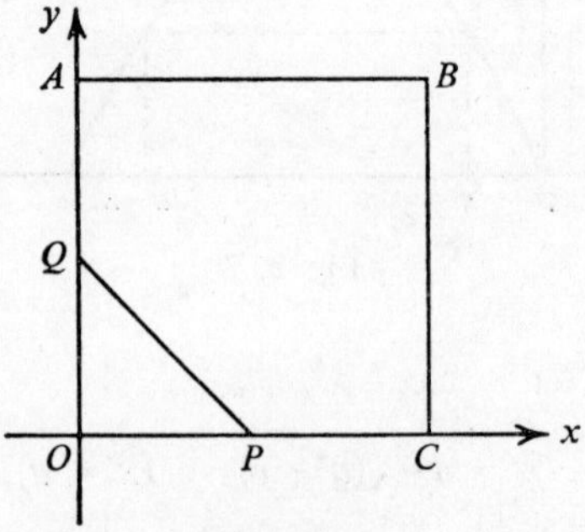

Fig. 25.9

for given z all possible points R lie on the line segment PQ, where P is the point $(z, 0)$. For $0 \leqslant z \leqslant 1$, $PQ \propto z$; for $1 < z \leqslant 2$, $PQ \propto (2-z)$ and thus the density function, f, of Z, whose value for given z is proportional to z, is triangular. In fact

$$f(x) = \begin{cases} 0 & \text{if } z < 0 \\ z & \text{if } 0 \leqslant z \leqslant 1 \\ (2-z) & \text{if } 1 < z \leqslant 2 \\ 0 & \text{if } z > 2. \end{cases}$$

Ex. 12. Find the distribution function, F, for the random variable Z of Example 5 and sketch its graph.

Ex. 13. If, with the notation of Example 5, the random variable W is defined by $w = x-y$, describe the distribution of W.

(ii) *The normal distribution*

Suppose that shots are fired at the centre, O, of a large circular target. Set up coordinate axes as shown in Figure 25.10 and let $P(x, y)$ be the point at which one of the shots hits the target. If we suppose that shots aimed at the centre are liable to errors, equally likely to occur above or below Ox and to the left or right of Oy, we may reasonably assume that the probability of x lying between x and $x+\delta x$, the 'x error', depends upon the

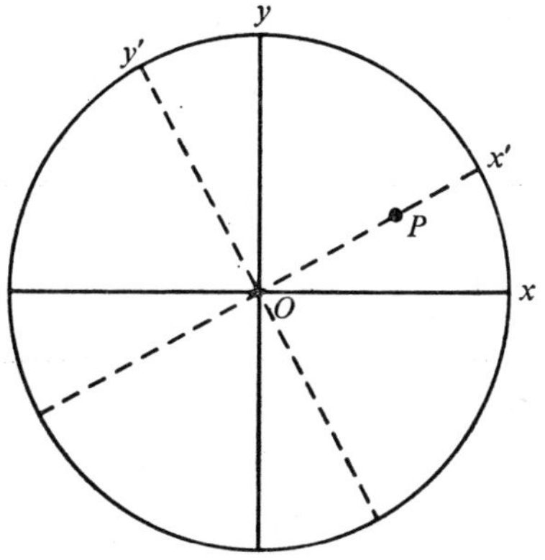

Fig. 25.10

numerical value of x; that is, the probability is of the form $\phi(x^2)\,\delta x$. On the assumption of the symmetry of the errors made above, the probability of the 'y error' is $\phi(y^2)\,\delta y$ and the probability of lying in a small area

$$\delta A = \delta x\,\delta y$$

around P is $\phi(x^2)\,\phi(y^2)\,\delta A$, on assuming that the errors occur independently. But on these assumptions the probability may also be written

$$\phi(x'^2)\,\phi(y'^2)\,\delta A, \qquad \text{(approx.)}$$

where new axes Ox', Oy' are taken with Ox' passing through P, as shown in Figure 25.10. Thus, since $x'^2 = x^2+y^2$ and $y' = 0$ at the point P,

$$\phi(x^2)\,\phi(y^2) = \phi(x'^2)\,\phi(y'^2) = \phi(x^2+y^2)\,\phi(0)$$

and ϕ is a function with values satisfying the equation

$$\phi(x^2)\,\phi(y^2) = k\phi(x^2+y^2).$$

A solution of this equation is $\phi(x^2) = k\mathrm{e}^{-\lambda x^2}$ where λ is taken as a positive constant, since it is reasonable to assume that the probability decreases the further we move from O.

The argument above (due to Thompson and Tait) suggests that the function $f(x) = k\mathrm{e}^{-\lambda x^2}$ could form a suitable model for the probability density of errors occurring in observations. This indeed turns out to be the case and we shall therefore make the following definition.

The random variable X is said to have a *normal distribution* if its density function, $\phi(x)$, is of the form

$$\phi(x) = \frac{1}{\surd(2\pi)}\mathrm{e}^{-\frac{1}{2}x^2}.$$

The letter ϕ is customarily used in place of f to denote the density function of the normal distribution; the corresponding distribution function is

$$\Phi(x) = \frac{1}{\surd(2\pi)}\int_{-\infty}^{x} \mathrm{e}^{-\frac{1}{2}x^2}\,\mathrm{d}x.$$

The factor $(2\pi)^{-\frac{1}{2}}$ appearing in the forms for ϕ and Φ ensures that the requirement

$$\int_{-\infty}^{+\infty} \phi(x)\,\mathrm{d}x = 1$$

is met—that is, that $\phi(x)$ is a genuine density function. This follows from the well-known integral

$$\int_{-\infty}^{+\infty} \mathrm{e}^{-\frac{1}{2}x^2}\,\mathrm{d}x = \surd(2\pi),$$

a result whose proof will be found in any sufficiently advanced book on calculus. (The reader should not be too discouraged by the remark, attributed to Sir William Thompson, about this integral: 'No-one can call himself a mathematician to whom this result is not obvious!')

The shape of the curve of the normal density function is shown in Figure 25.11. As is to be expected, the curve is symmetrical about $x = 0$, which clearly gives the *mean* of the distribution:

$$\mathscr{E}(X) = 0.$$

Furthermore, the variance of a normally distributed random variable X is 1, for

$$\begin{aligned}\mathscr{V}(X) = \sigma^2 &= \frac{1}{\surd(2\pi)}\int_{-\infty}^{+\infty} x^2\,\mathrm{e}^{-\frac{1}{2}x^2}\,\mathrm{d}x \\ &= -\frac{1}{\surd(2\pi)}\int_{-\infty}^{+\infty} x\,\mathrm{d}(\mathrm{e}^{-\frac{1}{2}x^2}) \\ &= \frac{1}{\surd(2\pi)}\int_{-\infty}^{\infty} \mathrm{e}^{-\frac{1}{2}x^2}\,\mathrm{d}x, \text{ on integrating by parts,} \\ &= 1, \text{ on quoting the integral result above.}\end{aligned}$$

Statistical tables (e.g. *The Cambridge Elementary Statistical Tables* by Lindley & Miller) give values of $\Phi(x)$ for values of x from 0 to about 4. (Since $\Phi(4) \approx 0{\cdot}99997$ values of $\Phi(x)$ for $x > 4$ are very rarely required.) By the symmetry of the curve

$$\phi(x) = \frac{1}{\sqrt{(2\pi)}} e^{-\frac{1}{2}x^2}$$

about the line $x = 0$, values of $\Phi(x)$ for $x < 0$ may be deduced immediately.

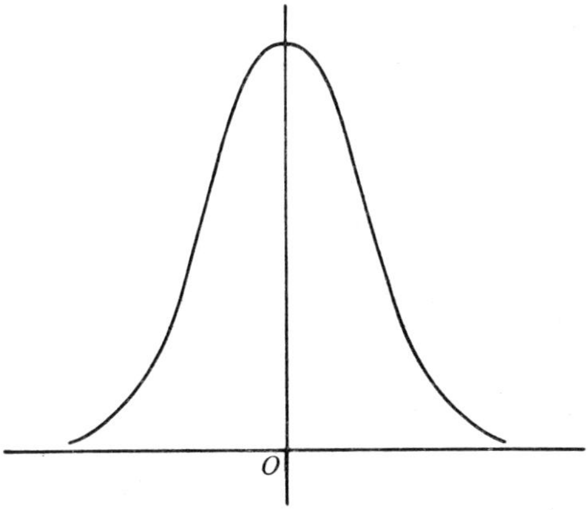

Fig. 25.11

The normal distribution is frequently used as a probability model. Suppose that the random variable X has mean μ and standard deviation σ; then the new random variable Y defined by

$$y = \frac{x-\mu}{\sigma}$$

has mean 0 and standard deviation 1. If Y is normally distributed it is customary to say that '*X is normally distributed about the mean μ with standard deviation σ*', sometimes written as '*X is distributed $N(\mu, \sigma)$*', and that '*X has been standardized to the normal random variable Y by the given substitution*'. For emphasis, Y is often referred to as 'the *standard* normal distribution $N(0, 1)$'.

Example 6. *If X is normally distributed $N(0, 1)$ find*

(i) $\Pr(x < 1{\cdot}4)$; (ii) $\Pr(0{\cdot}8 < x < 1{\cdot}4)$;
(iii) $\Pr(x < -1{\cdot}4)$; (iv) $\Pr(-1{\cdot}4 < x < 0{\cdot}8)$.

If Y is normally distributed $N(2, 0{\cdot}75)$, find

(v) $\Pr(y < 0)$; (vi) $\Pr(1 < y < 3)$.

From statistical tables,

(i) $\Pr(x < 1\cdot4) = \Phi(1\cdot4)$
$\approx 0\cdot9192$

(Figure 25.12).

(ii) $\Pr(0\cdot8 < x < 1\cdot4) = \Phi(1\cdot4) - \Phi(0\cdot8)$
$\approx 0\cdot9192 - 0\cdot7881$
$= 0\cdot1311$

(Figure 25.13).

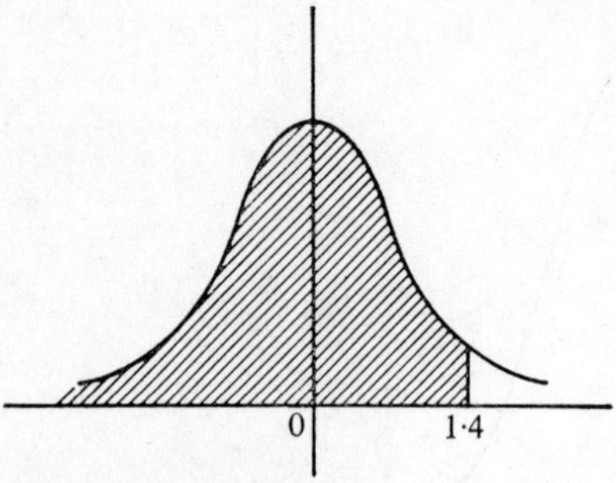

Fig. 25.12

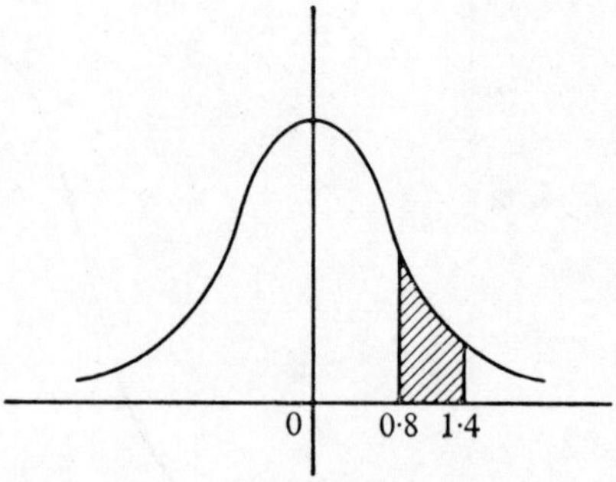

Fig. 25.13

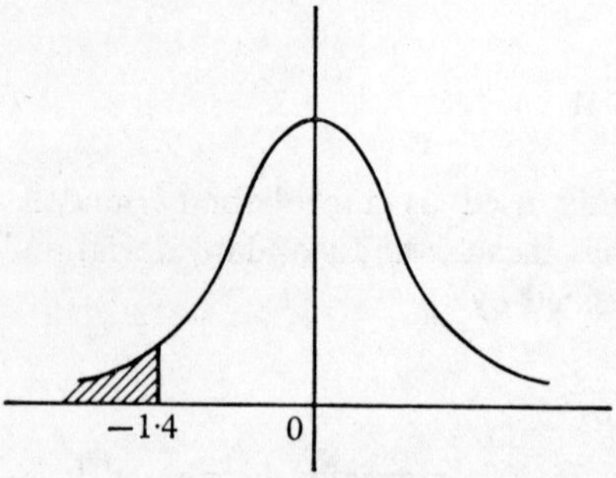

Fig. 25.14

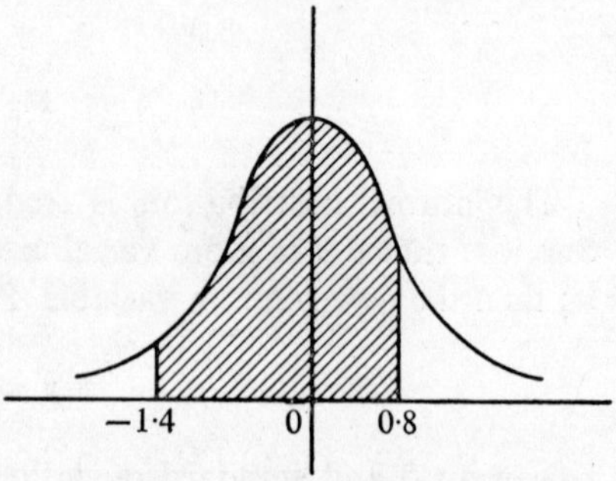

Fig. 25.15

(iii) $\Pr(x < -1\cdot4) = \Pr(x > 1\cdot4)$ by symmetry,
$= 1 - \Phi(1\cdot4)$
$\approx 1 - 0\cdot9192$
$= 0\cdot0808$

(Figure 25.14).

(iv) $\Pr(-1\cdot4 < x < 0\cdot8) = \Pr(x < 0\cdot8) - \Pr(x < -1\cdot4)$
$= \Phi(0\cdot8) - [1 - \Phi(1\cdot4)]$
$\approx 0\cdot7881 - 0\cdot0808$
$= 0\cdot7073$

(Figure 25.15).

(v) The random variable Z, where

$$z = \frac{y-2}{0\cdot 75}$$

has standardized normal distribution. Thus

$$\begin{aligned} \Pr(y < 0) &= \Pr(z < -2\cdot 67) \\ &= 1-\Phi(2\cdot 67) \\ &\approx 1-0\cdot 9962 \\ &= 0\cdot 0038. \end{aligned}$$

(vi) Again,

$$\begin{aligned} \Pr(1 < y < 3) &= \Pr(-1\cdot 33 < z < 1\cdot 33) \\ &= 2\Phi(1\cdot 33)-1 \\ &\approx 1\cdot 8164-1 \\ &= 0\cdot 8164. \end{aligned}$$

Example 7. The heights of a large number of schoolchildren are measured correct to the nearest centimetre and the mean and standard deviation of the resulting frequency distribution are calculated and found to have the values 122 cm *and* 5·2 cm *respectively. As a model of the situation it is assumed that the heights, x, of the children are distributed normally about a mean $\mu = 122$ with standard deviation $\sigma = 5\cdot 2$. Calculate the probabilities for each of the class intervals $x \leqslant 105$, $105 < x \leqslant 110$, $110 < x \leqslant 115, \ldots,$ $130 < x \leqslant 135$, $x > 135$.*

Since the heights are measured to the nearest centimetre, the upper limits for x for the class-intervals above are successively

$$105\cdot 5, \quad 110\cdot 5, \quad 115\cdot 5, \quad \ldots, \quad 135\cdot 5$$

(there is no upper limit, of course, for the last class).

We now standardize our variable, x, by the transformation

$$y = \frac{x-122}{5\cdot 2},$$

so that y has the standard normal distribution ($\mu = 0, \sigma = 1$) and using statistical tables, we find $\Phi(y)$ for the upper limit of each class-interval.

Finally, the probability associated with each class interval may be calculated

$$p = \Phi(y_2)-\Phi(y_1).$$

The working is set out in tabular form below:

Class	x	y	(y)	p
>135			1·0000	0·0047
130–135	135·5	2·60	0·9953	0·0458
125–130	130·5	1·64	0·9495	0·1978
120–125	125·5	0·68	0·7517	0·3658
115–120	120·5	−0·29	0·3859	0·2821
110–115	115·5	−1·26	0·1038	0·0906
105–110	110·5	−2·22	0·0132	0·0125
< 105	105·5	−3·18	0·0007	0·0007
				1·0000

It should be noticed that a continuous distribution has been used here as a mathematical model for a discrete situation (heights measured to the nearest centimetre). Again, it might be objected that, whereas heights could reasonably be expected to fall, say, within the interval 90–180 cm, the normal distribution is defined for all real values of x; however, as has been pointed out earlier, for the standard normal distribution,

$$\Pr(|y| > 4) \approx 0{\cdot}00006$$

and the two tails of the distribution, beyond $y = \pm 4$, may be neglected.

As well as appearing as a distribution in its own right, the normal distribution is often used as an approximation to the binomial distribution $B(n, p)$ for large n, provided neither p nor $1-p$ is too near zero. In Figure 25.16, the probabilities of obtaining the scores 0, 1, 2, ..., 20 in the

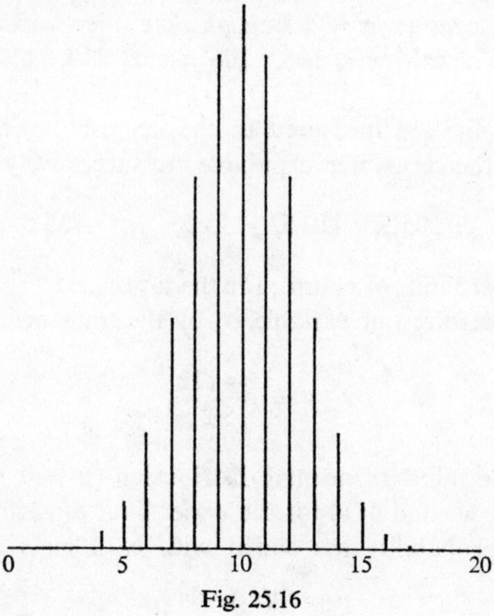

Fig. 25.16

binomial distribution $B(20, \frac{1}{2})$ are proportional to the lengths of the corresponding vertical lines. The outline shape is strongly reminiscent of the normal curve. Even in a skew case, where $p \neq \frac{1}{2}$, the outline is still approximately normal: see Figure 25.17, in which the length of the vertical lines are proportional to the probabilities of obtaining the scores 0, 1, 2, ..., 20 in the binomial distribution $B(20, \frac{2}{5})$.

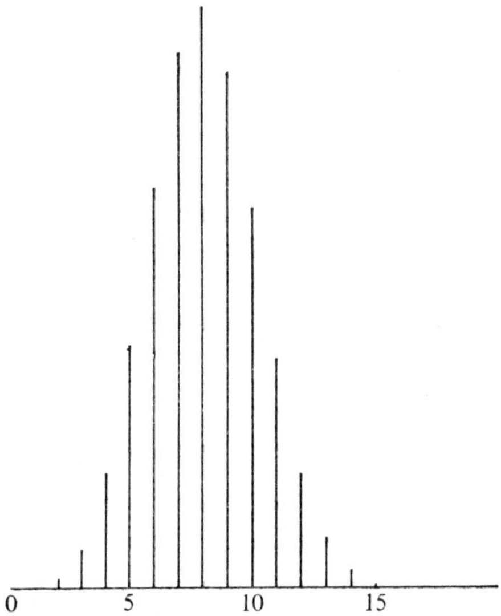

Fig. 25.17

In fact, it may be shown that, given a binomial distribution $B(n, p)$ where n is large and neither p nor $1-p$ is too near zero, then the probability of obtaining a score r, that is,

$$\binom{n}{r} p^r(1-p)^{n-r},$$

is approximately equal to σ^{-1} times the ordinate of the standard normal curve at the point

$$x = (r-\mu)/\sigma,$$

where μ, σ are the mean and standard deviation of the corresponding binomial distribution

$$\mu = np, \quad \sigma = \sqrt{(npq)}.$$

Furthermore (and more importantly for applications) we can relate areas under the normal curve to sums of successive terms of the given binomial distribution; in fact, the probability that the variable takes a value in the interval

$$r_1 \leqslant r \leqslant r_2$$

is approximately

$$\Phi\left(\frac{r_2+\frac{1}{2}-\mu}{\sigma}\right)-\Phi\left(\frac{r_1-\frac{1}{2}-\mu}{\sigma}\right)$$

where, as before, $\mu = np$, $\sigma = \sqrt{(npq)}$. (The limits r_1, r_2 are changed to $r_1-\frac{1}{2}$, $r_2+\frac{1}{2}$ respectively in order to avoid complications arising out of approximating to a discrete distribution by a continuous distribution; for example, we wish to associate a non-zero probability with a single value, r, of the variable.)

The reader is referred to one of the standard texts on probability for a proof of the results outlined above. As a working rule, the approximations give reasonable results if the lesser of the two numbers, *the mean number of successes* and *the mean number of failures* is greater than about five.

Example 8. *A coin is spun* 250 *times and turns up tails* 139 *times. Does this provide any evidence of bias?*

Take as the null hypothesis

$$H: \text{'the coin is unbiased'}.$$

We have to assess the probability of getting a result as bad as, or worse than, 139 tails on the assumption that H is true; that is, we have to determine

$$\Pr(139 \text{ or more tails} \mid H) + \Pr(139 \text{ or more heads} \mid H).$$

The calculation of this from the binomial distribution $B(250, \frac{1}{2})$ would be prohibitively laborious; we therefore use the normal approximation to the binomial distribution as our model where

$$\mu = 250 \times \tfrac{1}{2} = 125,$$

$$\sigma^2 = 125 \times \tfrac{1}{2} = 62{\cdot}5,$$

giving

$$\sigma = 7{\cdot}906.$$

We seek

$$2\Pr(x > 139),$$

where x is distributed $B(250, \frac{1}{2})$, and this is approximately

$$2\left[1-\Phi\left(\frac{138{\cdot}5-125}{7{\cdot}906}\right)\right]$$

$$\approx 2[1-\Phi(1{\cdot}71)]$$

$$\approx 0{\cdot}087.$$

There is thus insufficient evidence of bias at the 5 % level.

Note. Tables of the normal distribution show us that

$$\Pr(|y| \geqslant 1{\cdot}96) \approx 0{\cdot}05$$

and thus a *normal variable greater than* 1·96 *indicates significance at the* 5% *level.* In Figure 25.18 the two shaded areas each contain 2·5% of the total area under the curve.

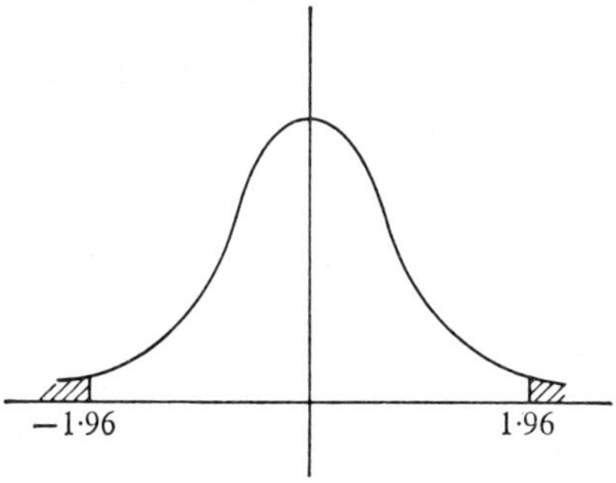

Fig. 25.18

One further application of the normal distribution should be mentioned: if $x_1, x_2, \ldots, x_n$ are independent observations of any random variable X and a new random variable Z is defined by

$$z = (x_1 + x_2 + \ldots + x_n)/n$$

then, if the mean and variance of X are respectively μ and σ^2, the distribution of Z will be approximately normal, with mean μ and variance σ^2/n. (*The Central Limit Theorem.*) The proof of this result is beyond the scope of this book.

(iii) *The exponential distribution*

The *exponential distribution* is defined by the density function

$$f(x) = \begin{cases} 0 & \text{if } x < 0, \\ \lambda e^{-\lambda x} & \text{if } x \geqslant 0, \end{cases}$$

where λ is a positive constant.

**Ex.* 14. Show that, for the exponential distribution,

$$\mathscr{E}(X) = 1/\lambda, \quad \mathscr{V}(X) = 1/\lambda^2.$$

The result of Ex. 14 suggests a resemblance between the exponential and Poisson distributions. We shall exhibit the relation between the two distributions in the following problem. (See Exs. 16–18.)

Consider the calls received at a telephone exchange, the rth call occurring at time t_r, where the time, t, is measured from $t = 0$. We shall assume that the number of calls constitutes a *purely random process*; that is, the number of calls received in any interval (t, t') is independent of anything that has occurred previously. Furthermore, we shall assume that the purely random process is a *stationary process*: that is, the number of calls received in the interval (t, t') depends only upon its length, not on its position.

Ex. 15. Discuss the validity of the assumptions made above in the light of what you imagine would be a typical exchange.

Suppose now that

$$\Pr\,(\textit{no calls are received in the interval}\ (0, t)) = p_0(t),$$

and let us assume that p_0 is a differentiable function. Since, by the definition of a purely random process, the probability of receiving no calls in the interval $(t, t+\delta t)$ is independent of receiving no calls in the interval $(0, t)$,

$$p_0(t+\delta t) = p_0(t)\,p_0(\delta t).$$

Put $t = \delta t = 0$:

$$p_0(0) = [p_0(0)]^2$$

and thus, since $p_0(0) \neq 0$, we must have $p_0(0) = 1$. Again

$$p_0(t+\delta t) = p_0(t)\,p_0(\delta t)$$

$$\Rightarrow \quad p_0(t+\delta t) - p_0(t) = p_0(t)\,[p_0(\delta t) - 1]$$

$$\Rightarrow \quad p_0(t+\delta t) - p_0(t) = p_0(t)\,[p_0(\delta t) - p_0(0)]$$

$$\Rightarrow \quad \frac{p_0(t+\delta t) - p_0(t)}{\delta t} = p_0(t)\frac{p_0(\delta t) - p_0(0)}{\delta t}.$$

Let $\delta t \to 0$:

$$p_0'(t) = p_0(t) . p_0'(0)$$

$$\Rightarrow \quad p_0(t) = A\mathrm{e}^{-\lambda t}, \tag{1}$$

where λ is a constant $(-p_0'(0))$ which must be positive since $p_0(t) \leqslant 1$ and $A = 1$ since $p_0(0) = 1$.

Result (1) now enables us to prove that if X is the random variable whose value is the length of the time interval up to the first call, then X has an exponential distribution. If F is the distribution for X then, for $x \geqslant 0$,

$$\begin{aligned} F(x) &= \Pr\,(\textit{length of the time interval up to the first call} \leqslant x) \\ &= \Pr\,(\textit{at least one call has been made by the time } x \textit{ has elapsed}) \\ &= 1 - \mathrm{e}^{-\lambda x} \end{aligned}$$

and thus

$$f(x) = \lambda \mathrm{e}^{-\lambda x} \quad \text{for} \quad x \geqslant 0.$$

Since $f(x) = 0$ for $x < 0$ it follows that X has an exponential distribution.

Ex. 16. If $p_n(t)$ is the probability that there are n telephone calls in the time interval $(0, t)$ and if δt is sufficiently small for there to be a negligible probability of more than one call in the interval $(t, t+\delta t)$ show that

$$p_n(t+\delta t) = (1-\lambda\delta t)\,p_n(t)+\lambda p_{n-1}(t)\,\delta t$$

and deduce that
$$\frac{\mathrm{d}p_n}{\mathrm{d}t} = \lambda(p_{n-1}-p_n).$$

Ex. 17. Prove by induction that the probability, $p_n(t)$, defined in Ex. 16, is given by

$$p_n(t) = \frac{\mathrm{e}^{-\lambda t}(\lambda t)^n}{n!}.$$

Ex. 18. Show that the distribution of the number, n, of calls received during a fixed time interval is Poisson and account for this in terms of the Poisson distribution being a limiting form of the binomial distribution.

Exercise 25(b)

1. The random variable, X, has uniform distribution in the interval $0 \leqslant x \leqslant 10$. Find $\Pr(X = x$, where $x^2-5x+4 > 0)$.

2. The line AB has length 10 cm. A point P is taken at random on AB, all points being equally likely; what is the probability that the area of the circle of radius AP will exceed 10 cm^2?

3. The line AB has length 10 cm. An interval of length 2 cm is marked at random on the line, the positions of the interval being uniformly distributed. What is the probability that it will contain the mid-point of AB?

4. A circular disc of radius 10 cm is placed on a table. Another disc, of radius 3 cm, is now placed on the table so that it is at least partially in contact with the first disc. On the assumption that the permissible positions of the smaller disc are uniformly distributed, what is the probability that it covers the centre of the larger disc?

5. A point P is taken at random on the side AB (and between A and B) of the square $ABCD$, the positions of P being uniformly distributed. If PC cuts BD at X, find the probability (i) that $BX < \frac{1}{2}BD$; (ii) that $BX < \frac{1}{4}BD$.

6. Figure 25.19 shows a square wooden frame $ABCD$ with a square hole $A'B'C'D'$ cut symmetrically in it. $AB = 50$ cm, $A'B' = 30$ cm. A ball of diameter 5 cm is dropped on to the frame. Assuming that its centre falls within the square $ABCD$ and that it is equally likely to meet the plane of the frame at any point within $ABCD$, what is the probability that the ball will pass straight through the hole without touching the frame?

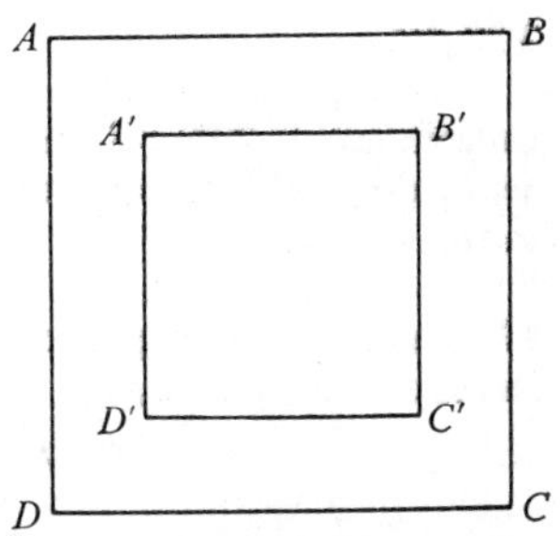

Fig. 25.19

7. A point A is marked on the circumference of a circle of radius r and a chord AP is drawn at random, the positions of the point P on the circumference being uniformly distributed. Find the expected length of the chord.

8. Determine the variance of the length AP of Question 7.

9. A point P is marked on the side AB of the square $ABCD$, the points within AB being uniformly distributed. Find the mean and variance of the area of the triangle APD.

10. ABC is an isosceles triangle right-angled at B. Through A a line is drawn at random to cut BC at P, the angle BAP being uniformly distributed between 0 and $\frac{1}{4}\pi$. If $AB = a$, show that the expected area of the triangle ABP is $(\ln 2)\, a^2/\pi$ and find its variance.

11. The intelligence quotients of 500 schoolchildren are assumed to be normally distributed with mean 105 and standard deviation 12. How many children may be expected:
(i) to have an intelligence quotient greater than 140;
(ii) to have an intelligence quotient less than 90;
(iii) to have an intelligence quotient between 100 and 110?

12. If an unbiased die is thrown 600 times, what is the probability of throwing a six less than 90 times?

13. X is normally distributed with mean 2 and standard deviation $\frac{1}{2}$. Find numbers p, q such that $\Pr(x > p) = 0{\cdot}2$ and $\Pr(p < x < q) = 0{\cdot}1$.

14. A multiple choice test has 100 questions, each question having written beside it five answers, only one of which is correct. If the pass mark is 30%, what is the probability that a student who makes a completely random guess at each answer will pass?

15. Rods are manufactured with a mean length of 20·2 cm and standard deviation 0·09 cm, the distribution being normal. If rods of length less than 20·1 cm are rejected, what is the probability that a rod that is retained has a length in excess of 20·3 cm?

16. Experience has shown that when a certain machine is functioning satisfactorily it produces capacitors with capacitances which are distributed normally with standard deviation 0·080 μF. Under these circumstances find the mean capacitance if 99% of the output has a capacitance of at least 2 μF. Find also what proportion of the output will then have a capacitance between 2·10 and 2·30 μF.

Tests on a large batch of capacitors just produced reveal that their mean capacitance is 2·20 μF and that 10 % of them have capacitance below 2 μF. What do you deduce about the variability of the capacitance of the capacitors in this batch? (M.E.I.)

17. The average proportion p of insects killed by administration of x units of insecticide is given by

$$p = \int_{-\infty}^{(x-\mu)/\sigma} (2\pi)^{-1/2}\, e^{-t^2/2}\, dt,$$

where μ and σ are constants.

When $x = 10$, $p = 0{\cdot}400$ and when $x = 15$, $p = 0{\cdot}900$. What dose will be lethal to 50% of the insect population, on average?

If a dose of 17·5 units is administered to each of 100 insects, how many will be expected to die? What is the probability that just two insects will survive? (M.E.I.)

18. Packets are advertised as containing 500 g of sugar. Tests carried out show that 6·7% of such packets contain a mass greater than 508 g, while 0·6 % have a mass less than 492 g. Estimate the average mass of the contents of a packet on the assumption that a normal distribution constitutes an acceptable mathematical model.

19. A firm advertises that their runner bean seeds give a 95% germination rate. Of 200 such seeds, 14 fail to germinate. Have you cause for complaint?

20. A man claims that he can forecast rainy or dry weather 48 hours ahead. Careful records are kept over 80 days and his forecast is found to be correct 47 times. Is his claim justified, or is he merely lucky?

21. A 'chance of failure' distribution is given for time t by the probability density function

$$p(t) = (1/a)\,\mathrm{e}^{-t/a} \quad (0 < t < \infty).$$

Show that a is the mean time of failure and that the variance is a^2.

Two components in a machine have failure time distributions corresponding to means a and $2a$ respectively. The machine will stop if either component fails and the failures of the two components are independent. Show that the chance of the machine continuing to operate for a time a from the start is $e^{-3/2}$. (O & C)

Miscellaneous Exercise 25

1. AB is a rod of length $2a$. A point P is taken on AB, the points being uniformly distributed, and the stick is broken at P. Find the expected value of AP^2+PB^2.

The two parts of the rod are placed on a table in such a way that A and B are $a\sqrt{3}$ apart. What is the probability that the triangle APB will be obtuse?

2. The *mean deviation about* μ for a distribution with density function f is defined by the equation

$$\eta = \int_{-\infty}^{+\infty} |x-\mu|\, f(x)\, \mathrm{d}x.$$

Prove that the mean deviation of a random variable distributed normally about μ with standard deviation σ is approximately $\frac{4}{5}\sigma$.

Rods are manufactured with a mean length of 18·4 cm and standard deviation 0·25 cm. If only those rods with a length greater than 18·4 cm are retained, what is their average length on the assumption that the rods were originally normally distributed about 18·4 cm with standard deviation 0·25 cm?

3. If X, Y are independently and uniformly distributed between 0 and 2, find $\Pr(xy < 1)$.

4. The side of a square is uniformly distributed between 0 and 1. Find the density and distribution functions for the area of the square and sketch their graphs.

Determine the expected value of the area and its variance.

5. The probability density function for the life x of a motor car tyre is given by

$$f(x) = \lambda e^{-\lambda x}$$

for $x \geqslant 0$, where $\lambda = 0{\cdot}04$ and x is measured in units of 1500 km. What is the probability that a single tyre will last more than 30 units?

A car has four tyres in use and they are all of the same age. What is the probability that all of them will need replacing before 30 000 km? (M.E.I.)

6. A random variable X has the probability density function

$$f(x) = \frac{\pi \exp\{-\pi x/(\sigma\sqrt{3})\}}{\sigma\sqrt{3}\{1+\exp[-\pi x/(\sigma\sqrt{3})]\}^2}$$

for all real x. Show that the distribution is symmetrical about $x = 0$. Determine the cumulative distribution $F(x) = \Pr\{X < x\}$.

You are given that the variance of X is σ^2. Compare the value of $F(\sigma)$, $F(2\sigma)$, $F(3\sigma)$ with the corresponding values for the normal distribution with zero mean and variance σ^2. (M.E.I.)

7. A random variable X is distributed normally with expectation μ and standard deviation σ. Find the mode and median of the distribution.

What is the proportion of the population lying between the points of inflexion of the curve of the probability density function? (M.E.I.)

8. The diameters of some machined components are distributed normally with mean 5·00 cm and standard deviation 0·05 cm. Find the expected proportion of the components which will be outside the range 4·925 to 5·075 cm and the ratio of the expected proportion in the range $5{\cdot}025 \leqslant d \leqslant 5{\cdot}050$ to the expected proportion in the range $5{\cdot}050 \leqslant d \leqslant 5{\cdot}075$.

It is desired to adjust the mean of the process so that there are, on average, twice as many components in the range 5·025–5·050 cm as in the range 5·050–5·075 cm. Show that this can be done if the mean is adjusted to a value between 5·00 cm and 4·95 cm, and find the value by trial to the nearest 0·01 cm. (M.E.I.)

9. The probability density function of a distribution is given by

$$f(x) = \frac{e^{-x}x^{\lambda-1}}{(\lambda-1)!} \quad (x \geqslant 0,\ \lambda \text{ an integer} > 0).$$

Find the expectation and variance of X. If $\mathscr{E}(X) = \mu$, find $\mathscr{E}\{(X-\mu)^3\}$.

Sketch $f(x)$ when $\lambda = 2$. (M.E.I.)

10. A grocer sells bread and can buy batches of 120 loaves. The number of daily customers for bread is distributed normally with mean 100 and variance 100. The net profit on the sale of a loaf is 2·5p and the net loss on an unsold loaf is 3·5p. What is the average daily net profit to the grocer? (M.E.I.)

11. The random variable X has normal distribution. If a new random variable Y is defined by

$$y = x^2$$

show that the distribution function, F, for Y is given by

$$F(Y) = \sqrt{\left(\frac{2}{\pi}\right)} \int_0^{\sqrt{y}} e^{-x^2/2}\, dx.$$

Deduce that Y has *a* χ^2 *distribution with one degree of freedom*, defined by the density function f, where

$$f(y) = \begin{cases} 0 & (y < 0) \\ \dfrac{1}{\sqrt{(2\pi)}} y^{-1/2} e^{-y/2} & (y \geqslant 0). \end{cases}$$

12. Two chords are drawn independently at random in a circle. What is the probability that they intersect?

13. Two men, A and B, are allowed a lunch-break of an hour. A is at liberty to leave the office at any time between 12.00 and 12.45, while B may leave the office between 12.45 and 2.00. What is the probability that they will both be out together, on the assumption that the permissible starting times for both A and B are uniformly distributed?

14. A needle is pivoted at the point $(0, a)$ and is rotated. When it comes to rest, the point P at which its axis (produced if necessary) cuts the x axis is marked. The random variable X is defined to have as its value the x coordinate of P. On the assumption that the angle that the needle makes with the y axis is distributed uniformly between $\pm\frac{1}{2}\pi$, prove that X has a *Cauchy distribution*, defined by the density function

$$f(x) = \frac{a}{\pi(a^2+x^2)} \quad (-\infty < x < \infty).$$

Discuss the existence of $\mathscr{E}(X)$ and $\mathscr{V}(X)$ for this distribution.
Sketch the graph of the density function for $a = 1$.

15. The *Laplace distribution* has probability density function

$$f(x) = Ce^{-\lambda|x|} \quad (-\infty < x < \infty)$$

where C is a constant and $|x|$ denotes the magnitude of x. Find $\Pr\{|x| \leqslant \lambda\}$ and the variance of X. (M.E.I.)

16. A famous early example of the 'Monte Carlo' method for solving problems by random numbers was Buffon's determination of π. Small rods of length L_1 were dropped at random on to a sheet of paper ruled with thin parallel lines a distance L_2 ($> L_1$) apart. A 'success' is recorded when any part of the rod touches a line. The ratio of 'successes' to the total number of tosses is recorded over several thousand tosses. How may π be estimated from this ratio? (C.S.)

17. Let X be a random variable uniformly (rectangularly) distributed over the interval $0 < x < 1$. Derive the probability density function of the following random variables: (*a*) $Y = X^2-1$, (*b*) $Z = \sin \pi X$. Find the mean and standard deviation of Y and Z. (C.S.)

18. The chance that the customer at the head of a queue completes service in any interval of length δt is $\mu\delta t$, and the chance that a new customer arrives in any interval is $\lambda\delta t$, arrivals and departures are independent and $\lambda < \mu$. The chance that at time t there are n customers in the queue (including the one being served) is denoted by $p_n(t)$. Show that

$$p_n(t) = \lambda\delta t p_{n-1}(t-\delta t)+(1-\lambda\delta t-\mu\delta t)p_n(t-\delta t)+\mu\delta t p_{n+1}(t-\delta t) \quad (n \geqslant 1)$$

and obtain the corresponding equation for $p_0(t)$. (over)

By considering the forms that these equations take under the steady-state condition $p_k(t) = p_k(t-\delta t)$, where k is zero or any positive integer, or by any other means, obtain p_n and the mean queue size (in terms if λ and μ) in the steady state. (C.S.)

19. X is a continuous random variable with mean μ and variance σ^2. λ is a positive constant. Prove Chebyshev's inequality

$$\Pr(|X-\mu| > \lambda\sigma) < 1/\lambda^2.$$

Show how this inequality may be sharpened, in the case $\lambda = 2$, if the distribution of X is assumed to be (i) normal, (ii) uniform.

20. Spacecraft land on a spherical planet of centre O. Each is able to transmit messages to, and receive messages from, any spacecraft on the half of the surface of the planet nearest to it.

(i) It is known that spacecraft have landed at points A and B on the surface of the planet. Show that the probability that a space craft, landing at random on the planet, will be able to communicate directly with the spacecraft at A and B is

$$(\pi-\theta)/2\pi$$

where θ is the angle AOB.

(ii) What is the probability that three spacecraft, all landing at random on the planet, will be in direct contact with each other? (C.S.)

21. Engine crankshafts are manufactured so that the diameters, in centimetres, form a normal distribution with mean 5 and standard deviation 0·03. Crankshafts with diameters less than 4·94 or greater than 5·06 are rejected. The accepted product is classified into three grades of size 4·940–4·988, 4·988–5·012, 5·012–5·060.

Show that:

(i) $\int x e^{-x^2/2}\, dx = -e^{-x^2/2}$,

(ii) $\int x^2 e^{-x^2/2}\, dx = -x\, e^{-x^2/2} + \int e^{-x^2/2}\, dx$.

Hence find the average diameter in each of the three grades and the ratio of the standard deviation in the middle grade to the standard deviation of the unclassified product before any rejection of under- and oversize shafts. (M.E.I.)

22. The police force in a certain district carries out tests on the brakes of automobiles chosen at random on the road. Each man is required to test 20 cars. Calculate the distribution of the number of cars with defective brakes in sets of 20 cars if the probability that a single car has defective brakes is 10 %.

Show the distribution graphically together with a plot of the normal distribution with the same mean and variance. Comment on the relation between the two distributions. (M.E.I.)

23. A population contains two strata I and II in proportions $\frac{1}{3}$, $\frac{2}{3}$ respectively. Both strata are exponentially distributed the first with mean 1 and probability density function

$$f(x_1) = \lambda_1 e^{-\lambda_1 x_1} \quad (x_1 \geqslant 0);$$

the second with mean 3 and probability density function

$$f(x_2) = \lambda_2 e^{-\lambda_2 x_2} \quad (x_2 \geqslant 0).$$

Find the mean and standard deviation of the population and the probability that a randomly chosen member of the population is greater than 2. (M.E.I.)

24. Initially a machine is in good running order but is subsequently liable to breakdown. As soon as a breakdown occurs repairs begin. If the machine is in good order at time t then the probability that a breakdown occurs in a small interval $(t, t+\mathrm{d}t)$ is $\alpha\,\mathrm{d}t$, and if it is under repair at time t the probability that the repair is completed in time $(t, t+\mathrm{d}t)$ is $\beta\,\mathrm{d}t$. Let $p(t)$ be the probability that the machine is under repair at time t.

Write down an equation relating $p(t+\mathrm{d}t)$ to $p(t)$ and hence show that $p(t)$ is

$$\frac{\alpha}{\alpha+\beta}\{1-\exp[-(\alpha+\beta)\,t]\}.$$ (C.S.)

26. *Numerical solution of equations*

1. ACCURACY

When equations arise in practical problems, their solutions are generally required only to some given order of accuracy. General algebraic methods of solution (for example, for the cubic equation), even if they are available, are often less suitable than approximate methods for deriving numerical solutions. Thus, to find the real root of the equation

$$x^3-2{\cdot}7x-5{\cdot}3 = 0$$

by Cardan's method would require burdensome calculations and no such method exists to solve, for example, the equation

$$x^5-2{\cdot}7x-5{\cdot}3 = 0.$$

Because of its great practical importance, the estimation of roots of numerical equations has been extensively studied; in this chapter we shall confine ourselves to some of the most elementary methods available: the reader who wishes to study the subject further should consult one of the standard texts on numerical methods (for example, Henrici, *Elements of Numerical Analysis*). It may be mentioned in passing that this subject allows more scope than most others in elementary mathematics for students to devise their own methods; although these may often prove less efficient than the standard techniques, much profit will be gained by developing them as far as possible—apart from the obvious enjoyment of producing something original.

Before considering the solution of equations it is worth pausing to consider some problems connected with the *accuracy* of calculations. In numerical calculations, errors occur through *rounding-off* to a given number of significant figures. If a number is to be rounded-off to N digits and the discarded digits form exactly half a unit, *round-off to the nearest even digit.* For example, to 3 significant figures,

$$3{\cdot}864 \approx 3{\cdot}86; \qquad 21990 \approx 2{\cdot}20\times 10^4;$$

$$0{\cdot}03765 \approx 3{\cdot}76\times 10^{-2}; \qquad 21{\cdot}55 \approx 2{\cdot}16\times 10.$$

**Ex.* 1. Suggest any advantage you see in adopting the convention above. Can you suggest any reason why even rather than odd digits have been selected?

Once numerical data have been rounded-off, their subsequent use in calculations introduces further errors which will be cumulative. It would be pleasant to be able to calculate exactly the error in any numerical answer. This is usually impossible, but we may be able to produce a positive '*acceptance*' for the answer; that is, a quantity which we are sure is more than the absolute error in the answer. For example, suppose we round-off a given angle θ, measured in degrees, to 3 significant figures to give 37·2°. The actual angle lies between $37{\cdot}2^\circ - 0{\cdot}05^\circ$ and $37{\cdot}2^\circ + 0{\cdot}05^\circ$. From four-figure tables, $\tan 37{\cdot}2^\circ = 0{\cdot}7590$ and the difference for $\pm 0{\cdot}05^\circ$ is

$$\tfrac{1}{4}(0{\cdot}7618 - 0{\cdot}7563) \approx 0{\cdot}0014.$$

Thus $$\tan\theta = 0{\cdot}7590 \pm 0{\cdot}0014;$$

in other words, $\tan\theta$ is 0·7590 with an acceptance 0·0014.

**Ex.* 2. If possible inaccuracies in the tangent tables are taken into account, show that $\tan\theta = 0{\cdot}7590 \pm 0{\cdot}0015$.

Now suppose y is a rational approximation to the number x; thus

$$x = y + \epsilon,$$

where ϵ is the error involved in writing y for x. The *absolute error* is $|\epsilon|$ while the *relative error* is $|\epsilon|/y$ (more correctly $|\epsilon|/x$, but the two forms are nearly equal for small relative errors and the form $|\epsilon|/y$ has the advantage of possessing a known denominator).

Usually we do not know the exact error, only an upper limit to the absolute error—the *maximum absolute error*. For example, if a number is rounded to 4 decimal places, the maximum absolute error is 5×10^{-5}. Similarly, we have a *maximum relative error*. In a calculation in which we know the maximum absolute errors in the given numbers it is important to be able to estimate the maximum absolute error in the answer.

Suppose that x_1 is written as y_1 with maximum absolute error $|\epsilon_1|$ (we use modulus signs to emphasize that maximum absolute errors are positive) and that x_2 is written as y_2 with maximum absolute error $|\epsilon_2|$. Then the maximum absolute error in writing $y_1 + y_2$ for $x_1 + x_2$ is $|\epsilon_1| + |\epsilon_2|$, while the maximum relative error is

$$(|\epsilon_1| + |\epsilon_2|)/(y_1 + y_2).$$

Ex. 3. Show that the maximum absolute and relative errors in writing $y_1 - y_2$ for $x_1 - x_2$ are respectively $|\epsilon_1| + |\epsilon_2|$ and $(|\epsilon_1| + |\epsilon_2|)/(y_1 - y_2)$.

**Ex.* 4. Neglecting terms such as $|\epsilon_1\epsilon_2|$, show that the maximum absolute and relative errors in writing $y_1 y_2$ for $x_1 x_2$ are respectively

$$|y_1\epsilon_2| + |y_2\epsilon_1| \quad \text{and} \quad |\epsilon_1/y_1| + |\epsilon_2/y_2|.$$

Show also that, writing y_1/y_2 for x_1/x_2 they are respectively

$$|\epsilon_1/y_2| + |y_1\epsilon_2/y_2^2| \quad \text{and} \quad |\epsilon_1/y_1| + |\epsilon_2/y_2|.$$

Ex. 5. If you have a calculating machine available, calculate:
(i) x_1+x_2, y_1+y_2; (ii) x_1-x_2, y_1-y_2; (iii) x_1x_2, y_1y_2;
(iv) x_1/x_2, y_1/y_2 (6 significant figures) where

$$x_1 = 2{\cdot}914, \quad x_2 = 0{\cdot}3472, \quad y_1 = 2{\cdot}9, \quad y_2 = 0{\cdot}35.$$

Ex. 6. What are the maximum absolute and relative errors in writing
(i) y_1^2 for x_1^2; (ii) $\sqrt{y_1}$ for $\sqrt{x_1}$; (iii) $\sqrt{(y_1+y_2)}$ for $\sqrt{(x_1+x_2)}$?

Before leaving the subject of errors, it is worth noting one particular case in which significant figures may be lost. Consider 3·144–3·097: both of these numbers are given to 4 significant figures, but their *difference*, 0·047, is correct to only 2 significant figures. It is sometimes possible to modify a calculation in such a way to minimize this loss of accuracy.

For example, if it is required to calculate the difference between the values of a function of x for two given values of x that are close together, it may help to use a Taylor expansion. Thus, using four figure tables,

$$\sin 31^\circ - \sin 30^\circ = 0{\cdot}0150,$$

but, by Taylor's expansion,

$$\begin{aligned} \sin 31^\circ &= \sin 30^\circ + (\pi/180)\cos 30^\circ - \tfrac{1}{2}(\pi/180)^2 \sin 30^\circ \ldots \\ &= 0{\cdot}5000 + 0{\cdot}01512 - 0{\cdot}00076 \end{aligned}$$

and $\sin 31^\circ - \sin 30^\circ \approx 0{\cdot}01504$; again using four figure tables and obtaining a result correct to 4 significant figures.

Again, suppose we have to solve the quadratic equation

$$x^2 - 16x - 1 = 0.$$

The roots are
$$\frac{16+\sqrt{260}}{2} \quad \text{and} \quad \frac{16-\sqrt{260}}{2}.$$

Direct evaluation of these two quantities gives

$$16{\cdot}06 \quad \text{and} \quad -0{\cdot}06$$

(using four figure tables). However, in obtaining the negative root, we lost significant figures by subtraction. A better approximation may be had by calculating the negative root using the more accurate positive root 16·06 and the known relation that the product of the roots is -1:

$$\alpha \approx -\frac{1}{16{\cdot}06} \approx -\,0{\cdot}0627.$$

Before leaving the subject of significant figures, it is worth noticing that a calculation can be carried out to a high degree of accuracy and yet not furnish a correct result even when corrected to 1 significant figure. For example, if the correct answer is 0·6499999 and our numerical answer is 0·6500001 then, to 1 significant figure our answer is 0·7 but the correct answer is 0·6 although our calculation was only 0·0000002 in error.

Example 1. *Given that* $\sqrt{31}=5{\cdot}568$ *and* $\sqrt{30}=5{\cdot}477$, *find* $\sqrt{31}-\sqrt{30}$.

Direct computation gives $\sqrt{31}-\sqrt{30}=0{\cdot}091$; but using

$$(\sqrt{31}-\sqrt{30})\,(\sqrt{31}+\sqrt{30})=1$$

we have

$$\sqrt{31}-\sqrt{30}=1/(5{\cdot}568+5{\cdot}477)$$

$$=0{\cdot}090539 \quad \text{(from 5 figure reciprocal tables)}$$

and this is in fact correct to 5 S.F.

2. LOCATING THE ROOTS OF THE EQUATION $f(x) = 0$

Before embarking upon the accurate estimation of a root of the equation $f(x) = 0$ it is usually necessary first to locate the root roughly.

One method of doing this is to draw the graph of $y = f(x)$ and find where it cuts the x axis. Indeed, it is possible, by accurately plotting a succession of graphs of the relevant regions of $f(x)$ on increasing scales, to obtain quite good approximations to roots of given equations. When using a graphical technique it is sometimes convenient to write the equation in a form other than $f(x) = 0$, e.g. $g(x) = x$. See, for example, Ex. 7.

**Ex*. 7. Verify from a freehand sketch that the equation

$$x = 2 \sin x$$

has a root lying between $\frac{1}{2}\pi$ and $\frac{3}{4}\pi$. Draw the graphs of $y = x$ and $y = 2 \sin x$ accurately on graph paper, taking values of x from $\frac{1}{2}\pi$ to $\frac{3}{4}\pi$ and hence obtain a better estimate of this root. Suggest how you could continue the process. If you have a table of sines of angles in radians, show how these may be used to obtain rapidly a good approximation.

The graphical method just mentioned suggests a simple analytic method for locating the presence of a root: if f is continuous and $f(a)$, $f(b)$ have opposite signs, then the equation $f(x) = 0$ has at least one root lying between $x = a$ and $x = b$.

Ex. 8. Show that the equation

$$x^5-x-1 = 0$$

has just one real root and that this root lies between $x = 1$ and $x = 2$.

Ex. 9. Show graphically that it is possible for roots of the polynomial equation $f(x) = 0$ to occur between $x = a$ and $x = b$ without $f(a)$, $f(b)$ being of different signs.

When a root has been located between $x = x_0$ and $x = x_1$, an approximate value for the root may be obtained by *linear interpolation*; that is, by approximating to the graph of $y = f(x)$ between $x = x_0$ and $x = x_1$ by a straight line (see Figure 26.1, in which $OA = x_0$, $OB = x_1$; we take OC as our approximate root, the exact root being OD). Since the equation of the line PQ, in the notation of Chapter 18 with $x_1 - x_0 = h$, is

$$h(y-f_0) = \Delta f_0(x-x_0)$$

we have, putting $y = 0$,

$$x = x_0 - \frac{hf_0}{\Delta f_0}$$

as an approximation to the required root.

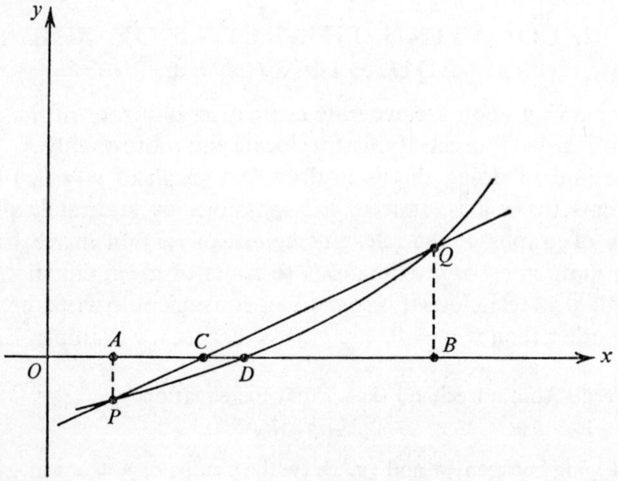

Fig. 26.1

**Ex.* 10. Show from graphical considerations how the computation of $\Delta^2 f_0$ enables one to tell whether the approximation by linear interpolation is likely to be too big or too small.

Example 2. *Show that the equation*

$$x^3+2x-4 = 0$$

has only one real root, and find its value, correct to 1 *decimal place, by linear interpolation.*

Write $f(x) \equiv x^3+2x-4$; then $f'(x) \equiv 3x^2+2$, and $f''(x) \equiv 6x$.

Thus we see immediately that, since the graph of f has no maximum or minimum, it can cut the x axis only once, giving just one real root (Figure 26.2).

By direct computation, $f(1) = -1$, $f(2) = 8$ and the real root lies between $x = 1$ and $x = 2$, probably nearer to $x = 1$.

By linear interpolation, a second approximation is given by

$$x = 1+\tfrac{1}{9}.$$

Since $f'(x) > 0$, $f''(x) > 0$ between $x = 1$ and $x = 2$ (see Figure 26.2) the root $x = 1\tfrac{1}{9}$ will be an underestimate.

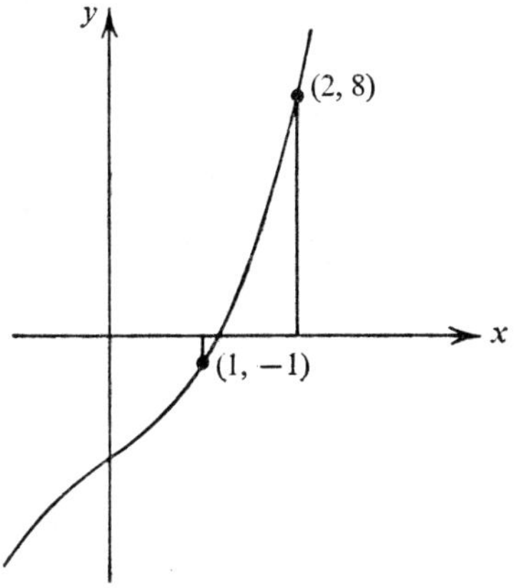

Fig. 26.2

By direct computation again we have

$$f(1{\cdot}1) = -0{\cdot}469, \quad f(1{\cdot}2) = +0{\cdot}128,$$

showing that the root lies between $x = 1{\cdot}1$ and $x = 1{\cdot}2$, and our next approximation to the root is

$$1{\cdot}1+\tfrac{469}{597}\times\tfrac{1}{10}.$$

As before, this is an underestimate and we can be confident, without further working, that $x = 1{\cdot}2$ is the value of the required root correct to 1 decimal place.

Another valuable method of locating roots of a polynomial equation is based on a consideration of the number of sign changes among the coefficients in the expression

$$a_0x^n+a_1x^{n-1}+\ldots+a_n.$$

Ex. 11. Prove that, if all of the coefficients are positive, then the equation

$$a_0x^n+a_1x^{n-1}+\ldots+a_n = 0$$

can have no positive roots.

Prove further that, if the coefficients of all odd powers of x are zero and the remaining coefficients are positive, then the equation can have no real roots.

The result of Ex. 11 may be extended to the following result, known as *Descartes's Rule of Signs*: the number of positive roots of a polynomial equation cannot exceed, and has the same parity as, the number of sign changes among the coefficients, reading from left to right. (Two numbers have the *same parity* if they are both even or both odd.)

For example, the equation

$$\begin{array}{cccccccc} 2x^9 & -7x^7 & -4x^6 & +3x^5 & +2x^3 & +x^2 & -5x & -4 = 0 \\ + & - & - & + & + & + & - & - \\ & \uparrow & & \uparrow & & & \uparrow & \end{array}$$

has three sign changes and thus either one or three positive roots. Again, the equation

$$\begin{array}{cccccc} x^8 & -6x^6 & -4x^5 & +3x^4 & -x^2 & +2 = 0 \\ + & - & - & + & - & + \\ & \uparrow & & \uparrow & \uparrow & \uparrow \end{array}$$

has four sign changes and thus has zero, two or four positive roots.

By writing $-x$ for x one may similarly find an upper limit to the number of negative roots.

Ex. 12. Show that the equation

$$x^7-3x^4-x^2-1 = 0$$

has just one real root, by considering separately the possible positive and negative roots.

The proof of Descartes's Rule is not difficult, but is complicated by the number of special cases that have to be considered. The reader who wishes to follow the proof through should try Exs. 13–17; in all these equations we adopt the notation

$$P(x) \equiv a_0x^n+a_1x^{n-1}+\ldots+a_{n-1}x+a_n,$$

where $a_n > 0$ and we suppose that $P(x)$ has k sign changes.

Ex. 13. Prove that $P'(x)$ has k sign changes if $a_{n-1} > 0$ and $(k-1)$ sign changes if $a_{n-1} < 0$.

Ex. 14. Prove graphically that, if $P'(0) > 0$, no roots of the equation $P(x) = 0$ lie between $x = 0$ and the least positive root of $P'(x) = 0$ and that if $P'(0) < 0$, at most one such root of $P(x) = 0$ exists.

Ex. 15. What happens in Exs. 13, 14 if $a_{n-1} = 0$?

Ex. 16. Complete the proof of Descartes's Rule of Signs using mathematical induction.

Exercise 26(a)

1. If x is rounded-off to 3 significant figures to give the numerical value 48·7, give the value and range of acceptance for (i) $\sin x^\circ$; (ii) $\tan x^\circ$, using 4-figure tables.

2. If x, y are rounded-off to 3 significant figures, their values are $x = 17{\cdot}2$, $y = 5{\cdot}16$; give the range of acceptance for (i) $x+y$; (ii) xy; (iii) x/y.

3. Compare the relative accuracy obtained from your square root tables for $x = \sqrt{3}-\sqrt{2}$ with that obtained for $x = (\sqrt{3}+\sqrt{2})^{-1}$.

If $\sqrt{2} = 1{\cdot}41421\ldots$ and $\sqrt{3} = 1{\cdot}73205\ldots$ find as accurately as you can a value for $\sqrt{3}-\sqrt{2}$.

4. Given that $\sqrt{130} = 11{\cdot}4018$ and $\sqrt{132} = 11{\cdot}4891$ find as accurately as you can a value for $\sqrt{132}-\sqrt{130}$.

5. The triangle ABC is right angled at B. AC, AB are measured to the nearest millimetre, their lengths being found to be 7·4 and 4·4 cm respectively. Use the theorem of Pythagoras to calculate the length of BC, stating what confidence you have in the reliability of your answer.

6. If $x = y+\epsilon$, use the Taylor expansion to find the absolute and relative error in taking $\tan x$ to be equal to $\tan y$.

7. With the notation of Question 6, find the absolute and relative error in taking $(1+x^2)^{-1/2}$ to be $(1+y^2)^{-1/2}$.

8. If $x = 4{\cdot}6 \pm 0{\cdot}05$ and $f(x) = x^3-2x+1$, and if x is taken to be 4·6, find the relative errors in (i) x; (ii) $f(x)$. Comment upon your result.

9. Find graphically, using the method of enlarging scales, the least positive root of the equation

$$\cot\left(\frac{\pi x}{2}\right) = 1+x$$

correct to 2 decimal places.

10. Solve graphically the equation $2x = 1+\ln x$.

11. Solve graphically $x^3 = 10$.

12. Find $2^{1/5}$, correct to 2 decimal places, by a graphical method.

13. Locate the three roots of the equation

$$2x^3-6x-3 = 0$$

and find their approximate values, using linear interpolation.

14. Locate the two real roots of the equation

$$x^4+x^2+10x-24 = 0$$

and find their approximate values, using linear interpolation.

15. Use Descartes's Rule of Signs to prove that the equations

(i) $x^5-2x^3-3 = 0$; (ii) $x^4-x^3+5x^2+2 = 0$; (iii) $x^6+x^4-4x^3+5x+2 = 0$

each have at least two complex roots.

16. Use Descartes's Rule of Signs to show that the equation

$$x^5+x+1 = 0$$

has only one real root, and that this root is negative.

Show that this result can also be obtained from simple graphical considerations.

17. What can you say about the reality of the roots of the equation

$$x^n+x^2+2 = 0$$

where n is an integer greater than 2?

3. ITERATIVE PROCESSES FOR SOLVING EQUATIONS

In this section we consider the application of *iterative methods* for finding numerical solutions of equations; that is, methods which develop successive approximations to a root of a given equation by a simple repetitive process depending upon a recurrence relation.

As a first example, consider the equation

$$x^2-5x-5 = 0.$$

(A quadratic equation is chosen for simplicity, but the method to be outlined below is applicable to any polynomial equation.)

The given equation has roots lying between -1 and 0 and between 5 and 6. We shall denote successive approximations to one of the roots by $x_0, x_1, x_2, \ldots$.

We first seek the negative root: take $x_0 = -1$. Since the equation may be rewritten

$$x = \tfrac{1}{5}x^2-1$$

we try, as a plausible attempt at developing successive approximations, the recurrence relation

$$x_{r+1} = \tfrac{1}{5}x_r^2-1.$$

Starting with $x_0 = -1$, this gives successive approximations

$$-1,\ -0{\cdot}8,\ -0{\cdot}872,\ -0{\cdot}847,\ -0{\cdot}857.$$

Thus $x_4 = -0{\cdot}857$, and the process is seen to be converging (albeit rather slowly) towards the negative root $-0{\cdot}8541 \ldots$.

Now suppose we adopt the same recurrence relation to find the root lying between 5 and 6:

$$x_0 = 6 \quad \text{gives} \quad 6{\cdot}2,\ 6{\cdot}69, \text{ etc.}$$

diverging.

$$x_0 = 5 \quad \text{gives} \quad 4,\ 2{\cdot}2,\ -0{\cdot}32,\ -0{\cdot}980,\ -0{\cdot}808, \text{ etc.}$$

and we are clearly converging towards the wrong root!

To see what has gone wrong let us write

$$\alpha = x_r + \epsilon_r,$$

where α is the exact root of the given equation and ϵ_r is the error in the estimate x_r.

$$x_{r+1} = \tfrac{1}{5}x_r^2 - 1$$

gives
$$\alpha - \epsilon_{r+1} = \tfrac{1}{5}(\alpha - \epsilon_r)^2 - 1$$

which reduces to
$$\epsilon_{r+1} = \tfrac{2}{5}\alpha\epsilon_r,$$

on neglecting ϵ_r^2 and recalling that $\alpha = \frac{1}{5}\alpha^2 - 1$. Thus $|\epsilon_{r+1}| < |\epsilon_r|$ only if $|\alpha| < 2{\cdot}5$; in other words, the absolute errors diminish in this iterative process only if the root to which we are approximating is less than 2·5 in magnitude.

Graphically, the solution of

$$x = \tfrac{1}{5}x^2 - 1$$

means finding the x coordinate of the intersection of the straight line

$$y = x$$

with the parabola
$$y = \tfrac{1}{5}x^2 - 1.$$

Our iterative process consists in starting with a given value of x, ($x_0 = -1$) finding the corresponding point on the parabola, moving from there to the line $y = x$ ($x_1 = \frac{1}{5}x_0^2 - 1$), and thence to the parabola, then on to the line $y = x$, again ($x_2 = \frac{1}{5}x_1^2 - 1$), and so on, spiralling in to the root $x = -0{\cdot}854...$ in the cobweb pattern shown in Figure 26.3.

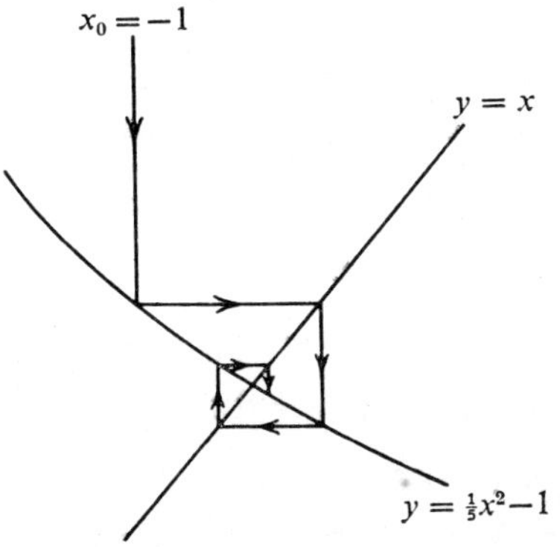

Fig. 26.3

However, if we start at $x_0 = 5$, a quite different pattern emerges. The reader should draw the graph and illustrate the various stages of the iterative process: he will see that the path taken moves rapidly across the graph to the negative root.

**Ex.* 17. Write the equation

$$x^2-5x-5 = 0$$

in the form

$$x = 5+5/x$$

and obtain the recurrence relation

$$x_{r+1} = 5+5/x_r.$$

Show that, with $x_0 = 6$, this gives $x_5 = 5{\cdot}8541$ which is correct to four decimal places.

Show further that, in this case,

$$|\epsilon_{r+1}| \approx \left|\frac{5\epsilon_r}{\alpha^2}\right|.$$

**Ex.* 18. Illustrate graphically the iterative process of Ex. 17 by sketching the graphs

$$y = x \quad \text{and} \quad y = 5+5/x.$$

Ex. 19. Explain geometrically why convergence is more rapid for finding the positive root from $x_{r+1} = 5+5/x_r$ than for finding the negative root from

$$x_{r+1} = \tfrac{1}{5}x_r^2-1.$$

Ex. 20. Explain why, in both the iterative processes considered so far, the successive approximations oscillate from side to side of the exact root.

Now consider a polynomial equation rewritten in the form

$$x = f(x).$$

(We have already seen that there are various ways of doing this: our object now is to choose the best one for a given root.)

Suppose we take

$$x_{r+1} = f(x_r),$$

where x_r, x_{r+1} are successive approximations to the exact root α, with errors ϵ_r, ϵ_{r+1}:

$$\alpha = x_r+\epsilon_r, \quad \alpha = x_{r+1}+\epsilon_{r+1}.$$

Since α is a root of the given equation,

$$\alpha = f(\alpha).$$

But

$$x_{r+1} = f(x_r),$$

giving

$$\alpha-\epsilon_{r+1} = f(\alpha-\epsilon_r)$$

$$\approx f(\alpha)-\epsilon_r f'(\alpha), \quad \text{on using Taylor's expansion.}$$

Thus

$$\epsilon_{r+1} \approx \epsilon_r f'(\alpha).$$

Successive applications of this result give

$$\epsilon_{r+1} \approx \epsilon_r f'(\alpha) \approx \epsilon_{r-1}[f'(\alpha)]^2 \approx \ldots \approx \epsilon_0[f'(\alpha)]^{r+1};$$

thus, provided $|f'(\alpha)| < 1$, $|\epsilon_{r+1}| < |\epsilon_0|$

and ϵ_r decreases in magnitude as r increases. If, however, $|f'(\alpha)| > 1$, the error increases as r increases. The reader should now reconsider the two iterative processes for the quadratic equation considered previously, in the light of this analysis.

Ex. 21. Explain graphically what happens to the iterative process if $f'(\alpha) = 1$.

**Ex.* 22. Illustrate the significance of the condition $|f'(\alpha)| < 1$ for the convergence of the iterative process

$$x_{r+1} = f(x_r)$$

by drawing four graphs with:

(i) $f'(\alpha) < 1, f''(\alpha) < 0$; (ii) $f'(\alpha) < 1, f''(\alpha) > 0$;
(iii) $f'(\alpha) > 1, f''(\alpha) < 0$; (iv) $f'(\alpha) > 1, f''(\alpha) > 0$.

As we have already seen, the iterative procedure outlined above converges to the required root rather slowly. A more powerful iterative procedure is provided by the *Newton–Raphson process*. Consider the equation

$$f(x) = 0$$

and suppose, as before, that α is an exact root, with x_0 the first approximation and error ϵ_0; that is

$$\alpha = x_0 + \epsilon_0.$$

We then have

$$f(\alpha) = 0,$$

$$\Leftrightarrow f(x_0+\epsilon_0) = 0,$$

and thus, on using a Taylor expansion and regarding ϵ_0 as being sufficiently small for us to be able to ignore ϵ_0^2 and higher powers,

$$f(x_0) + \epsilon_0 f'(x_0) \approx 0.$$

This yields an expression for the error term

$$\epsilon_0 \approx -\frac{f(x_0)}{f'(x_0)} \quad (\text{provided } f'(x_0) \neq 0),$$

and we may take as our next approximation

$$x_1 = x_0 - \frac{f(x_0)}{f'(x_0)}.$$

Repetition of this process leads us to the recurrence relation

$$x_{r+1} = x_r - \frac{f(x_r)}{f'(x_r)}.$$

Example 3. *Show that the cubic equation*

$$x^3 - 5x - 8 = 0$$

has just one real root, and find its value, correct to 2 *decimal places.*

Writing $$f(x) \equiv x^3-5x-8,$$
we have $$f'(x) \equiv 3x^2-5,$$
and the curve $y = f(x)$ is seen to have two stationary points, a maximum at $x \approx -\sqrt{1\cdot 7}$, and a minimum at $x \approx +\sqrt{1\cdot 7}$. But $-\sqrt{1\cdot 7} \approx -1\cdot 3$ and $f(-1\cdot 3) < 0$ therefore the curve cuts the x axis just once and the equation thus has only one real root (which is clearly positive). (See Figure 26.4.)

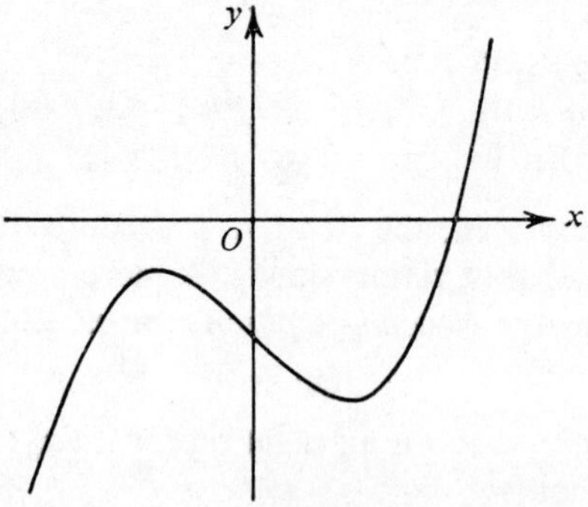

Fig. 26.4

The next step in the solution is to locate the root: to do this we substitute integral values for x until we discover a sign change. In order to write down the differences Δf, $\Delta^2 f$, ... we tabulate our working, writing the values of f in a vertical line:

x	f	Δf	$\Delta^2 f$	$\Delta^3 f$
0	-8			
		-4		
1	-12		6	
		2		6
2	-10		12	
		14		
3	4			

The root we seek lies between 2 and 3; using linear interpolation we take as our first approximation

$$x_0 = 2+\tfrac{10}{14}\times 1 \approx 2\cdot 7.$$

Since $\Delta f(2) > 0$, the curve $y = f(x)$ is increasing between $x = 2$ and $x = 3$ and since $\Delta^2 f(2) > 0$, it is increasing at an increasing rate: we deduce that 2·7 is an underestimate of the exact root (see Figure 26.5, which has not been drawn to scale).

Since $f'(x) \equiv 3x^2-5$,

$$x_1 = 2{\cdot}7 - \frac{f(2{\cdot}7)}{f'(2{\cdot}7)}$$

$$= 2{\cdot}7 + \frac{1{\cdot}817}{16{\cdot}87}$$

$$\approx 2{\cdot}808.$$

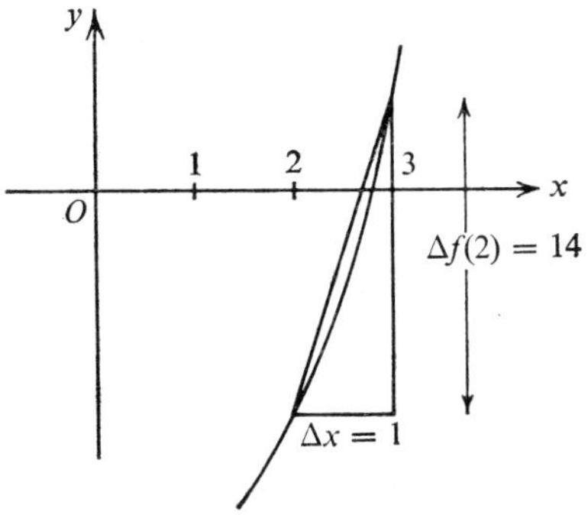

Fig. 26.5

The process is now repeated, taking, $x_1 = 2{\cdot}808$ as our second approximation. (The process is much facilitated by the use of a hand-calculating machine: if one is available, recall the method of nested-multiplication for the evaluation of polynomials—see Chapter 18.)

$$x_2 = 2{\cdot}808 - \frac{f(2{\cdot}808)}{f'(2{\cdot}808)}$$

$$= 2{\cdot}808 - \frac{0{\cdot}1}{18{\cdot}65}$$

$$= 2{\cdot}802.$$

Since we may regard this as a more accurate approximation to the required root than 2·808, we take

$$x = 2{\cdot}80$$

as the root to 2 decimal places.

Ex. 23. Use a calculating machine to obtain the root of the equation

$$x^3-5x-8 = 0$$

correct to 4 decimal places.

A feature of the Newton–Raphson process is that, since at each stage the value of $f(x_r)$ is calculated, a running check may be kept on the *residuals*;

that is, the values obtained by substituting our successive approximations into the polynomial. (We want the residuals to be zero eventually.)

We now investigate, as before, how rapidly the process converges. Suppose that the exact root of the equation $f(x) = 0$ that we seek is α and that we obtain a sequence of approximations $x_0, x_1, x_2, \ldots$, where

$$\alpha = x_r + \epsilon_r.$$

From the recurrence relation

$$x_{r+1} = x_r - \frac{f(x_r)}{f'(x_r)},$$

we thus have
$$\alpha - \epsilon_{r+1} = \alpha - \epsilon_r - \frac{f(\alpha-\epsilon_r)}{f'(\alpha-\epsilon_r)},$$

giving
$$\epsilon_{r+1} = \epsilon_r + \frac{f(\alpha-\epsilon_r)}{f'(\alpha-\epsilon_r)}.$$

Now write
$$g(x) = \frac{f(x)}{f'(x)}, \quad \text{for} \quad f'(x) \neq 0;$$

then
$$g(\alpha) = 0, \quad \text{provided } f'(\alpha) \neq 0$$

and also, since
$$g(x) f'(x) = f(x),$$

we have
$$g'(x) f'(x) + g(x) f''(x) = f'(x)$$

and
$$g''(x) f'(x) + 2g'(x) f''(x) + g(x) f'''(x) = f''(x).$$

Thus
$$g'(\alpha) = 1, \quad g''(\alpha) = -f''(\alpha) / f'(\alpha).$$

From the equation
$$\epsilon_{r+1} = \epsilon_r + g(\alpha - \epsilon_r)$$

we now have, using Taylor's theorem and ignoring powers of ϵ_r higher than the second,

$$\epsilon_{r+1} \approx \epsilon_r + g(\alpha) - \epsilon_r g'(\alpha) + \tfrac{1}{2}\epsilon_r^2 g''(\alpha)$$
$$= -\frac{f''(\alpha)}{2f'(\alpha)} \epsilon_r^2.$$

The analysis above shows us that each error is proportional to the *square* of the error in the preceding term: we deduce that convergence is more rapid in the Newton–Raphson process than in the first iterative method discussed in this Section. (The Newton–Raphson method is a *second-order process*.)

Geometrically, the Newton–Raphson process is equivalent to drawing a sequence of tangents to the curve $y = f(x)$.

Let $y_0 = f(x_0)$; then the tangent at the point (x_0, y_0) on the curve $y = f(x)$ has equation

$$y - y_0 = f'(x_0)(x - x_0)$$

and this meets the x axis at

$$x_1 = x_0 - \frac{y_0}{f'(x_0)}$$

$$= x_0 - \frac{f(x_0)}{f'(x_0)} \quad \text{(see Figure 26.6).}$$

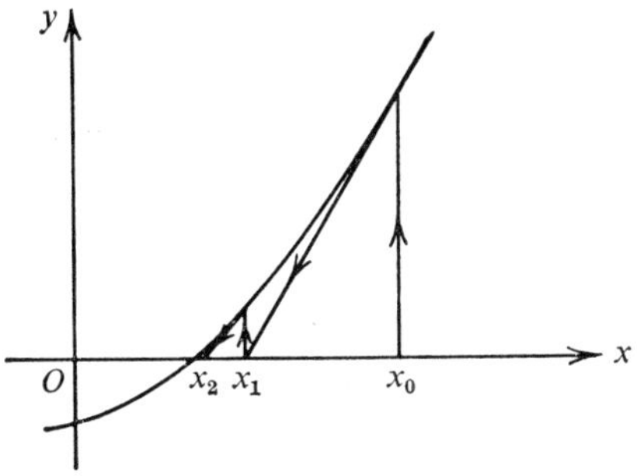

Fig. 26.6

Ex. 24. Show by graphical considerations, that, once the process starts to converge to a root, it always does so from one side. Try to produce an analytical argument to support this assertion.

Ex. 25. Explain by a graphical argument how inaccuracies may arise in the neighbourhood of two nearly equal roots.

The Newton–Raphson process involves division by the awkward number $f'(x_r)$. A simplification is effected by using the *von Mises iteration*:

$$x_{r+1} = x_r - \frac{f(x_r)}{f'(x_0)},$$

where the variable denominator $f'(x_r)$ is replaced by the constant $f'(x_0)$. The convergence is less rapid but nevertheless fairly good.

Ex. 26 Find the negative root of the equations

$$x^2 - 2x - 2 = 0$$

by Newton–Raphson and von Mises's iteration, taking $x_0 = -1$.

Ex. 27. Interpret the von Mises process graphically. Explain why a very efficient procedure is to use the Newton method strictly for a suitable number of stages and then stick with a constant value of f', e.g. $f'(x_2)$ after two stages.

We conclude this chapter with an example of how an iterative process of a required order may be developed—in this case determining the reciprocal of a number.

Example 4. *If* $1/a$ *is calculated from the recurrence relation*

$$x_{r+1} = 2x_r - ax_r^2$$

prove that $$\epsilon_{r+1} = a\epsilon_r^2.$$

Find also the connection between ϵ_r, ϵ_{r+1} *if* $1/a$ *is calculated from the recurrence relation*

$$x_{r+1} = 3x_r - 3ax_r^2 - a^2x_r^3$$

and suggest a fourth-order iterative process for finding $1/a$.

If $$x_{r+1} = 2x_r - ax_r^2$$

we have $$x_r + \epsilon_r = \frac{1}{a}$$

$$\Rightarrow \quad \epsilon_r = \frac{1}{a} - x_r$$

$$\Rightarrow \quad \epsilon_r^2 = \frac{1}{a^2} - \frac{2x_r}{a} + x_r^2$$

$$\Rightarrow \quad a\epsilon_r^2 = \frac{1}{a} - (2x_r - ax_r^2)$$

$$\Rightarrow \quad x_{r+1} + a\epsilon_r^2 = \frac{1}{a}$$

$$\Rightarrow \quad \epsilon_{r+1} = a\epsilon_r^2.$$

Similarly, if $x_{r+1} = 3x_r - 3ax_r^2 + a^2x_r^3$,

$$x_r + \epsilon_r = \frac{1}{a}$$

$$\Rightarrow \quad \epsilon_r^3 = \frac{1}{a^3} - \frac{3x_r}{a^2} + \frac{3x_r^2}{a} - x_r^3$$

$$\Rightarrow \quad a^2\epsilon_r^3 = \frac{1}{a} - (3x_r - 3ax_r^2 + a^2x_r^3)$$

$$\Rightarrow \quad \epsilon_{r+1} = a^2\epsilon_r^3.$$

Now suppose we seek a recurrence relation such that $\epsilon_{r+1} \propto \epsilon_r^4$.

$$x_r + \epsilon_r = \frac{1}{a}$$

$$\Rightarrow \quad \epsilon_r^4 = \frac{1}{a^4} - \frac{1}{a^3}(4x_r - 6ax_r^2 + 4a^2x_r^3 - a^3x_r^4)$$

$$\Rightarrow \quad a^3\epsilon_r^4 = \frac{1}{a} - (4x_r - 6ax_r^2 + 4a^2x_r^3 - a^3x_r^4).$$

Thus, if we set $\quad x_{r+1} = 4x_r - 6ax_r^2 + 4a^2x_r^3 - a^3x_r^4,$

it follows that $\quad \epsilon_{r+1} = a^3\epsilon_r^4$

and we have a fourth-order iterative process for finding $1/a$.

Exercise 26(b)

Find the real roots of the equations 1–8, *using any suitable iterative process. Use linear interpolation to find the first approximation and give your final answer to* 3 *significant figures.*

1. $x^3 - 3 = 0.$

2. $x^3 - 100 = 0.$

3. $x^5 - 5 = 0.$

4. $x^2 - 3x - 11 = 0.$

5. $x^3 - 6x^2 + 10x - 9 = 0$ (1 root).

6. $x^3 - 3x^2 - 3x - 7 = 0$ (1 root).

7. $x^4 - 7x - 12 = 0$ (2 roots).

8. $x^3 + 3x^2 - 9x - 16 = 0$ (3 roots).

9. Find to 4 significant figures the least positive root of the equation

$$x^4 - 13x^2 - 18x - 5 = 0.$$

10. Show that the equation

$$2x^5 - 10x^3 + 10x - 1 = 0$$

has two roots between 1 and 2 and find their numerical values, correct to 4 significant figures.

11. Use the Newton–Raphson method to find an approximate value for the least positive root of the equation

$$3 \tan x = 4x \quad \text{(4 decimal places).}$$

12. Find an approximate value of x such that $x + e^x = 3$.

13. Solve, correct to 2 decimal places, the equation

$$\sin\frac{\pi x}{2} = 3x - 1.$$

14. Show that the equation

$$(2k+1)\,x^3 - k(x+1) = 0,$$

where k is large and positive, has a root near to 1.

Find the equation of the tangent to the curve

$$y = (2k+1)\,x^3 - k(x+1)$$

at the point $x = 1$. From the equation of this tangent find a better approximation to the indicated root of the original equation. (O & C)

15. Find the root of the equation

$$\sin x = x^2$$

other than $x = 0$, to 3 decimal places.

16. Establish Newton's formula for obtaining a closer approximation to a real root of the equation $f(x) = 0$.

Use this method to find, correct to 3 significant figures, the positive root of the equation

$$4 \cos x - 2x - 1 = 0. \qquad \text{(L.)}$$

17. By using Newton's method of approximation, or any other method, find the value of x correct to 3 decimal places for which the expression

$$\frac{x+1}{\ln x}$$

has a stationary value. (L.)

Miscellaneous Exercise 26

1. Find the greatest root of the equation

$$x^3 - 3x + 1 = 0,$$

correct to 3 decimal places.

2. The roots of the quadratic equation $ax^2 + bx - 1 = 0$ are calculated from the recurrence relation

$$x_{r+1} = \frac{1}{ax_r + b}.$$

Interpret this process geometrically, and prove that

$$\epsilon_{r+1} = -aX^2\epsilon_r,$$

where X is the exact value of the root being calculated and $X = x_r + \epsilon_r$.

With the help of reciprocal tables, use this method to solve the equation

$$5x^2 - 3x - 1 = 0.$$

3. Give a sketch showing the general shape of the graph of $y = \sec x$ for values of x from $x = 0$ to $x = \frac{9}{2}\pi$.

Deduce from the graph that large roots of the equation

$$x \cos x = 2$$

are approximately equal to $(n + \frac{1}{2})\pi$, where n is a large integer; and prove that closer approximations are given by

$$x = (n + \tfrac{1}{2})\pi \pm \frac{4}{(2n+1)\pi},$$

where the positive sign is taken when n is odd, and the negative sign when n is even. (O & C)

4. A root of the equation

$$\sin^3 \tfrac{1}{2}x^\circ + \cos x^\circ - \tfrac{249}{400} = 0$$

is close to 60. Find the value of the root, correct to 0·1. (O & C)

5. Draw an accurate graph of $y = \frac{1}{2}\pi \sin x$ between $x = 0$ and $x = \pi$. Draw in the same diagram the lines $y = mx$ for $m = \frac{1}{5}, \frac{2}{5}, \frac{3}{5}, \frac{4}{5}, 1$. Determine the values of x where these lines cut the graph of $y = \frac{1}{2}\pi \sin x$, giving your answers in the form $k\pi$, where k is correct to 2 decimal places.

Use the values obtained to draw a separate graph of $y = \frac{1}{2}\pi \sin x/x$ between $x = \frac{1}{2}\pi$ and $x = \pi$. (O & C)

6. Verify that $x = \frac{1}{3}\pi$ is an approximate solution of the equation $\cos x = \frac{1}{2}x$, and show that a better approximation is 1·03. (O & C)

7. If $(x_r + \epsilon_r)^2 = a$, prove that

$$\sqrt{a} = \frac{1}{2}\left(\frac{a}{x_r} + x_r\right) - \frac{\epsilon_r^2}{2x_r}$$

and deduce the recurrence relation

$$x_{r+1} = \frac{1}{2}\left(\frac{a}{x_r} + x_r\right)$$

for finding $\sqrt{a}$.

Suggest an intuitive argument leading to this recurrence relation and show that the same relation is obtained by applying the Newton–Raphson process to the equation $x^2 - a = 0$.

8. $\sqrt{a}$ is calculated from the recurrence relation

$$x_{r+1} = \frac{a^2 + 6ax_r^2 + x_r^4}{4x_r(a + x_r^2)}.$$

Show that

$$\epsilon_{r+1} = -\frac{\epsilon_r^4}{4x_r(a + x_r^2)}$$

and find $\sqrt{11}$ correct to 6 decimal places.

9. Develop a recurrence relation for finding $a^{1/3}$ in which $\epsilon_{r+1} = k\epsilon_r^2$ (where k depends upon r) and hence find $\sqrt[3]{6}$ to 5 decimal places.

10. Prove graphically, or otherwise, that the equation

$$\cos x = mx \quad (m \neq 0)$$

has one and only one root in the interval $-\frac{1}{2}\pi < x < \frac{1}{2}\pi$.

The angle α is defined as that root of the equation

$$\cot \alpha = -\alpha$$

which lies between $\frac{1}{2}\pi$ and π. Prove that, when m lies in the range

$$-\sin \alpha < m < 0,$$

the equation $\cos x = mx$ has three and only three roots in the interval

$$-\tfrac{1}{2}\pi < x < \tfrac{3}{2}\pi.$$

By means of careful graphs of the functions $\cot x$ and $-x$, or otherwise, obtain the value of α, giving your answer in radians to 2 significant figures.

11. If a root of the equation $f(x) = 0$ is obtained by an iterative process in which $\epsilon_{r+1} \approx k\epsilon_r$ (k constant) show that the expression

$$x_r - \frac{(x_r - x_{r-1})^2}{x_r - 2x_{r-1} + x_{r-2}}$$

may be taken as an approximation to the required root.

12. Find approximately the root of the equation

$$x^4+3x^3-7x^2+18x-18 = 0$$

which lies between 1 and 2.

13. Prove, graphically or otherwise, that the equation

$$f(x) \equiv 1+\frac{1}{x^2}-\tan x = 0$$

has a root near $x = \frac{1}{4}\pi+k\pi$, where k is any large integer.

Denoting $(k+\frac{1}{4})\,\pi$ by K, prove that a better approximation to the root is given by $K+\lambda$, where

$$f(K)+\lambda f'(K) = 0,$$

and that

$$\lambda \approx \frac{1}{2K^2}.$$

Find an approximation to the root correct to the term of order K^{-4}. (O & C)

14. The equation $x^2-ax+b = 0$ has roots α, β. Form the equation with roots α^2, β^2 and, if the equation with roots $\alpha^{2^n}, \beta^{2^n}$ is $x^2-a_nx+b_n = 0$, write down the equation with roots $\alpha^{2^{n+1}}, \beta^{2^{n+1}}$.

If $\alpha = \lambda\beta$ where $|\lambda| < 1$, show how the formation of a sequence of such equations may be used to give successive approximations to β and hence α.

Under what circumstances does this method provide a very satisfactory method for solving quadratic equations? Discuss the magnitude of the errors involved and consider the cases of (i) equal, (ii) complex, roots.

Discuss the application of this method of solution to equations of degree higher than the second.

15. Prove that the equation $e^x = \lambda x$ has a real solution if and only if $\lambda \geqslant e$, but that $xe^x = \mu$ has a real solution for all real values of μ.

Solve the equation $xe^x = 1$ to 5 decimal places.

16. x_r, x_{r+1}, x_{r+2} are successive approximations to a root of the equation $f(x) = 0$ obtained by the Newton–Raphson process. Writing C_r for the correction $-f(x_1)/f'(x_r)$, so that $C_{r+1} = -f(x_r+C_r)/f'(x_r+C_r)$ show that

$$\frac{C_{r+1}}{C_r^2} \approx -\frac{1}{2}\frac{f''(x_r)}{f'(x_r)}$$

by using Taylor's result,

$$f(x+h) \approx f(x)+hf'(x)+\tfrac{1}{2}h^2f''(x),$$

where h is small.

Deduce that

$$x_{r+2}-x_{r+1} \approx -\frac{[f(x_r)]^2f''(x_r)}{2[f'(x_r)]^3}.$$

17. It is required to find the nth root of the number a by the Newton–Raphson process. Show that this can be done by solving the equation $x^{n+k}-ax^k = 0$.

Using the analysis of Question 16, show that most rapid convergence is obtained by choosing $k = \frac{1}{2}(1-n)$. Use this method to obtain a value for $\sqrt[3]{2}$, carrying out two iterations.

27. *The ellipse and hyperbola*

1. CONICS: FOCI AND DIRECTRICES

In Chapter 22 we considered the geometry of the parabola and rectangular hyperbola; both these curves are examples of a type of plane curve called a *conic*, which we shall now define. Given a point S and a line l not containing S, a conic is the set of all points P in the plane of l and S such that the ratio of the distance of P from S to the distance of P from l is a fixed number.

If we call the foot of the perpendicular from P to l, M, and the fixed ratio e (not to be confused with the base (e) of the natural logarithms) then a conic is the set of points P in the plane containing S and l, such that

$$\left|\frac{SP}{PM}\right| = e \quad \text{(Figure 27.1).}$$

The point S is called a *focus* of the conic; l is a *directrix*. The number e is called the *eccentricity* of the conic. The type of conic is determined by

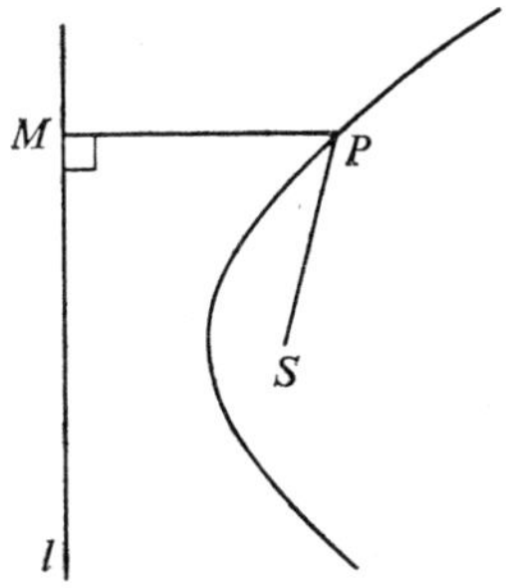

Fig. 27.1

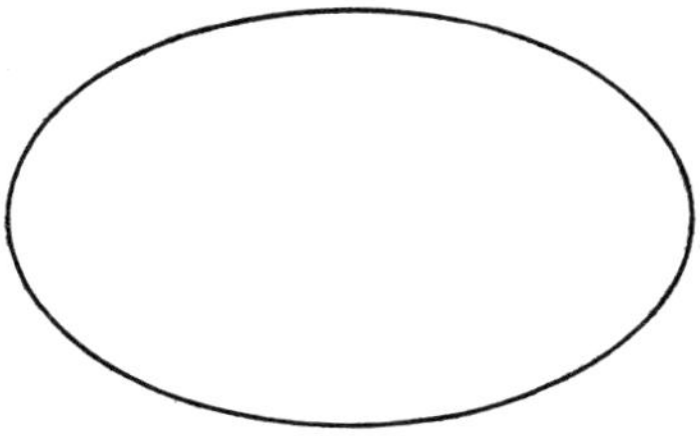

Fig. 27.2

the value of e. The reader will no doubt already have noticed that, with $e = 1$, the conic defined is a parabola. More generally, we define the following types of conic:

(i) if $0 < e < 1$, the conic is called an *ellipse*;
(ii) if $e = 1$, the conic is called a *parabola*;
(iii) if $e > 1$ the conic is called a *hyperbola*.

The shape of the three types of conic are shown in Figures 27.2 (ellipse), 27.3 (parabola) and 27.4 (hyperbola). (The dotted lines in Figure 27.4 are not part of the hyperbola: they are, in fact, asymptotes and are included as an aid towards drawing the curve.)

Ex. 1. Explain why the set of points

$$\{(x, y): x^2 = 4[(x-1)^2+(y-1)^2]\}$$

represents an ellipse.

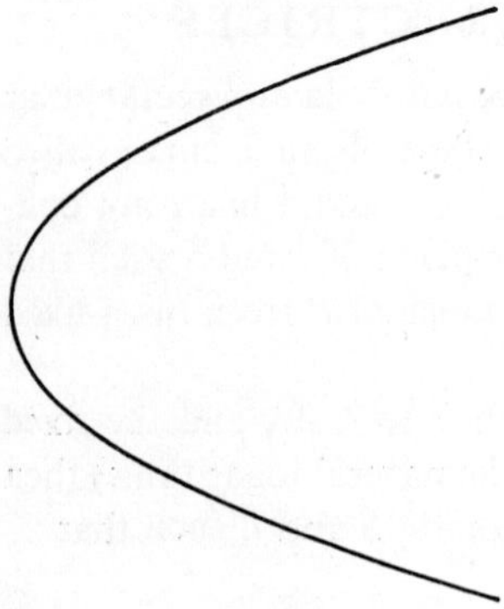

Fig. 27.3

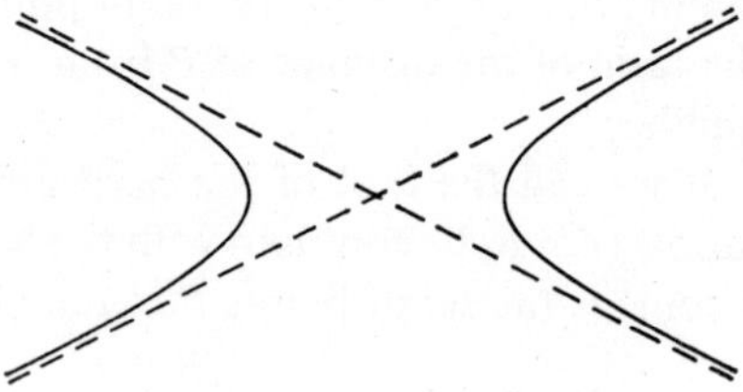

Fig. 27.4

2. THE ELLIPSE

The ellipse has been defined in Section 1 as a conic with eccentricity $e < 1$. Before attempting to obtain the Cartesian equation of an ellipse, it is worthwhile to consider a little of the geometry of the curve, in order that we may be able to choose the most suitable coordinate axes. Suppose that S is the given focus and l the corresponding directrix and let K be the foot of the perpendicular from S on to l.

If e is the given eccentricity, then there are two points A, A' on the line SK which belong to the ellipse—namely the points dividing SK internally and externally in the ratio $e:1$ (see Figure 27.5 and notice that A' lies on KS produced, since $e < 1$). Let O be the mid-point of AA' and set $OS = s$, $OK = k$. Then, if $AA' = 2a$,

$$SA = a-s, \quad A'S = a+s,$$
$$AK = k-a, \quad A'K = k+a,$$

and thus, by the definition of A, A' as points of the ellipse,

$$a-s = e(k-a)$$
$$a+s = e(k+a).$$

Solving these two equations we have

$$s = ae, \quad k = a/e.$$

Let us now take O as the origin, OK as the x axis and the perpendicular to this line through O as the y axis. Then, by what we have just shown, S is the point $(ae, 0)$ and l is the line $x-a/e = 0$. Let $P(x, y)$ be any point of the ellipse; by the definition of the ellipse

$$SP^2 = e^2PM^2,$$

where M is the foot of the perpendicular from P on to l (Figure 27.6).

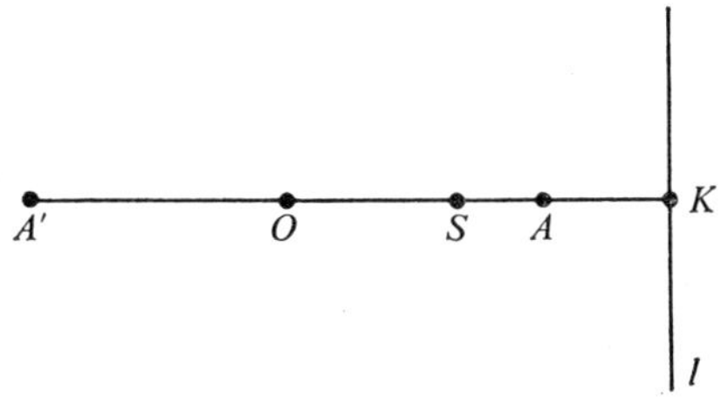

Fig. 27.5

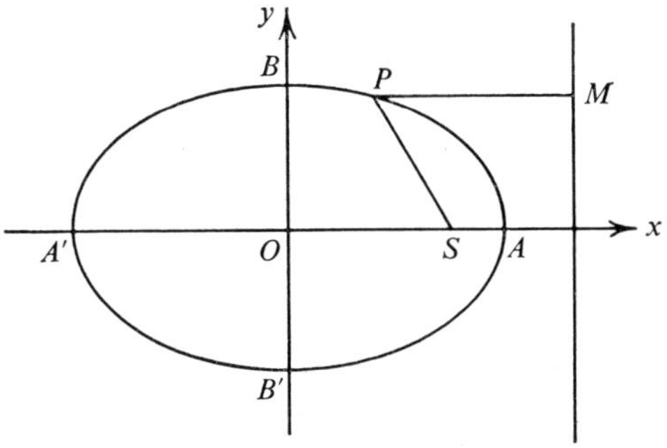

Fig. 27.6

Thus, the ellipse is the set of points

$$\{(x, y): (x-ae)^2+y^2 = e^2(a/e-x)^2\}.$$

The equation may be rewritten

$$x^2-2aex+a^2e^2+y^2 = a^2-2aex+e^2x^2$$

$$\Rightarrow \quad x^2(1-e^2)+y^2 = a^2(1-e^2)$$

$$\Rightarrow \quad \frac{x^2}{a^2}+\frac{y^2}{a^2(1-e^2)} = 1.$$

If we write $$b^2 = a^2(1-e^2),$$

the equation becomes

$$\frac{x^2}{a^2}+\frac{y^2}{b^2} = 1,$$

which is called *the standard form of equation for the ellipse.*

**Ex.* 2. Show that the ellipse

$$b^2x^2+a^2y^2 = a^2b^2$$

is symmetrical about the x and the y axes and that (with the notation of Figure 27.6) B is the point $(0, b)$ and B' is the point $(0, -b)$.

**Ex.* 3. Deduce, by an appeal to symmetry, the existence of a second focus S' $(-ae, 0)$ and second corresponding directrix $x+a/e = 0$.

**Ex.* 4. O is called the *centre* of the ellipse. Show that every chord PP' of the ellipse which passes through O is bisected at O.

AA', BB' are respectively the *major* and *minor axes* of the ellipse; a is the length of the *major semi-axis*, b is the length of the *minor semi-axis* ($b < a$, hence the word 'minor'). From the original definition of b, we have the important equation connecting a and b:

$$b^2 = a^2(1-e^2).$$

Ex. 5. If a circle is regarded as a special case of an ellipse with equal major and minor axes, show that it corresponds to a conic with zero eccentricity.

Ex. 6. A chord drawn through a focus perpendicular to the major axis of an ellipse is called a *latus rectum* of the ellipse. Show that the length of the latus rectum is $2b^2/a$.

Ex. 7. Show that the eccentricity of the ellipse

$$\frac{x^2}{25}+\frac{y^2}{16} = 1$$

is $\frac{3}{5}$ and deduce the coordinates of its foci.

Ex. 8. Show, by translating the coordinate axes, that the equation

$$\frac{(x-1)^2}{3}+\frac{(y+2)^2}{2} = 1$$

represents an ellipse with centre $(1, -2)$. Find the coordinate of its foci and show the eccentricity to be $\frac{1}{3}\sqrt{3}$.

The ellipse arises in nature as the typical orbit of objects moving under the action of an attractive force varying inversely as the square of their distance from a fixed point. For example, planets describe ellipses (with slight variations and discrepancies due to the attractions exerted on each other) with the sun at one focus—hence the use of the letter S for the focus. Similarly, for the motion of satellites around the Earth, the centre of the Earth being situated at a focus (again with a slight discrepancy, this time because the Earth is neither exactly spherical, nor uniform). The aesthetically pleasing shape of the ellipse has long been admired: it can be seen, for example, in the elliptic arches of some bridges which with their reflection in still water yield complete ellipses.

Ex. 9. The eccentricity of the Earth's orbit is approximately $\frac{1}{60}$ while that of the planet Pluto is $\frac{1}{4}$. Compare the shapes of their orbits.

Ex. 10. An astronomical unit (a.u.) is the mean distance of the Earth from the Sun. (It is approximately 500 light-seconds.) At its nearest approach to the Sun, the planet Mercury is distant about 0·308 a.u. from the Sun, while its maximum distance is about 0·466 a.u. Calculate the approximate eccentricity of the orbit.

**Ex.* 11. Show that $x = a\cos\theta$, $y = b\sin\theta$ gives a parametric representation for points of the ellipse $b^2x^2+a^2y^2 = a^2b^2$.

A rod has three points P, Q, R marked on it, where $PQ = a$, $QR = b$. If P, Q are constrained to move along two fixed perpendicular lines, show that R moves along the arc of an ellipse. (This is the engineer's paper trammel method for drawing ellipses.)

**Ex.* 12. Draw in the same diagram the ellipse

$$b^2x^2+a^2y^2 = a^2b^2$$

and the circle

$$x^2+y^2 = a^2$$

(the *auxiliary circle* of the ellipse). Given a point P of the ellipse, let the perpendicular through P to the major axis of the ellipse cut the auxiliary circle at Q. If OQ makes an angle θ with the major axis, show that Q is the point $(a\cos\theta, a\sin\theta)$ and that P is the point $(a\cos\theta, b\sin\theta)$. θ is called the *eccentric angle* of PQ.

The relationship between the ellipse

$$b^2x^2+a^2y^2 = a^2b^2$$

and its auxiliary circle

$$x^2+y^2 = a^2$$

outlined in Ex. 11 and Ex. 12 is worth a little further study.

It will be recalled (see Chapter 13) that linear transformations of the plane into itself map straight lines into straight lines and in particular, parallel straight lines into parallel lines. They also map ratios of lengths on a line into the same ratio of corresponding lengths on the image line; in particular they map the mid-point of a line segment into the mid-point of the image line segment. We shall now consider the particular linear transformation, T, defined by the equation

$$x' = x, \quad y' = by/a.$$

Ex. 13. Show that perpendicular lines do *not* map into perpendicular lines under T, unless $a^2 = b^2$. Identify the two linear transformations for which $a = b$ and $a = -b$, and explain why these do preserve perpendicularity.

**Ex.* 14. Show that the image of the circle

$$\{(x, y): x^2+y^2 = a^2\}$$

under T is the ellipse

$$\{(x', y'): b^2x'^2+a^2y'^2 = a^2b^2\}.$$

It follows from the preceding remarks and the result of Ex. 14 that properties of parallel chords of circles, mid-points of chords of circles, etc. are preserved by T and thus correspond to identical properties of the ellipse. For example, since the mid-points of parallel chords of a circle lie on a straight line through the centre, the same is true for an ellipse. The

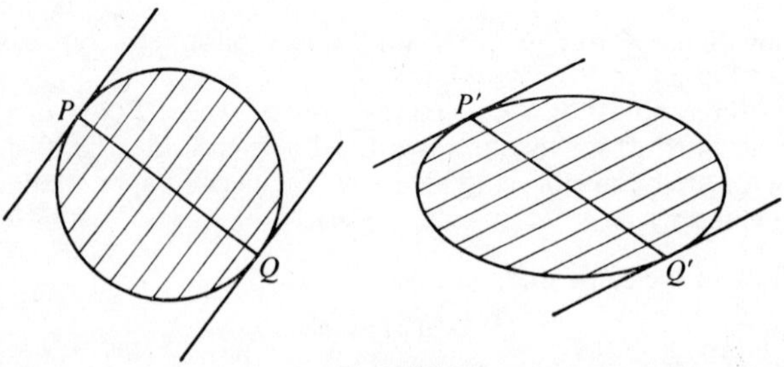

Fig. 27.7

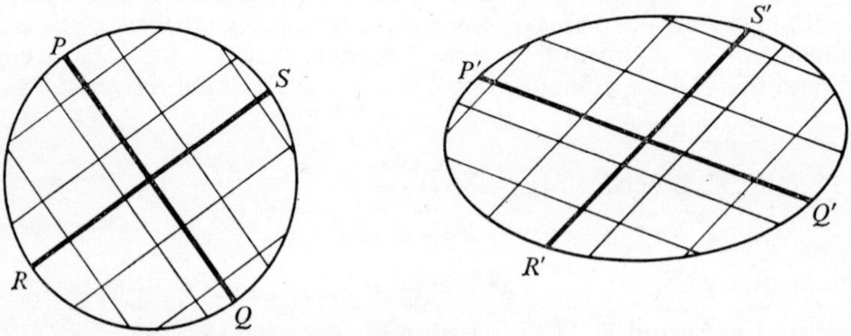

Fig. 27.8

line obtained is called a *diameter* of the ellipse. Furthermore, the two lines parallel to the given chords passing through the intersections P, Q of the corresponding diameter with the circle are tangents: the same is true of the ellipse (see Figure 27.7).

Again, in the circle, the centres of chords parallel to PQ define another diameter, RS, such that all chords parallel to RS are bisected by PQ. The same property is thus true of the ellipse: $P'Q'$ and $R'S'$ are called *conjugate diameters*; they have the property that each bisects all chords parallel to the other (see Figure 27.8).

The geometrical interpretation of T is as follows. Given a point P, draw

PX perpendicular to the x axis to meet it at X (*PX* is the *ordinate* of X); then P' is the point on *PX* such that

$$P'X = (b/a)\,PX.$$

In particular, using the language of Ex. 12, the image of a point P on the auxiliary circle lies on the ellipse (see Figure 27.9).

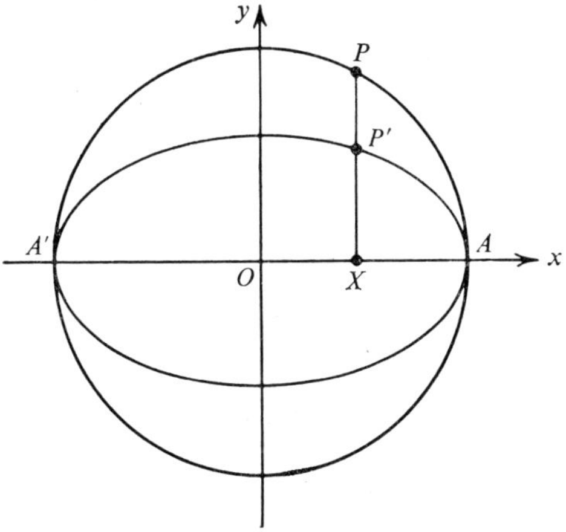

Fig. 27.9

**Ex.* 15. Show that T is a non-singular linear transformation, that is, that T^{-1} exists.

Example 1. *LP is the tangent at P to an ellipse, centre O, and PR is the chord parallel to LO. Show that, if RQ is a diameter of the ellipse, then LQ is a tangent.*

Suppose that the ellipse is taken in standard form and let us denote image elements under T^{-1} (see Ex. 15) by dashes. The corresponding property of the circle (see Figure 27.10) is immediately obvious, either by an appeal to symmetry, or, equivalently, by proving the triangles $L'P'O$, $L'Q'O$ congruent. Thus, since the properties with which we are concerned remain invariant under the linear transformation T (under which the image of P' is P, etc.) the result is also true for the ellipse (see Figure 27.11).

Notice that no appeal can be made to symmetry in the figure for the ellipse, but since the final result is couched in terms of tangents, diameters, and parallel lines only, the results for ellipse and circle correspond completely.

In solving problems on the ellipse it is sometimes necessary to obtain the equations of particular lines, notably the tangent and normal. The quickest way to arrive at these equations is to use calculus, although alternative methods are available.

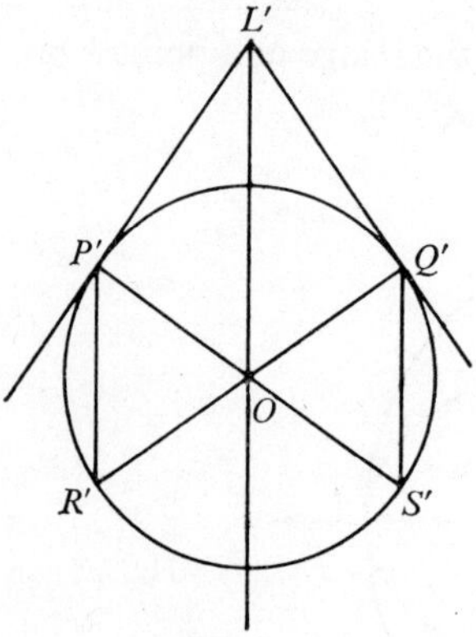

Fig. 27.10

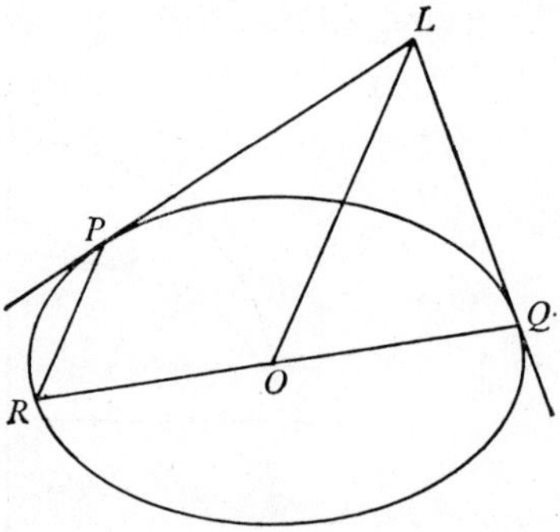

Fig. 27.11

**Ex*. 16. Show that, if

$$b^2x^2+a^2y^2 = a^2b^2,$$

then

$$\frac{dy}{dx} = -\frac{b^2x}{a^2y} \quad (y \neq 0).$$

Deduce that the equation of the tangent at the point (x_1, y_1) on the ellipse is

$$\frac{x-x_1}{a^2y_1} = \frac{y-y_1}{-b^2x_1} = \lambda$$

and show that this may be rewritten in the form

$$\frac{xx_1}{a^2}+\frac{yy_1}{b^2} = 1.$$

**Ex*. 17. Show that the equation of the tangent to the ellipse at the point $(a \cos \theta, b \sin \theta)$ is

$$\frac{x \cos \theta}{a}+\frac{y \sin \theta}{b} = 1,$$

and that the equation of the normal is

$$ax \sin \theta - by \cos \theta = (a^2-b^2) \cos \theta \sin \theta.$$

**Ex*. 18. Show that the line L, with equation

$$\frac{x-x_1}{l} = \frac{y-y_1}{m} = \lambda,$$

cuts the ellipse

$$b^2x^2+a^2y^2 = a^2b^2$$

at points with parameters λ_1, λ_2 which are roots of the quadratic equation

$$\lambda^2\left(\frac{l^2}{a^2}+\frac{m^2}{b^2}\right)+2\lambda\left(\frac{lx_1}{a^2}+\frac{my_1}{b^2}\right)+\left(\frac{x_1^2}{a^2}+\frac{y_1^2}{b^2}-1\right) = 0.$$

Deduce that the point $P_1(x_1, y_1)$ is the mid-point of the chord L if

$$\frac{lx_1}{a^2}+\frac{my_1}{b^2} = 0$$

and that this gives the equation of the chord with mid-point P_1 as

$$\frac{x-x_1}{a^2y_1} = \frac{y-y_1}{-b^2x_1} = \mu.$$

Show how to deduce from this the equation of the tangent at the point $P_2(x_2, y_2)$ of the ellipse.

The ellipse, like the parabola, has a rich geometry. We shall now prove two of its most famous properties. (Since both are focal properties, neither is derivable from the circle.)

(i) *Focal distance property*

For any point P on the ellipse with foci S, S' and major axis $2a$,

$$SP+S'P = 2a.$$

Take the ellipse in the standard form

$$b^2x^2+a^2y^2 = a^2b^2$$

and let M, M' be the feet of the perpendiculars from P on to the two directrices (see Figure 27.12).

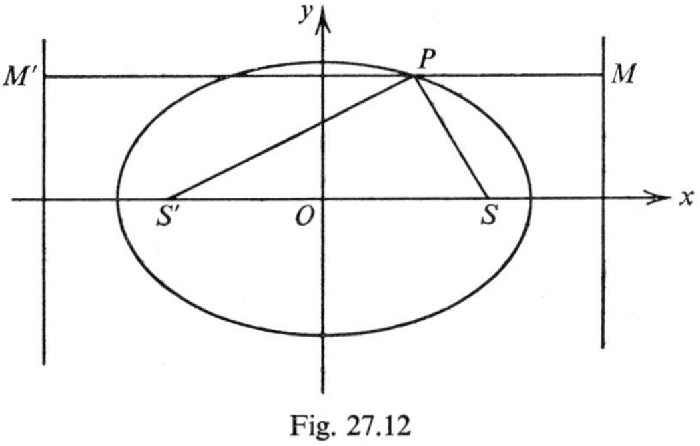

Fig. 27.12

Let $P(h, k)$ be any point on the ellipse; then, by the focus–directrix definition of the ellipse

$$\begin{aligned} SP+S'P &= e(PM+M'P) \\ &= e\left[\left(\frac{a}{e}-k\right)+\left(\frac{a}{e}+k\right)\right] \\ &= 2a. \end{aligned}$$

**Ex.* 19. Prove that $SQ+S'Q < 2a$ for all points Q within the ellipse and that $SR+S'R > 2a$ for all points R outside the ellipse.

**Ex.* 20. Prove a converse of the above result, namely that if S, S' are fixed points and P is any point such that $SP+S'P$ is constant and greater than SS', then the locus of P is an ellipse with S, S' as foci. (If the constant distance is taken as $2a$, show that a unique ellipse exists with S, S' as foci and major axis of length $2a$; then use Ex. 19.)

**Ex.* 21. If a, b are complex numbers such that $|a-b| < c$, where c is a real positive constant, describe the set of points

$$\{z \in C: |z-a|+|z-b| \leqslant c\}$$

in the Argand diagram.

Ex. 22. Explain the theory underlying the following well-known mechanical construction for an ellipse. Two drawing pins are stuck in a sheet of paper and a loop of cotton is placed loosely around the two pins. The loop is made taut by the point of a pencil which is then made to trace out a curve on the paper, keeping the loop taut at all stages of the construction.

(ii) *Reflection property*

The tangent and normal at any point P of an ellipse bisect the angle SPS'.

Take the ellipse in standard form and let P be the point

$$(a \cos \theta, b \sin \theta).$$

With the notation of Figure 27.13, the equation of the normal, PG, is (see Ex. 17)

$$ax \sin \theta - by \cos \theta = (a^2-b^2) \cos \theta \sin \theta,$$

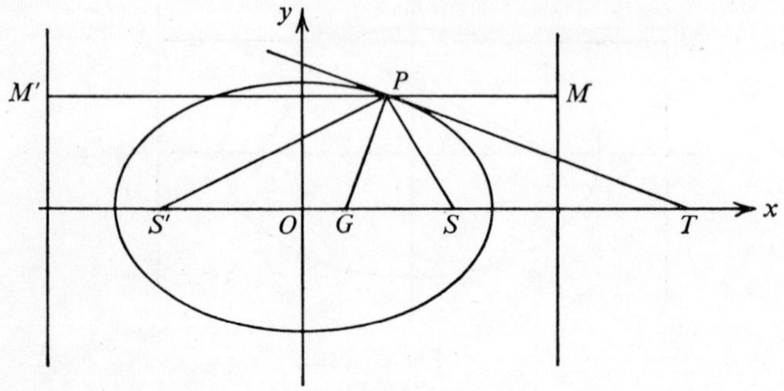

Fig. 27.13

and thus G is the point

$$([(a^2-b^2) \cos \theta]/a, 0).$$

Using $b^2 = a^2(1-e^2)$, this reduces to

$$(ae^2 \cos \theta, 0).$$

Thus $$S'G = ae+ae^2\cos\theta, \quad GS = ae^2\cos\theta$$

and $$\frac{GS}{S'G} = \frac{ae-ae^2\cos\theta}{ae+ae^2\cos\theta} = \frac{a/e-a\cos\theta}{a/e+a\cos\theta} = \frac{PM}{M'P} = \frac{PS}{PS'}$$

and so, by the angle bisector theorem for a triangle, PG bisects $\angle SPS'$ internally. Since $\angle GPT$ is a right angle, it follows immediately that PT bisects $\angle SPS'$ externally.

Ex. 23. ABC is any triangle and P is any point on the external bisector of the angle BAC. Prove that
$$BP+CP \geqslant BA+CA,$$
with equality only if P coincides with A.

Example 2. *Prove that the chord of contact of tangents from the point* $C(h, k)$ *to the ellipse*
$$\frac{x^2}{a^2}+\frac{y^2}{b^2} = 1$$
is the line
$$\frac{hx}{a^2}+\frac{ky}{b^2} = 1.$$

Deduce that, if the tangents at the extremities of a variable chord through the fixed point $D(\alpha, \beta)$ *meet at* R, *then* R *lies on the straight line*
$$\frac{\alpha x}{a^2}+\frac{\beta y}{b^2} = 1.$$

From Ex. 16, we know that the equation of the tangent to the ellipse at the point $P_1(x_1, y_1)$ is
$$\frac{xx_1}{a^2}+\frac{yy_1}{b^2} = 1.$$

Now suppose that $P_1(x_1, y_1)$, $P_2(x_2, y_2)$ are the points of contact of tangents from $C(h, k)$ to the ellipse. Then the coordinates (h, k) satisfy the equations
$$\frac{xx_1}{a^2}+\frac{yy_1}{b^2} = 1 \quad\text{and}\quad \frac{xx_2}{a^2}+\frac{yy_2}{b^2} = 1$$
and we have
$$\frac{hx_1}{a^2}+\frac{ky_1}{b^2} = 1 \quad\text{and}\quad \frac{hx_2}{a^2}+\frac{ky_2}{b^2} = 1.$$
Thus $P_1(x_1, y_1)$ and $P_2(x_2, y_2)$ lie on the line
$$\frac{hx}{a^2}+\frac{ky}{b^2} = 1,$$
which must represent the required chord of contact.

If $D(\alpha, \beta)$ lies on this chord of contact,
$$\frac{h\alpha}{a^2}+\frac{k\beta}{b^2} = 1$$

for all positions of the point (h, k). Thus (h, k) lies on the line

$$\frac{\alpha x}{a^2}+\frac{\beta y}{b^2} = 1.$$

Example 3. *The normal at the point P to the ellipse Σ with equation*

$$b^2x^2+a^2y^2 = a^2b^2$$

meets the x axis at H and the y axis at K, and Q is the fourth vertex of the rectangle $OHQK$. Prove that the locus of Q is a concentric similar ellipse Σ', but with the minor axis of Σ' lying along the major axis of Σ.

[Locus questions are usually best attempted using the parametric form $(a \cos \theta, b \sin \theta)$: the coordinates of the point whose locus is sought are then obtained in terms of the parameter θ, which may be eliminated by using the identity

$$\cos^2 \theta+\sin^2 \theta = 1$$

to yield the equation of the locus.]

Let P be the point

$$(a \cos \theta, b \sin \theta).$$

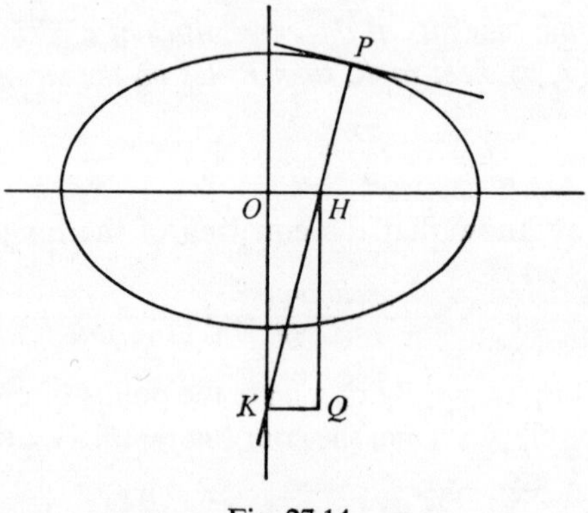

Fig. 27.14

Then the normal at P has equation $ax \sin \theta - by \cos \theta = (a^2-b^2) \cos \theta \sin \theta$ (see Ex. 17). Thus we may write down the coordinates of the points H $(y = 0)$ and K $(x = 0)$ and hence of Q (see Figure 27.14):

$$H: \left(\frac{a^2-b^2}{a} \cos \theta, 0\right),$$

$$K: \left(0, -\frac{a^2-b^2}{b} \sin \theta\right),$$

$$Q: \left(\frac{a^2-b^2}{a} \cos \theta, -\frac{a^2-b^2}{b} \sin \theta\right).$$

If we call the coordinates of $Q(x, y)$ we have

$$\cos\theta = \frac{ax}{a^2-b^2}, \quad -\sin\theta = \frac{by}{a^2-b^2}$$

and the equation of the locus of Q is

$$\frac{a^2x^2}{(a^2-b^2)^2}+\frac{b^2y^2}{(a^2-b^2)^2} = 1,$$

which, by comparison with the equation

$$\frac{x^2}{A^2}+\frac{y^2}{B^2} = 1,$$

is seen to be an ellipse, Σ', with

$$A = \frac{a^2-b^2}{a}; \quad B = \frac{a^2-b^2}{b}.$$

The centre of Σ' is the origin and $A:B = b:a$; the ellipses Σ and Σ' are thus concentric and similar. Furthermore, if $a > b$, then $A < B$ and the minor axis of Σ' lies along the x axis, that is, the major axis of Σ.

(Strictly speaking, the first part of our solution is incomplete until we have verified that *every* point of Σ' is a point of the locus. The reader may care to supply the details.)

Exercise 27(a)

1. Find the eccentricities of the following ellipses:

(i) $\frac{x^2}{8}+\frac{y^2}{6} = 1$; (ii) $\frac{x^2}{4}+y^2 = 1$; (iii) $3x^2+5y^2 = 1$.

2. Draw a rough sketch of the ellipse

$$\frac{x^2}{5}+\frac{y^2}{9} = 1.$$

Find its eccentricity, the coordinates of its foci and the equations of the corresponding directrices.

3. Show that the equation

$$3(x-2)^2+4(y+1)^2 = 36$$

represents an ellipse and find its eccentricity and the coordinates of its foci.

4. Show that the equation

$$x^2+2y^2+6x-4y+9 = 0$$

represents an ellipse and find its eccentricity and the coordinates of its foci.

5. Show that the equation

$$4x^2+3y^2-16x+12y+16 = 0$$

represents an ellipse and find its eccentricity and the coordinates of its foci.

6. The arch of an elliptic bridge is in the form of half of an ellipse. The span is 12·5 m and the maximum height of the arch is 1·75 m. Find the eccentricity of the ellipse.

7. Find the equation of the tangent to the ellipse

$$\frac{x^2}{21}+\frac{y^2}{7} = 1$$

at the point $(3, -2)$.

8. Find the equation of the tangent to the ellipse

$$\frac{x^2}{3}+\frac{y^2}{6} = 1$$

at the point $(1, 2)$.

Prove that the line $x-y = 3$ also touches the ellipse.

9. Prove that the line $x-3y+7 = 0$

touches the ellipse $2x^2+3y^2 = 14$,

and find the point of contact.

Where does the line $3x-y+7 = 0$

meet the ellipse. $3x^2+2y^2 = 14$?

10. An ellipse has foci S, S', minor axis BB' and major axis of length $2a$. Prove that $BSB'S'$ is a rhombus of side a.

11. P is a point on the ellipse

$$b^2x^2+a^2y^2 = a^2b^2$$

and N is the foot of the perpendicular from P to AA', where A is the point $(a, 0)$ and A' the point $(-a, 0)$. Prove that

$$\frac{PN^2}{A'N.NA} = \frac{b^2}{a^2}.$$

If the tangent at P meets the directrix corresponding to the focus S at T, prove that $\angle PST$ is a right angle.

12. The feet of the perpendiculars from the foci S, S' of the ellipse

$$b^2x^2+a^2y^2 = a^2b^2$$

to the tangent at the point P are Y, Y'. Prove that Y, Y' lie on the auxiliary circle of the ellipse and that

$$SY.S'Y' = b^2.$$

13. P is a variable point of an ellipse, focus S. Prove that the locus of the mid-point of PS is an ellipse and locate its centre.

14. The tangents at the points P, Q of an ellipse, centre O, meet at the point T. Prove that OT bisects PQ.

15. The parallel chords P_1Q_1, P_2Q_2 of an ellipse are bisected by the diameter UV. Prove that P_1Q_2 and P_2Q_1 meet on UV, as also do P_1P_2 and Q_1Q_2.

16. UV is a diameter of an ellipse and P is any point on the ellipse. Prove that the diameters parallel to PU, PV are conjugate.

17. With the notation of Question 16, the tangent at U meets PV at T and the tangent at P meets TU at M. Prove that M is the mid-point of TU.

18. The diameter UV of an ellipse bisects the chord PQ. UP and VQ meet at X, UQ and VP meet at Y; prove that XY and PQ are parallel.

19. What is the locus of the mid-points of chords of an ellipse which pass through a common point?

20. UV is a diameter of an ellipse and the tangents at U, V are u, v respectively; the tangent at any point P meets u, v at X and Y. Prove that X and Y lie on conjugate diameters of the ellipse.

21. Prove that tangents at the extremities of a focal chord of an ellipse meet on the corresponding directrix.

22. The tangent to the ellipse

$$\frac{x^2}{a^2}+\frac{y^2}{b^2} = 1$$

at the point P meets the axes at Q and R. Find the locus of the mid-point of QR.

23. The perpendicular from the centre of an ellipse with focus S to the tangent at a point P meets SP produced at Q. Prove that the locus of Q is a circle, and find its centre and radius.

24. Define geometrically the eccentric angle ϕ of a point P on the ellipse

$$\frac{x^2}{a^2}+\frac{y^2}{b^2} = 1 \quad (a > b)$$

and express the coordinates of P in terms of ϕ.

Prove that the equation of the normal at P is

$$\frac{ax}{\cos\phi}-\frac{by}{\sin\phi} = a^2-b^2.$$

O is the centre of the ellipse and QP is the ordinate of P, the normal at P cuts the x axis at N. Show that

$$NQ = \frac{b^2}{a}\cos\phi.$$

If the normal at P bisects the angle OPQ prove that the eccentricity e satisfies the equation

$$e^2(1+\sin^2\phi) = 1,$$

that $OP = ae$, and that

$$\tfrac{1}{2}\sqrt{2} < e < 1.$$

(O & C)

25. S, S' are the foci of the ellipse

$$\frac{x^2}{a^2}+\frac{y^2}{b^2} = 1$$

and $P(a\cos\phi, b\sin\phi)$ is a point on the curve. Calculate the lengths PS, PS' in terms of a, e and ϕ, where e is the eccentricity of the ellipse and verify that $PS+PS'$ is constant.

The tangents to the ellipse at P cuts the x axis at T, and the normal at P cuts the x axis at N. Prove that

(i) $OT.ON = a^2e^2$; (ii) $PT/PN = \tan\phi/(1-e^2)$

where O is the centre of the ellipse. (O & C)

26. Show that the coordinates of any point P on the ellipse

$$\frac{x^2}{a^2}+\frac{y^2}{b^2} = 1$$

can be expressed as $(a\cos\phi, b\sin\phi)$. Prove that the equation of the tangent to the ellipse at P is

$$\frac{x\cos\phi}{a}+\frac{y\sin\phi}{b} = 1.$$

P and Q are two points on the ellipse, such that ϕ has the value ϕ_1 at P and ϕ has the value ϕ_2 at Q. If the tangents to the ellipse at P and Q meet on the line $ay = bx$, prove that

$$\phi_1+\phi_2 = \tfrac{1}{2}\pi \quad \text{or} \quad \tfrac{5}{2}\pi.$$ (O & C)

27. Prove that, if $a^2l^2+b^2m^2 = n^2$, then the line

$$lx+my+n = 0$$

touches the ellipse

$$\frac{x^2}{a^2}+\frac{y^2}{b^2} = 1$$

and find the coordinates of the point of contact.

Find the equations of the common tangents to the two ellipses

$$\frac{x^2}{14}+\frac{y^2}{4} = 1, \quad \frac{x^2}{23}+\frac{y^2}{3} = 1.$$

28. If Z is the foot of the perpendicular from the centre of the ellipse

$$b^2x^2+a^2y^2 = a^2b^2$$

to the tangent at a variable point P, prove that the locus of Z is the curve

$$(x^2+y^2)^2 = b^2y^2+a^2x^2.$$

3. THE HYPERBOLA

A hyperbola is a conic with eccentricity $e > 1$. The analysis of the hyperbola follows closely that for the ellipse and there are many striking similarities between the geometry of the two curves. To obtain the standard form of the equation of the hyperbola we first derive certain geometrical results, just as we did for the ellipse in Section 2. Suppose that S is the given focus and l the corresponding directrix and let K be the foot of the perpendicular from S on to l. There are two points, A, A', on the line SK which belong to the hyperbola—namely, the points dividing SK internally and externally in the ratio $e:1$. Since, for the hyperbola, $e > 1$, l will lie between A and A' (see Figure 27.15, and compare with Figure 27.5). Again we take O as the mid-point of AA' and set $OS = s$, $OK = k$. Then, if $AA' = 2a$,

$$SA = s-a, \quad A'S = s+a,$$

$$KA = a-k, \quad A'K = a+k,$$

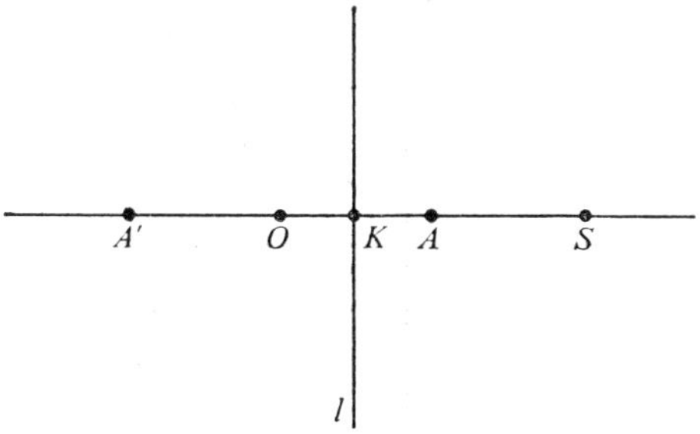

Fig. 27.15

and thus, by the definition of A, A',

$$s-a = e(a-k),$$

$$s+a = e(a+k).$$

Solving these two equations we have

$$s = ae, \quad k = a/e.$$

(Compare these results with the corresponding results obtained for the ellipse in Section 2.)

Following closely the corresponding analysis for the ellipse, we take O as the origin, OK as the x axis, and the perpendicular to this line through O as the y axis. Then, exactly as for the ellipse, S is the point $(ae, 0)$ and l is the line $x-a/e = 0$. If $P(x, y)$ is any point of the hyperbola, by the focus–directrix definition we have

$$SP^2 = e^2PM^2,$$

where M is the foot of the perpendicular from P on to l.

Thus, the hyperbola is the set of points

$$\{(x, y): (x-ae)^2+y^2 = e^2(x-a/e)^2\}.$$

The equation may be written

$$x^2-2aex+a^2e^2+y^2 = e^2x^2-2aex+a^2$$

$$\Rightarrow \quad x^2(e^2-1)-y^2 = a^2(e^2-1)$$

$$\Rightarrow \quad \frac{x^2}{a^2}-\frac{y^2}{a^2(e^2-1)} = 1.$$

Since $e^2 > 1$, $a^2(e^2-1) > 0$ and we may write

$$b^2 = a^2(e^2-1).$$

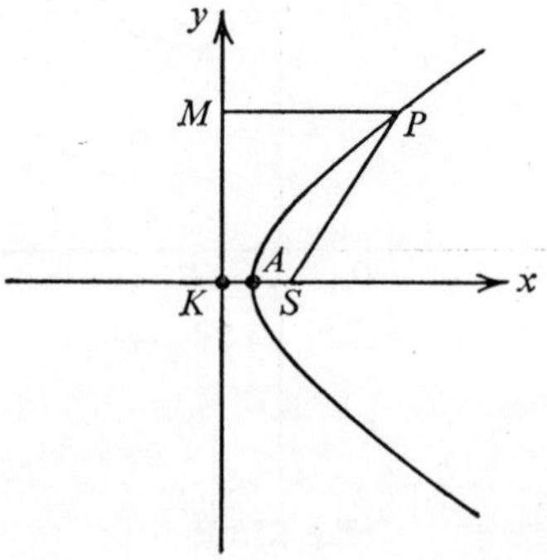

Fig. 27.16

(Notice that this is the first slight point of departure from the corresponding analysis for the ellipse. Notice also that, for a hyperbola, b can be greater than a.) With this notation the equation may be written as

$$\frac{x^2}{a^2}-\frac{y^2}{b^2} = 1,$$

which is called *the standard form of equation for the hyperbola.*

**Ex.* 24. Show that the hyperbola is symmetrical about both the x and y axes and deduce by an appeal to symmetry, the existence of a second focus $S'(-ae, 0)$ and second corresponding directrix.

**Ex.* 25. Prove that all chords through the centre, O, are bisected at O and that the length of the latus rectum (see Ex. 6) is $2b^2/a$.

The form of the complete curve is shown in Figure 27.17. Notice that, if $P(x, y)$ lies on the hyperbola,

$$x^2 = a^2\left(1+\frac{y^2}{b^2}\right) \geqslant a^2,$$

and thus no part of the curve lies within the interval $-a < x < a$. It follows that the points $B(0, b)$ and $B'(0, -b)$ do *not* lie on the curve in contrast to the corresponding points of the ellipse. $A\ (a, 0)$ and $A'(-a, 0)$ are called the *vertices* of the hyperbola; AA' is the *transverse axis* of the hyperbola, while BB' is the *conjugate axis* (see Ex. 26).

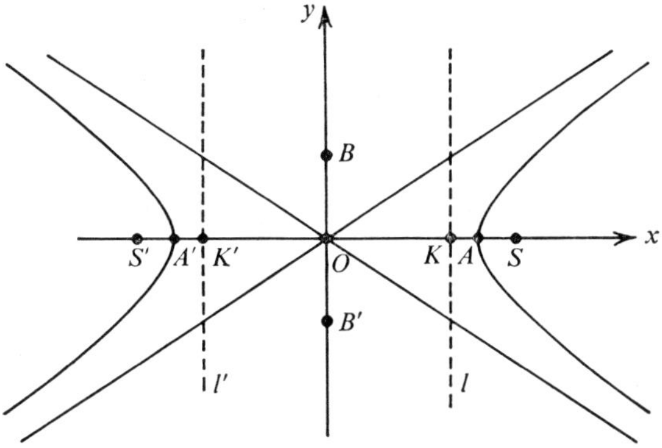

Fig. 27.17

The equation of the hyperbola may be written in the form

$$\left(\frac{x}{a}-\frac{y}{b}\right)\left(\frac{x}{a}+\frac{y}{b}\right) = 1,$$

that is

$$\frac{x}{a}-\frac{y}{b} = \frac{1}{\dfrac{x}{a}+\dfrac{y}{b}}.$$

As x, y both tend to ∞, $\left(\frac{x}{a}-\frac{y}{b}\right)$ tends to zero and the equation of the curve approximates to

$$\frac{x}{a}-\frac{y}{b} = 0.$$

This line is thus an asymptote to the hyperbola; by a similar argument so also is the line

$$\frac{x}{a}+\frac{y}{b} = 0.$$

**Ex.* 26. Show that the equations of the asymptotes may be written in the form

$$\frac{x^2}{a^2}-\frac{y^2}{b^2} = 0$$

and prove that the acute angle between them is 2 arctan (b/a).

**Ex*. 27. Sketch in the same diagram the two hyperbolas

$$\frac{x^2}{a^2}-\frac{y^2}{b^2}=1 \quad\text{and}\quad \frac{x^2}{a^2}-\frac{y^2}{b^2}=-1$$

and mark in the points $A(a, 0)$, $A'(-a, 0)$, $B(0, b)$, $B'(0, -b)$. Two hyperbolas of this form are said to be *conjugate*; show that conjugate hyperbolas have the same asymptotes.

A rectangular hyperbola was defined in Chapter 22 as the curve which, with a suitable choice of axes, has equation

$$xy = c^2.$$

To justify the use of the word hyperbola in this definition, that is, to show that such a curve has the required focus–directrix property, consider the linear transformation, T, of the plane into itself defined by the equations

$$x' = \frac{1}{\sqrt{2}}(x+y),$$

$$y' = \frac{1}{\sqrt{2}}(-x+y).$$

It is not difficult to see that T preserves distance; for, if the matrix of T is $\mathbf{A}$, then

$$\mathbf{A} = \begin{pmatrix} \frac{1}{\sqrt{2}} & \frac{1}{\sqrt{2}} \\ -\frac{1}{\sqrt{2}} & \frac{1}{\sqrt{2}} \end{pmatrix}$$

$$= \begin{pmatrix} \cos(-\frac{1}{4}\pi) & -\sin(-\frac{1}{4}\pi) \\ \sin(-\frac{1}{4}\pi) & \cos(-\frac{1}{4}\pi) \end{pmatrix}$$

which represents a *rotation* through an angle $-\frac{1}{4}\pi$ (see p. 243, vol. 1).

It follows that, under T, the image of a curve C has precisely the same appearance and geometrical properties as the curve C itself, but is rotated through an angle of $-\frac{1}{4}\pi$. But, since

$$x = \frac{1}{\sqrt{2}}(x'-y'),$$

$$y = \frac{1}{\sqrt{2}}(x'+y'),$$

we see that any point of the set

$$\{(x, y): xy = c^2\}$$

maps into a point of the set

$$\{(x, y): x'^2-y'^2 = 2c^2\}.$$

This latter set clearly represents a hyperbola, with $a^2 = b^2 = 2c^2$ and asymptotes

$$\{(x', y'): x' \pm y' = 0\}.$$

Thus, the equation

$$xy = c^2$$

represents a hyperbola with perpendicular asymptotes: we may therefore redefine a *rectangular hyperbola* as a hyperbola with perpendicular asymptotes.

Ex. 28. Prove that the eccentricity of a rectangular hyperbola is $\sqrt{2}$ and that the latus rectum is equal to the distance between the vertices of the curve. (A rectangular hyperbola is sometimes called an *equilateral hyperbola*. All rectangular hyperbolas are similar to one another.)

Ex. 29. Show that the foci of the rectangular hyperbola

$$xy = c^2$$

are the points $(\pm c\sqrt{2}, \pm c\sqrt{2})$ and find the equation of the corresponding directrices.

Ex. 30. A parametric form for a point on the curve

$$xy = \tfrac{1}{2}a^2$$

is $(at\sqrt{2}, a/(t\sqrt{2}))$; by considering the linear transformation T, deduce that the parametric form for a point on the curve

$$x^2 - y^2 = a^2$$

is $(a(1/t+t)/2, a(1/t-t)/2)$. Examine how this point moves along the curve as t varies between $-\infty$ and $+\infty$; in particular, explain what happens as t approaches the value zero from below and from above.

Ex. 31. Show that $x = a \sec \theta$, $y = b \tan \theta$ gives a parametric representation for all points of the hyperbola

$$b^2x^2 - a^2y^2 = a^2b^2$$

and explain how $P(a \sec \theta, b \tan \theta)$ moves on the hyperbola as θ varies from 0 to 2π. In particular, what happens as θ approaches the values $\frac{1}{2}\pi$, $\frac{3}{2}\pi$ from below and from above?

Ex. 32. The circle.

$$x^2 + y^2 = a^2$$

is called the *auxiliary circle* of the hyperbola

$$b^2x^2 - a^2y^2 = a^2b^2$$

(see Ex. 12). Sketch in the same diagram a hyperbola and its auxiliary circle. P is a point of the hyperbola with positive coordinates, N is the foot of the perpendicular from P to the transverse axis and NQ is a tangent to the auxiliary circle at the point Q with positive coordinates. The angle $NOQ = \theta$. Show that Q is the point $(a \cos \theta, a \sin \theta)$ and that P is the point $(a \sec \theta, b \tan \theta)$.

Ex. 33. Show that the equation of the tangent to the hyperbola

$$\frac{x^2}{a^2}-\frac{y^2}{b^2}=1$$

at the point $P(a\sec\theta, b\tan\theta)$ has equation

$$\frac{x\sec\theta}{a}-\frac{y\tan\theta}{b}=1.$$

Corresponding to the property $SP+S'P = 2a$ for points of an ellipse we have the following property for the hyperbola.

Focal distance property

If P is any point on a hyperbola, foci S, S', then $|SP-S'P| = 2a$.

Take the hyperbola in the standard form

$$b^2x^2-a^2y^2 = a^2b^2$$

and let M, M' be the feet of the perpendiculars from P on to the two directrices (see Figure 27.18). Let P be the point (h, k); then, by the definition of the hyperbola

$$SP = ePM, \quad S'P = ePM'$$

and

$$|SP-S'P| = e|MM'| = 2a.$$

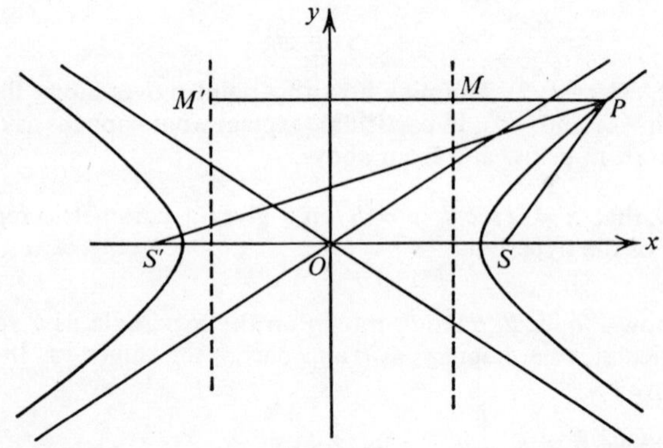

Fig. 27.18

It is not difficult to see that $SP-S'P = -2a$ if P lies on the branch of the hyperbola enclosing S, and that $SP-S'P = +2a$ if P lies on the opposite branch.

**Ex.* 34. Prove that, if S, S' are fixed points and P moves in such a way that $SP-S'P$ is constant, then the locus of P is a branch of a hyperbola.

**Ex*. 35. Interpret the conditions $|SP-S'P| < 2a$ and $|SP-S'P| > 2a$ geometrically.

Ex. 36. Devise a mechanical construction for a branch of a hyperbola with given foci and length of transverse axis, based upon the relation $SP-S'P = 2a$.

**Ex*. 37. By adapting the method used for the ellipse, prove that the tangent and normal at the point P of a hyperbola with foci S, S' bisect the angle SPS'.

Hyperbolic orbits arise under the action of forces of repulsion varying inversely as the square of the distance from the centre of force, such as arise in the case of electrically charged particles. They also arise under attractive forces, such as gravity, when the energy content of the orbit is too great for it to be an ellipse, such as, for example, comets which orbit the sun only once before retreating into outer space. (Comets such as Halley's comet, which reappear, obviously follow elliptic orbits: the orbits are usually highly eccentric, that is, $e \approx 1$.) The property $|SP-S'P| = 2a$ is used in range-finding. If a gun at P is fired and the times taken for the sound to reach two listening posts, S and S', recorded, $|SP-S'P| = 2a$ may be determined and P lies on the unique hyperbola with foci S, S' and transverse axis $2a$. If the same experiment is conducted from S and a further listening post S'', a second hyperbola is obtained and P lies at the intersection of the two hyperbolas. A more sophisticated application of the same idea is employed in navigation. [See the article 'Sound Ranging' in No. 195 of the *Mathematical Gazette* (July 1928) by W. Hope-Jones.]

The ellipse and hyperbola have very similar geometrical properties—most of the properties peculiar to the hyperbola are associated with its asymptotes. Both the ellipse and hyperbola possess a centre, that is, a point at which all chords are bisected and, for this reason, they are called *central conics*; the parabola is not a central conic. If we are concerned with a property common to all central conics, we may take the equation of a typical conic in the form

$$\alpha x^2+\beta y^2 = 1$$

(where α, β are not both negative).

Example 4. *Prove that the line*

$$y = mx+c$$

touches the central conic

$$\alpha x^2+\beta y^2 = 1 \quad (\alpha, \beta \neq 0)$$

if and only if

$$c^2 = (\alpha+\beta m^2)/(\alpha\beta).$$

Prove that perpendicular tangents to a central conic meet, in general, on a circle, the director circle of the conic, and point out what exceptional cases may arise.

The given line and conic meet in points with x coordinates given by

$$\alpha x^2+\beta(mx+c)^2 = 1,$$

that is, by $$(\alpha+\beta m^2)\,x^2+2\beta mcx+(\beta c^2-1) = 0.$$

The line $y = mx+c$ touches the conic $\alpha x^2+\beta y^2 = 1$

$\Leftrightarrow$ $\beta^2m^2c^2 = (\alpha+\beta m^2)(\beta c^2-1)$ (the condition for double roots of the quadratic equation above)

$\Leftrightarrow$ $\alpha\beta c^2 = \alpha+\beta m^2$

$\Leftrightarrow$ $c^2 = (\alpha+\beta m^2)/(\alpha\beta)$, since $\alpha, \beta \neq 0$.

Now let $y = mx+c$ be a tangent through the point (h, k). Then we have

$$k = mh+c$$

and thus, using the condition proved above for the line to be a tangent,

$$\alpha\beta(k-mh)^2 = \alpha+\beta m^2,$$

which reduces to $m^2(\alpha\beta h^2-\beta)-2\alpha\beta hkm+(\alpha\beta k^2-\alpha) = 0$. This is a quadratic equation in m (showing that *two* tangents can, in general, be drawn through the point (h, k)). Let the roots be m_1, m_2: then

$$m_1m_2 = (\alpha\beta k^2-\alpha)/(\alpha\beta h^2-\beta).$$

Thus, (h, k) lies on perpendicular tangents if

$$\alpha\beta h^2-\beta = -(\alpha\beta k^2-\alpha),$$

that is, if $$h^2+k^2 = \frac{1}{\alpha}+\frac{1}{\beta},$$

and the locus of (h, k) is thus

$$x^2+y^2 = \frac{1}{\alpha}+\frac{1}{\beta}.$$

This is a circle provided $(1/\alpha)+(1/\beta) > 0$. This is certainly always true for an ellipse. For a rectangular hyperbola, $(1/\alpha)+(1/\beta) = 0$ and the director circle reduces to a point (the origin); for any other hyperbola, the circle exists provided $(1/\alpha) > -(1/\beta)$, that is, in the usual notation, provided $a^2 > b^2$: a hyperbola in which the conjugate axis is greater than the transverse axis has no director circle. Since the angle between the asymptotes is $2\arctan(b/a)$, a hyperbola has no director circle if the angle between those parts of the asymptotes containing the curve is greater than $\frac{1}{2}\pi$.

Ex. 38. An elliptic lamina moves in a plane in such a way that it touches each of two fixed perpendicular lines. What is the locus of its centre?

Exercise 27(b)

1. Find the eccentricities of the following hyperbolas:
(i) $\frac{1}{4}x^2-\frac{1}{5}y^2 = 1$; (ii) $2x^2-y^2 = 1$; (iii) $(x-1)^2-2(y-2)^2 = 1$.

2. Sketch roughly the hyperbola
$$\tfrac{1}{6}x^2-\tfrac{1}{2}y^2 = 1$$
and find its eccentricity, the coordinates of its foci and the equations of the corresponding directrices.

3. Sketch roughly the hyperbola
$$\tfrac{1}{16}(x-1)^2-\tfrac{1}{9}(y+1)^2 = 1,$$
find its eccentricity, the coordinates of its foci and the equations of the corresponding directrices.

4. Show that the equation
$$x^2-4y^2+6x+8y+1 = 0$$
represents a hyperbola, and find its eccentricity, the coordinates of its foci and the equations of the corresponding directrices.

5. Find the equations of the asymptotes of the hyperbolas
(i) $9x^2-4y^2 = 12$; (ii) $3x^2-y^2 = 1$.

6. Find the equations of the asymptotes of the hyperbolas
(i) $x^2-4y^2+2x+8y-5 = 0$; (ii) $2x^2-4y^2-8y-5 = 0$.

7. Find the equation of the hyperbola with asymptotes
$$2x-y = 0, \quad 2x+y = 0,$$
which passes through the point (2, 3).

8. Find the equation of the hyperbola with asymptotes
$$3x-2y-9 = 0, \quad 3x+2y+3 = 0,$$
which passes through the point (2, −4).

9. Find the equation of the tangent to the hyperbola $2x^2-3y^2 = 6$ at the point (3, −2) and prove that the line $x+y+1 = 0$ also touches the curve.

10. Find the equation of the normal to the hyperbola $3x^2-y^2 = 2$ at the point (1, 1) and find the x coordinate of the point where it meets the curve again.

11. Find the equation of the tangent to the rectangular hyperbola
$$x^2-y^2 = 3$$
which passes through the point (3, −3).

12. Prove that a hyperbola intersects a line parallel to one of its asymptotes at just one point.

13. Find the equation of the rectangular hyperbola which has the points (4, 0), (−4, 0) as foci.

14. If the eccentricity of a hyperbola is e, prove that the eccentricity of the conjugate hyperbola is $e/\sqrt{(e^2-1)}$.

15. Prove the results of Exercise 27(*a*) 11 for the hyperbola

$$b^2x^2 - a^2y^2 = a^2b^2.$$

16. Prove the result of Exercise 27(*a*) 12 for the hyperbola

$$b^2x^2 - a^2y^2 = a^2b^2.$$

17. If P is a variable point on a hyperbola with vertex A, show that the locus of the mid-point of PA is another hyperbola and find its centre and eccentricity.

18. Find the equation of the hyperbola $(x^2/a^2) - (y^2/b^2) = 1$ when the origin is changed to the point whose coordinates are (h, k), and the new axes are parallel to the original axes.

The coordinates of a point are given by

$$x = 2\cos^2\theta/(2\cos^2\theta - 1), \quad y = 2\tan 2\theta.$$

Show that the locus of the point is a hyperbola, and find the coordinates of the centre.

Find the equation of the tangent to the curve at the point $\theta = \frac{1}{8}\pi$. (O & C)

19. Obtain the equations of the tangent and normal to the hyperbola

$$4x^2 - y^2 = 36$$

at the point $P(5, 8)$.

The tangent at P meets the y axis at Q and the normal at P meets the x axis at R. Prove that the area of the triangle PQR is 145 square units.

Obtain the coordinates of the point S such that $PQRS$ is a rectangle. (O & C)

20. Prove that the normal to the hyperbola

$$\frac{x^2}{a^2} - \frac{y^2}{b^2} = 1$$

at the point $(a\sec\phi, b\tan\phi)$ is given by the equation

$$ax\sin\phi + by = (a^2+b^2)\tan\phi.$$

The normal at a variable point P on the hyperbola meets the axes at X and Y; the mid-point of XY is Z. Prove that, if O is the centre of the hyperbola:

(i) the length of OZ is not less than $(a^2+b^2)/2a$;

(ii) OZ is inversely proportional to the perpendicular distance from O to the tangent at P;

(iii) if $a = b$, then Z always coincides with P. (O & C)

21. The tangent at the point P of the rectangular hyperbola

$$xy = c^2$$

meets the asymptotes at Q, R. Prove that $PQ = RP$.

By considering a linear transformation T with matrix of the form

$$\begin{pmatrix} \alpha & \alpha \\ \beta & -\beta \end{pmatrix}$$

show that the result above is true also for the hyperbola

$$b^2x^2 - a^2y^2 = a^2b^2.$$

Repeat the above process to prove the following more general result: if a chord PQ of a hyperbola meets the asymptotes at P', Q', then $PP' = QQ'$.

Suggest a further generalization of this result by considering a second rectangular hyperbola $xy = d^2$ in place of the asymptotes.

4. POLAR EQUATIONS OF CONICS

Given coordinate axes Ox, Oy, the position of a point is uniquely determined by its Cartesian coordinates (x, y). However, other systems of coordinates are available; in Figure 27.19 the position of the point P is determined if, given the point O and the line OX, we know the length $r = OP$ and the angle $\theta = XOP$. (r, θ) are called the *polar coordinates* of P relative to the *origin* or *pole O* and the *initial line OX*. We adopt the usual convention that positive values of θ are measured in the anticlockwise sense.

The polar coordinates of a point are not uniquely determined. For example, consider the point P with Cartesian coordinates $(1, -\sqrt{3})$ (see Figure 27.20). Taking O as origin and Ox as initial line, the polar coordinates of P can be taken in any of the alternative forms

$$(2, -\tfrac{1}{3}\pi), \quad (2, \tfrac{5}{3}\pi), \quad (-2, \tfrac{2}{3}\pi), \quad \text{etc.}$$

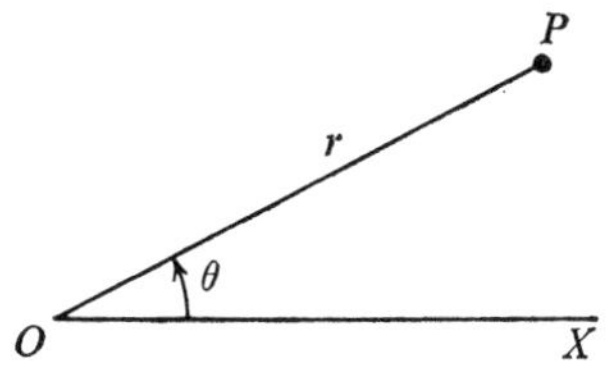

Fig. 27.19

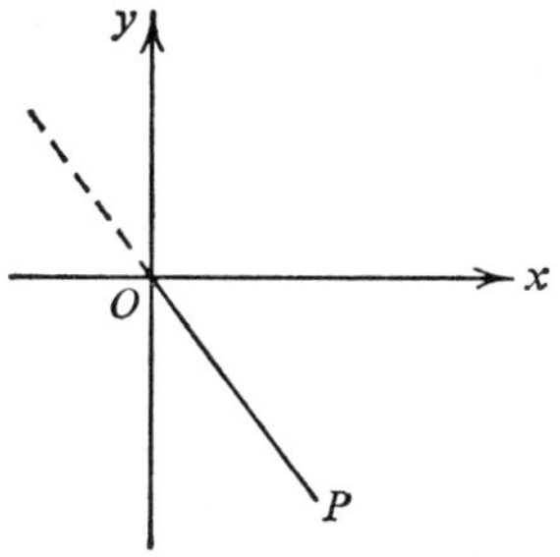

Fig. 27.20

If unique polar coordinates are desired, we define the *principal polar coordinates* of a point to be the ones for which

$$r > 0, \quad -\pi < \theta \leqslant \pi.$$

Generally, however, we allow r to take both positive and negative values.

In the same way that a set of points in a plane could be defined by an equation or inequality involving Cartesian coordinates, sets may be defined using polar coordinates. For example, the set

$$\{(r, \theta): r = a\}$$

represents a circle, centre O and radius a; the set

$$\{(r, \theta): r\cos\theta = a\}$$

represents a straight line (traversed twice) perpendicular to the initial line and at a distance a from O; the set

$$\{(r, \theta): a < r < b\}$$

represents the annulus defined by two concentric circles of radii a and b.

Ex. 39. Describe in words and draw sketches of the following sets of points:
(i) $\{(r, \theta): r\sin\theta = a\}$; (ii) $\{(r, \theta): \theta = c\}$;
(iii) $\{(r, \theta): r = a\sin\alpha\operatorname{cosec}(\theta-\alpha)\}$.
(*Hint: rewrite the defining equation and think of the Sine Rule.*)
(iv) $\{(r, \theta): a < r < b,\ \ 0 < \theta < \tfrac{1}{2}\pi\}$; (v) $\{(r, \theta): r\sec\theta < 2a\}$.

Ex. 40. Find the polar equation of:
(i) the line passing through the points with polar coordinates $(1, 0)$ and $(1, \tfrac{1}{2}\pi)$;
(ii) the circle through the points with polar coordinates $(2, \tfrac{1}{2}\pi)$ and touching the initial line at the origin;
(iii) the circle of radius 1 with centre at the point with polar coordinates $(\sqrt{2}, \tfrac{1}{4}\pi)$.

Using the relation

$$x = r\cos\theta, \quad y = r\sin\theta$$

it is a simple matter to rewrite the defining equation of a set of points in the plane in terms of polar coordinates. For example, the ellipse Γ may be written in the alternative forms

$$\begin{aligned}\Gamma &= \{(x, y): b^2x^2+a^2y^2 = a^2b^2\}\\ &= \{(r, \theta): r^2(b^2\cos^2\theta+a^2\sin^2\theta) = a^2b^2\}.\end{aligned}$$

However, it may well happen that a curve is best expressed in polar form without recourse to Cartesian coordinates. Suppose, for example, that we are given a conic of eccentricity e and semi-latus rectum l. The focus–directrix definition suggests that it would be reasonable to take the origin, O, at a focus and the initial line along the major axis.

In Figure 27.21, O is the focus of the conic, MN is the corresponding directrix and $OL = l$ is the semi-latus rectum.

Then, since

$$OP = ePM,$$

we have,

$$r = ePM.$$

But

$$\begin{aligned}PM &= LN-r\cos\theta\\ &= l/e-r\cos\theta,\end{aligned}$$

giving us

$$r(1+e\cos\theta) = l.$$

This equation is the *standard form for the equation of a conic in polar coordinates* with the origin at a focus and the initial line along the major axis.

Ex. 41. Sketch the parabola $r(1+\cos\theta) = 2$.

Ex. 42. Show that the equation $r(2+\cos\theta) = 2$ represents an ellipse and sketch the curve.

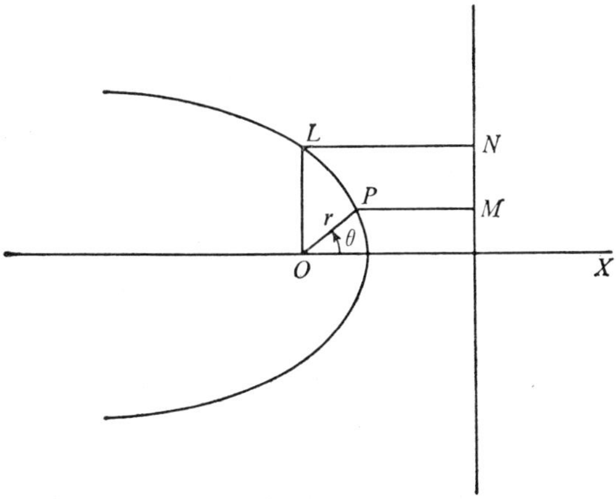

Fig. 27.21

The polar form for a conic is of great value in dealing with properties of focal chords. We conclude this section with an example illustrating its use in this context.

Example 5. *PQ, UV are perpendicular focal chords of a rectangular hyperbola, focus S. Prove that:*

(i) $|PS.SQ| = |US.SV|$, (ii) $PQ = UV$.

Taking S as origin and the major axis SX as initial line, the polar equation of the rectangular hyperbola is

$$r(1+\sqrt{2}\cos\theta) = l.$$

Let P, Q, U, V have polar coordinates

$$(r_1, \theta),\quad (r_2, \theta+\pi),\quad (r_3, \theta+\tfrac{1}{2}\pi),\quad (r_4, \theta+\tfrac{3}{2}\pi).$$

Then, from the equation of the conic we have

$$r_1 = l(1+\sqrt{2}\cos\theta)^{-1},$$
$$r_2 = l(1-\sqrt{2}\cos\theta)^{-1},$$
$$r_3 = l(1-\sqrt{2}\sin\theta)^{-1},$$
$$r_4 = l(1+\sqrt{2}\sin\theta)^{-1}.$$

(Notice that θ cannot be an odd multiple of $\frac{1}{4}\pi$ if the chords PQ, UV are to exist.) Taking sense along each chord into account, this gives

$$PS = l(1+\surd 2\cos\theta)^{-1},$$
$$SQ = l(1-\surd 2\cos\theta)^{-1},$$
$$US = l(1-\surd 2\sin\theta)^{-1},$$
$$SV = l(1+\surd 2\sin\theta)^{-1}.$$

(i) $$PS.SQ = l^2(1-2\cos^2\theta)^{-1} = -l^2\sec 2\theta,$$
$$US.SV = l^2(1-2\sin^2\theta)^{-1} = l^2\sec 2\theta.$$

$$\therefore\ PS.SQ+US.SV = 0.$$

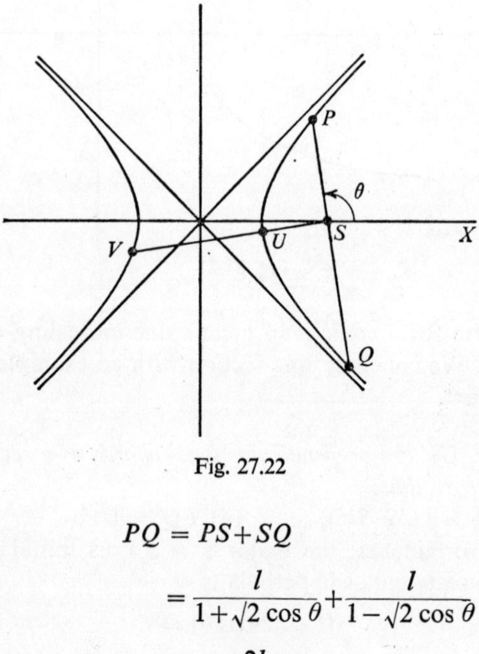

Fig. 27.22

(ii) $$PQ = PS+SQ$$
$$= \frac{l}{1+\surd 2\cos\theta}+\frac{l}{1-\surd 2\cos\theta}$$
$$= \frac{2l}{1-2\cos^2\theta},$$
$$|PQ| = |2l\sec 2\theta|,$$

and similarly $$|UV| = |2l\sec 2\theta|.$$

Ex. 43. PQ is a focal chord through the focus S of a conic of semi-latus rectum l. Prove that

$$\frac{1}{PS}+\frac{1}{SQ} = \frac{2}{l}.$$

Ex. 44. PP', QQ' are perpendicular focal chords of a conic; prove that

$$\frac{1}{PP'}+\frac{1}{QQ'}$$

is constant.

Ex. 45. Show that the polar equation of the directrix corresponding to the focus O is

$$er\cos\theta = l.$$

Show also that a polar equation of the form

$$1/r = a\cos\theta+b\sin\theta$$

represents a straight line.

Deduce (with the help of Exercise 27(*a*), question 11 (second part)), that the equation

$$\frac{l}{r} = \cos(\theta-\alpha)+e\cos\theta$$

is the equation of the tangent to the standard conic at the point with vectorial angle α.

Prove that the tangents at the ends of a focal chord of a conic meet on the corresponding directrix.

5. SECTIONS OF A CONE

This section is included for the attention of readers interested in the historical development of the geometry of the conics.

The conics, or conic sections, were first studied, as their name suggests, as sections of a right circular cone. Their history extends back to the time of Ancient Greece: they were first extensively studied by Apollonius of Perga (247–205 B.C.) who wrote a treatise on their properties. The focus–directrix property, which we have made the basis of our definition, was not discovered until later.

The various types of conic arise as sections of a cone made by planes making various angles with the axis of the cone. For the purposes of the following definitions, the cone is taken to extend infinitely in both directions from the vertex. (See Figure 27.23.)

First observe that any plane perpendicular to the axis of the cone (and not through the vertex) cuts the cone in a *circle*, which may therefore be regarded as a particular type of conic.

If the plane is oblique, not parallel to one of the generating lines of the cone and cuts only one half of the cone, the resulting section is an *ellipse* (Figure 27.24).

If the plane is parallel to one of the generating lines of the cone, the resulting section is a *parabola* (Figure 27.25).

If the plane is oblique and cuts both halves of the cone, but does not pass through the vertex of the cone, the section is a hyperbola (Figure 27.26).

Ex. 46. Show how a pair of straight lines arises as a conic section.

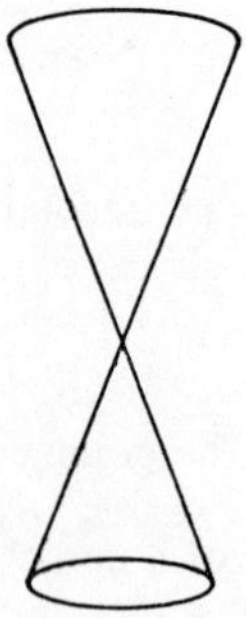

Fig. 27.23

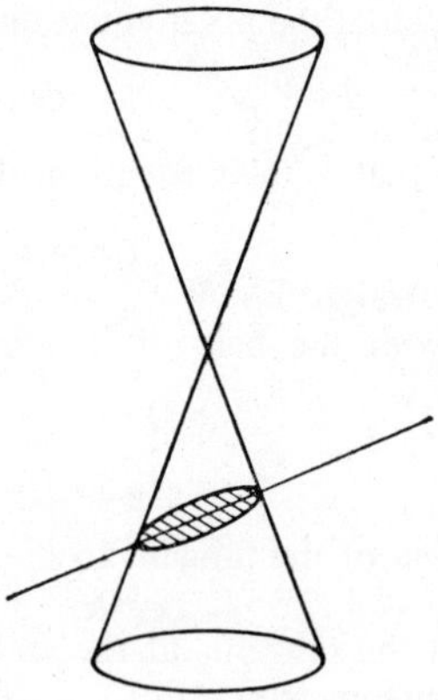

Fig. 27.24

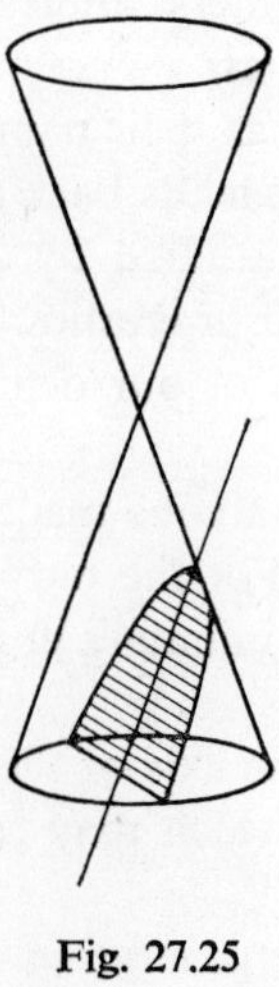

Fig. 27.25

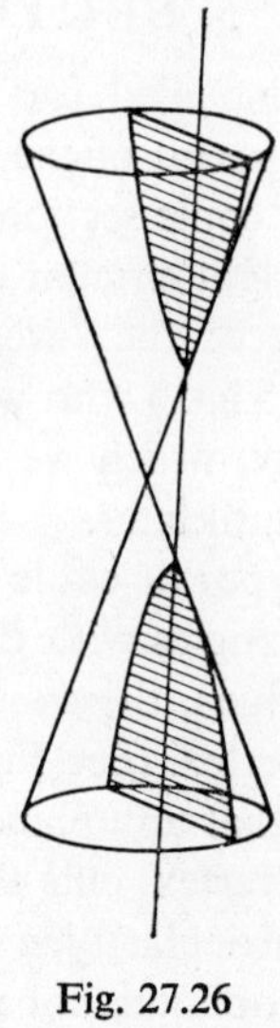

Fig. 27.26

The connection between the conic section and the focus–directrix definitions is exhibited in Exs. 47–52. The notation refers to Figure 27.27 (for convenience of drawing, we take an elliptic section and show only one half of the cone).

V is the vertex of a right circular cone with its axis vertical and we consider the section of the cone by the plane Π. A sphere may be drawn to touch the plane Π at S and also to touch the cone in a circle lying in a horizontal plane Π'. The vertical plane containing V and S cuts Π and the cone at A and A' and Π' and the cone at C and C'. l is the line of intersection of Π and Π' and $A'A$ meets l at K. P is any point on the cone lying in the plane Π and VP touches the sphere at P'. M is the point of l such that PM and AK are parallel, NPQ is a horizontal section, N lying on AA' and Q lying on VA.

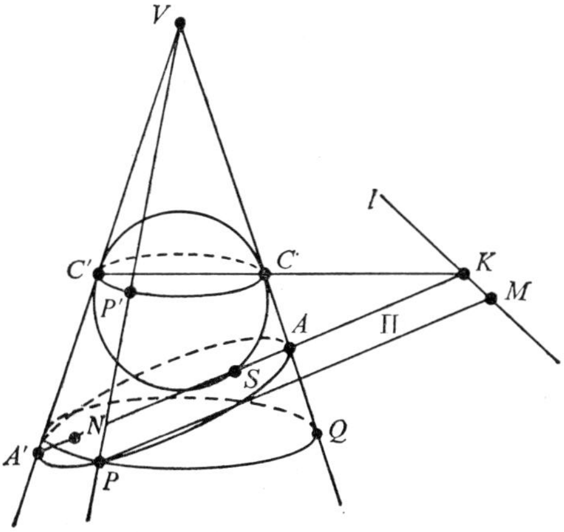

Fig. 27.27

Ex. 47. Show that $PNKM$ is a rectangle.

Ex. 48. Show that $SP = QC$ and $PM = NK$.

Ex. 49. Show that $QC/NK = AC/AK$.

Ex. 50. Deduce that SP/PM is fixed for a given plane Π, and locate a focus and its corresponding directrix for the resulting conic section.

Ex. 51. Show how the parabolic cross-section arises.

Ex. 52. Show that, both for elliptic and for hyperbolic sections, a second sphere may be drawn to touch the cone and the plane Π, giving rise to a second focus and a second directrix.

Miscellaneous Exercise 27

1. Show that the equation

$$x^2+2y^2-2x+12y+8 = 0$$

represents an ellipse, and find its eccentricity and the coordinates of its centre.

Prove that

$$3x-2y+2 = 0$$

is a tangent to the ellipse and find the point of contact.

2. The tangents drawn from a point $P(x_1, y_1)$ to the circle $x^2+y^2 = r^2$ touch the circle at L, M. Show that the equation of LM is $xx_1+yy_1 = r^2$.

If P moves on a hyperbola of eccentricity e with its centre at the origin, show that LM touches a hyperbola of eccentricity e', where

$$\frac{1}{e^2}+\frac{1}{e'^2} = 1.$$

Draw a figure for the case when $e < \sqrt{2}$ and the circle touches the given hyperbola. (London)

3. Show that the equation of the tangent at $P(a \cos \alpha, b \sin \alpha)$ to the ellipse

$$b^2x^2+a^2y^2 = a^2b^2$$

is $bx \cos \alpha+ay \sin \alpha = ab$.

The tangent at P cuts the x and y axes at A and B respectively and the normal at P cuts the x and y axes at C and D respectively. Find the ratio $PC:PD$.

If AD and BC meet at E, prove that BE is perpendicular to AD and hence, or otherwise, find the equation of the circle through A, B, E. (London)

4. Prove that the conics

$$\frac{x^2}{a^2}+\frac{y^2}{b^2} = 1 \quad \text{and} \quad \frac{x^2}{a^2+\lambda}+\frac{y^2}{b^2+\lambda} = 1$$

have the same foci, whatever value λ takes (provided $\lambda \neq -a^2$ or $-b^2$).

Find the equation of the rectangular hyperbola whose foci coincide with those of the ellipse $b^2x^2+a^2y^2 = a^2b^2$.

Show further that two ellipses which have the same foci cannot intersect in real points but that if an ellipse and a hyperbola have the same foci, they intersect orthogonally at four real points.

5. P is any point on a rectangular hyperbola, centre O and foci S, S'. Prove that $OP^2 = SP.S'P$.

6. P is the point $(a \cos \theta, b \sin \theta)$ of the ellipse

$$b^2x^2+a^2y^2 = a^2b^2$$

and $Q(a \cos \theta, a \sin \theta)$ is the corresponding point of the auxiliary circle. Prove that the perpendicular distance from S to the tangent at Q to the circle is equal to SP.

7. Write down the equation of

(i) an ellipse which has its minor axis along the y axis and touches the x axis;

(ii) an ellipse which has its minor axis along the x axis and touches the y axis.

Two such ellipses are given. Assuming that they meet in four (real) points, write down the equation of any conic through these four points, and prove that its centre lies on a certain rectangular hyperbola. (O & C)

8. A tangent to the ellipse

$$b^2x^2+a^2y^2 = a^2b^2$$

$(a, b > 0)$ meets the ellipse

$$bx^2+ay^2 = ab(a+b)$$

at the points P, Q. Prove that the tangents to the second ellipse at P and Q meet on its director circle.

9. Two hyperbolas, S and S', have the equations

$$\frac{x^2}{a^2}-\frac{y^2}{b^2}=1,$$

$$\frac{x^2}{a^2}-\frac{y^2}{b^2}=\lambda,$$

where $\lambda < 1$. Prove that the tangent to S at any point P meets S' in two points Q and R which are equidistant from P.

If $\lambda = -1$, prove that the tangents to S' at Q and R meet on S at P', the reflection of P in the origin. (O & C)

10. The asymptotes of the hyperbola

$$\frac{x^2}{a^2}-\frac{y^2}{b^2}=1$$

are l and l', and P is a point on the hyperbola. The perpendicular from P to l meets l and l' at X and Y, and the perpendicular from P to l' meets l' and l at X' and Y'. By expressing the coordinates of P in parametric form, or otherwise, prove that, for all positions of P,

(i) $PX.PX'=\dfrac{a^2b^2}{a^2+b^2}$;

(ii) $PX.PY=\dfrac{a^2b^2}{a^2-b^2}$;

(iii) $PY.PY'=\dfrac{a^2b^2(a^2+b^2)}{(a^2-b^2)^2}$. (O & C)

11. Show that, if the point $(x_1+r\cos\theta, y_1+r\sin\theta)$ lies on the central conic $ax^2+by^2=1$, then r satisfies the quadratic equation

$$r^2(a\cos^2\theta+b\sin^2\theta)+2r(ax_1\cos\theta+by_1\sin\theta)+ax_1^2+by_1^2-1=0.$$

Deduce that, if (x_1, y_1) is the mid-point of the chord PQ, then PQ has gradient $-ax_1/by_1$.

Prove that the locus of mid-points of parallel chords of a central conic is a diameter of the conic.

12. Use the analysis of Question 11 to prove Newton's Theorem for a central conic: if PQ, RS are two chords intersecting at X, then the ratio

$$\frac{PX.XQ}{RX.XS}$$

depends only upon the directions of the chords PQ and RS, and not on their positions.

13. Given the outline of an ellipse, show how you would construct the centre, the axes, the foci and the directrices.

14. Prove that, if tangents are drawn to a hyperbola from any point of the conjugate hyperbola, their chord of contact touches the opposite branch of the conjugate hyperbola and is bisected by it.

15. Find the equation of the perpendicular bisector of the line joining the points (x_1, y_1), (x_2, y_2).

A fixed circle has centre C and radius $2a$. A is a fixed point inside the circle and P is a variable point on the circumference. Prove that the perpendicular bisector of AP touches the ellipse whose foci are at C and A, and whose major axis is of length $2a$. (C.S.)

16. A point P is taken at random inside an ellipse of eccentricity e. Calculate the probability (in terms of e) that the sum of the focal distances of P should be not greater than the distance from a focus to the opposite end of the major axis. (C.S.)

(*Note: the area of an ellipse with major and minor semi-axes of lengths a and b respectively is πab—a result easily deduced by the usual calculus methods, or alternatively by considering the effect on areas of the linear transformation T of Section* 2 *of this Chapter.*)

17. Σ_1 is a circle, centre O, radius b, Σ_2 is a circle, centre A radius a $(a < b)$ which touches Σ_1 internally. Describe the locus of the centre, P, of a variable circle, Σ, which touches Σ_1 internally and Σ_2 externally.

If Σ_1 is a straight line, Σ_2 is a circle, centre A and radius a which touches Σ_1 and P the centre of a variable circle Σ which has Σ_1 as a tangent and touches Σ_2 externally, describe the locus of P.

18. Prove that at most four normals may be drawn from a point A to a central conic with centre O.

Prove further that, if the normals at the points P_1, P_2, P_3, P_4 on a central conic intersect at A, then $P_1P_2P_3P_4$ lie on a rectangular hyperbola which passes through O and A and has its asymptotes parallel to the axes of the given conic. (The rectangular hyperbola of this question is known as the *hyperbola of Apollonius*.)

19. Prove that the vector equation (referred to the vertex as origin) of the tangent to the parabola $y^2 = 4ax$ at the point $\mathbf{p} = at^2\mathbf{i}+2at\mathbf{j}$ is $\mathbf{r} = \mathbf{p}+\lambda\mathbf{u}$ where $\mathbf{u}$ is the unit vector $(t\mathbf{i}+\mathbf{j})/(1+t^2)^{\frac{1}{2}}$.

Prove that the locus of the meets of tangents to the parabola $y^2 = 4ax$ which cut at a fixed angle α is a hyperbola of eccentricity $\sec\alpha$.

20. Prove that, if the chord PQ of any conic with focus S, when produced, cuts the corresponding directrix at R, then SR is a bisector of the angle PSQ. Deduce that, if the tangent at P cuts the directrix at T, then the angle PST is a right angle.

21. If X is any point on the tangent at the point P of any conic with focus S, and if H, K are respectively the feet of the perpendiculars from X to SP and the directrix corresponding to S, prove that $SH = eXK$, where e is the eccentricity.

28. *Further matrices*

Throughout this chapter it is to be assumed, unless explicitly stated otherwise, that the matrix **A** *of a linear transformation T is referred to the base vectors* **i**, **j** (*or* **i**, **j**, **k** *in three dimensions*).

1. EIGENVALUES AND EIGENVECTORS FOR 2×2 MATRICES

Let us write the matrix of the linear transformation, $T: R^2 \to R^2$ as **A** where

$$\mathbf{A} = \begin{pmatrix} a_1 & b_1 \\ a_2 & b_2 \end{pmatrix}.$$

Whatever form T, and therefore **A**, has, the origin O is mapped into itself. In general, no other point remains fixed under T but, if there is a point P, distinct from O, such that $T(P) = P$, then *all* points of the line OP are mapped into themselves.

Ex. 1. Prove the assertion that if T maps a point P (other than the origin) into itself, then it maps every point of the line OP into itself.

Ex. 2. Prove that, if T maps two points, P and Q, into themselves, where O, P, Q are distinct and not collinear, then T is the identity transformation, that is, the transformation with matrix

$$\begin{pmatrix} 1 & 0 \\ 0 & 1 \end{pmatrix}.$$

We may now investigate the answer to the question: Even if T does not map any point other than the origin into itself, is there a line l through the origin such that every point of l is mapped into a point of l? If such a line exists, we say that l is an *invariant line* and that l *maps into itself* under T.

Ex. 3. If S is a point such that $T(S) \in OS$, prove that every point of the line OS is mapped into a point of OS.

Suppose that S is a point other than the origin, with the property that $T(S) \in OS$; write $\mathbf{OS} = \mathbf{s}$. Then $\mathbf{As} = \lambda\mathbf{s}$ and thus

$$(\mathbf{A} - \lambda\mathbf{I})\,\mathbf{s} = \mathbf{0}.$$

Since $\mathbf{s} \neq \mathbf{0}$, we must have $\det(\mathbf{A} - \lambda\mathbf{I}) = 0$ (see p. 294). This is a quadratic equation in λ, *the characteristic equation of the matrix* **A**, with, in general, two distinct roots, λ_1 and λ_2, called the *eigenvalues* of the matrix **A**. Any non-zero solution, $\mathbf{s}_1$, of the homogeneous equation

$$(\mathbf{A}-\lambda_1\mathbf{I})\,\mathbf{s} = \mathbf{0}$$

is called an *eigenvector corresponding to the eigenvalue* λ_1; similarly, we can find eigenvectors corresponding to the eigenvalue λ_2.

Example 1. *Find the eigenvalues of the matrix*

$$\mathbf{A} = \begin{pmatrix} 4 & 2 \\ 6 & 5 \end{pmatrix}$$

and determine their corresponding unit eigenvectors.

The characteristic equation

$$\det\begin{pmatrix} 4-\lambda & 2 \\ 6 & 5-\lambda \end{pmatrix} = 0$$

$$\Leftrightarrow \qquad \lambda^2-9\lambda+8 = 0$$

$$\Leftrightarrow \quad \lambda = 1 \quad \text{or} \quad \lambda = 8.$$

Consider first the eigenvalue $\lambda = 1$: if $\mathbf{s}_1$ is any corresponding eigenvector then, writing

$$\mathbf{s}_1 = \begin{pmatrix} x \\ y \end{pmatrix},$$

we have

$$(\mathbf{A}-\mathbf{I})\,\mathbf{s}_1 = \mathbf{0},$$

that is

$$\begin{pmatrix} 3 & 2 \\ 6 & 4 \end{pmatrix}\begin{pmatrix} x \\ y \end{pmatrix} = \begin{pmatrix} 0 \\ 0 \end{pmatrix}.$$

On solving these equations we obtain $x = 2\mu$, $y = -3\mu$, giving as an eigenvector

$$\mathbf{s}_1 = \begin{pmatrix} 2\mu \\ -3\mu \end{pmatrix}.$$

In particular, a unit eigenvector is given by

$$\mathbf{e}_1 = \begin{pmatrix} 2/\sqrt{13} \\ -3/\sqrt{13} \end{pmatrix}.$$

All the points of the line $\quad \mathbf{r} = k_1\mathbf{e}_1,$

that is, all points of the line $\quad 3x+2y = 0,$

map into points of the same line (in fact, in this case, into themselves since $\lambda = 1$): $3x+2y = 0$ is an *invariant line* under the transformation.

Now consider the eigenvalue $\lambda = 8$: if $\mathbf{s}_2$ is any corresponding eigenvector we have

$$(\mathbf{A}-8\mathbf{I})\,\mathbf{s}_2 = \mathbf{0},$$

that is,
$$\begin{pmatrix} -4 & 2 \\ 6 & -3 \end{pmatrix} \begin{pmatrix} x \\ y \end{pmatrix} = \begin{pmatrix} 0 \\ 0 \end{pmatrix},$$

giving
$$\mathbf{s}_2 = \begin{pmatrix} \nu \\ 2\nu \end{pmatrix}.$$

In particular, a unit eigenvector is given by

$$\mathbf{e}_2 = \begin{pmatrix} 1/\sqrt{5} \\ 2/\sqrt{5} \end{pmatrix}.$$

All points of the line
$$\mathbf{r} = k_2\mathbf{e}_2,$$

that is, of the line
$$2x - y = 0,$$

map into points of the same line (but, apart from O, *not* this time into themselves): $2x-y = 0$ is a second invariant line under the transformation.

We shall now analyse the linear transformation represented by the matrix $\mathbf{A}$ of Example 1 in greater detail. The matrix

$$\mathbf{P} = \begin{pmatrix} 2 & 1 \\ -3 & 2 \end{pmatrix}$$

formed by taking as its columns the eigenvectors

$$\mathbf{s}_1 = \begin{pmatrix} 2 \\ -3 \end{pmatrix} \quad \text{and} \quad \mathbf{s}_2 = \begin{pmatrix} 1 \\ 2 \end{pmatrix}$$

(these eigenvectors being chosen for their simplicity, although any non-zero multiples of $\mathbf{s}_1$ and $\mathbf{s}_2$ would do just as well) has an important property. Since

$$\mathbf{A}\mathbf{s}_1 = \mathbf{s}_1 \quad \text{and} \quad \mathbf{A}\mathbf{s}_2 = 8\mathbf{s}_2$$

it follows that the product $\mathbf{AP}$ is a 2×2 matrix with $\mathbf{s}_1$ as its first column and $8\mathbf{s}_2$ as its second column. Thus

$$\mathbf{AP} = \mathbf{P}\begin{pmatrix} 1 & 0 \\ 0 & 8 \end{pmatrix}.$$

But, since $\det \mathbf{P} = 7$, $\mathbf{P}$ is non-singular and we may multiply both sides of the above equation on the left by $\mathbf{P}^{-1}$ to obtain

$$\mathbf{P}^{-1}\mathbf{AP} = \begin{pmatrix} 1 & 0 \\ 0 & 8 \end{pmatrix}.$$

The matrix $\mathbf{A}$ has been *reduced to diagonal-form.* Notice that the elements of the diagonal are precisely the eigenvalues of $\mathbf{A}$.

The linear transformation T, with matrix $\mathbf{A}$ relative to $\mathbf{i}, \mathbf{j}$ as base vectors, maps the point with position vector $\alpha\mathbf{s}_1 + \beta\mathbf{s}_2$ where $\mathbf{s}_1, \mathbf{s}_2$ are defined above, into the point with position vector $\alpha\mathbf{s}_1 + 8\beta\mathbf{s}_2$; it follows that, if we take $\mathbf{s}_1, \mathbf{s}_2$ as base vectors, the matrix of T is

$$\begin{pmatrix} 1 & 0 \\ 0 & 8 \end{pmatrix};$$

for

$$\begin{pmatrix} 1 & 0 \\ 0 & 8 \end{pmatrix}\begin{pmatrix} \alpha \\ \beta \end{pmatrix} = \begin{pmatrix} \alpha \\ 8\beta \end{pmatrix}$$

as required.

Thus, when considering the transformation T, it is simpler to express vectors in terms of the eigenvectors $\mathbf{s}_1, \mathbf{s}_2$ rather than in terms of the more usual base vectors $\mathbf{i}, \mathbf{j}$. For example, the point Q, where

$$\mathbf{q} = 2\mathbf{s}_1 + \tfrac{1}{2}\mathbf{s}_2,$$

maps into the point Q', where

$$\mathbf{q}' = 2\mathbf{s}_1 + 4\mathbf{s}_2.$$

Figure 28.1 shows the effect of the transformation T upon Q:

$$\mathbf{OQ} = 2\mathbf{s}_1 + \tfrac{1}{2}\mathbf{s}_2 = \mathbf{OP} + \mathbf{PQ},$$

$$\mathbf{OQ}' = 2\mathbf{s}_1 + 4\mathbf{s}_2 = \mathbf{OP}' + \mathbf{P}'\mathbf{Q}',$$

where $OP'Q'P$ is a parallelogram.

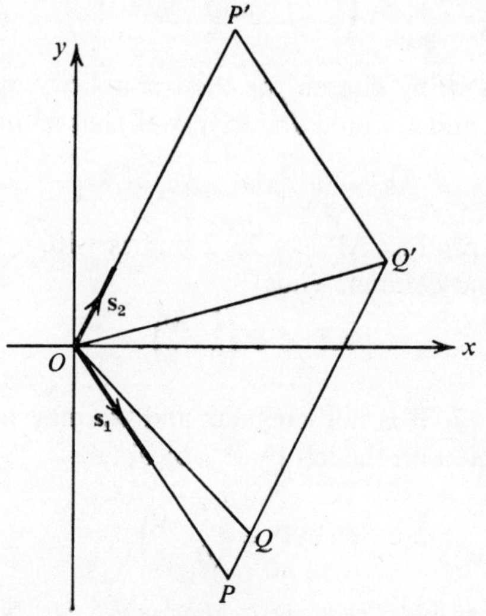

Fig. 28.1

Ex. 4. Calculate the eigenvalues and corresponding unit eigenvectors of the matrix

$$\mathbf{B} = \begin{pmatrix} 0 & -3 \\ -2 & 1 \end{pmatrix}.$$

Hence describe geometrically the linear transformation T which has matrix $\mathbf{B}$ relative to the base vectors $\mathbf{i}, \mathbf{j}$.

Reverting to Example 1, the reduction to diagonal form enables us to calculate powers of $\mathbf{A}$. For example,

$$\begin{aligned}\begin{pmatrix} 1 & 0 \\ 0 & 8^3 \end{pmatrix} = (\mathbf{P}^{-1}\mathbf{A}\mathbf{P})^3 &= (\mathbf{P}^{-1}\mathbf{A}\mathbf{P})(\mathbf{P}^{-1}\mathbf{A}\mathbf{P})(\mathbf{P}^{-1}\mathbf{A}\mathbf{P}) \\ &= \mathbf{P}^{-1}\mathbf{A}(\mathbf{P}\mathbf{P}^{-1})\mathbf{A}(\mathbf{P}\mathbf{P}^{-1})\mathbf{A}\mathbf{P} \\ &= \mathbf{P}^{-1}\mathbf{A}^3\mathbf{P}\end{aligned}$$

$$\begin{aligned}\therefore \quad \mathbf{A}^3 &= \mathbf{P}\begin{pmatrix} 1 & 0 \\ 0 & 8^3 \end{pmatrix}\mathbf{P}^{-1} \\ &= \frac{1}{7}\begin{pmatrix} 2 & 1 \\ -3 & 2 \end{pmatrix}\begin{pmatrix} 1 & 0 \\ 0 & 512 \end{pmatrix}\begin{pmatrix} 2 & -1 \\ 3 & 2 \end{pmatrix} \\ &= \frac{1}{7}\begin{pmatrix} 2 & 512 \\ -3 & 1024 \end{pmatrix}\begin{pmatrix} 2 & -1 \\ 3 & 2 \end{pmatrix} \\ &= \frac{1}{7}\begin{pmatrix} 1540 & 1022 \\ 3066 & 2051 \end{pmatrix} \\ &= \begin{pmatrix} 220 & 146 \\ 438 & 293 \end{pmatrix}.\end{aligned}$$

More generally we have

$$\begin{aligned}\mathbf{A}^n &= \frac{1}{7}\begin{pmatrix} 2 & 1 \\ -3 & 2 \end{pmatrix}\begin{pmatrix} 1 & 0 \\ 0 & 8^n \end{pmatrix}\begin{pmatrix} 2 & -1 \\ 3 & 2 \end{pmatrix} \\ &= \frac{1}{7}\begin{pmatrix} 4+3.8^n & -2+2.8^n \\ -6+6.8^n & 3+4.8^n \end{pmatrix}.\end{aligned}$$

Ex. 5. Verify by direct calculation the form given for $\mathbf{A}^n$ above and also prove the result by mathematical induction.

**Ex.* 6. If $\mathbf{s}_1$ and $\mathbf{s}_2$ are eigenvectors corresponding to the distinct real eigenvalues λ_1, λ_2 of a matrix $\mathbf{A}$ and if $\mathbf{P}$ is the matrix with first column $\mathbf{s}_1$ and second column $\mathbf{s}_2$, and assuming that $\mathbf{P}$ is non-singular, show that

$$\mathbf{P}^{-1}\mathbf{A}\mathbf{P} = \begin{pmatrix} \lambda_1 & 0 \\ 0 & \lambda_2 \end{pmatrix}.$$

Eigenvalues need not be real. For example, if

$$\mathbf{A} = \begin{pmatrix} 1 & 1 \\ -1 & 1 \end{pmatrix}$$

the characteristic equation of **A** is

$$(\lambda-1)^2+1 = 0$$

giving eigenvalues $\lambda_1 = 1+\mathrm{j}$, $\lambda_2 = 1-\mathrm{j}$, with corresponding eigenvectors

$$\begin{pmatrix}1\\ \mathrm{j}\end{pmatrix},\ \begin{pmatrix}1\\ -\mathrm{j}\end{pmatrix}.$$

In this case, the invariant lines are imaginary.

Of the linear transformations of the plane into itself, those which preserve distance are of particular importance—that is, linear transformations T with the property that, if $T(P) = P_T$,† then $OP = OP_T$ for all points P of the plane. In Theorem 28.1 we shall show that the matrix **A** of such a transformation is of a special type; to this end we make the following definition (which holds for $n \times n$ matrices, although we are considering here only 2×2 matrices): **A** is called an *orthogonal matrix* if $\mathbf{AA}' = \mathbf{I}$.

Ex.* 7. Show that, if **A is an orthogonal matrix then $\det \mathbf{A} = \pm 1$, **A** is non-singular, and $\mathbf{A}^{-1} = \mathbf{A}'$.

Ex. 8. In the course of the proof of Theorem 28.1 we shall require the fact that the transpose of a product **AB** of two matrices is the product of the transposes of **A** and **B** in reverse order; that is

$$(\mathbf{AB})' = \mathbf{B}'\mathbf{A}'.$$

Prove this result

(i) when **A** is 2×2 and **B** is 2×1;

(ii) when **A** is 2×2 and **B** is 2×2.

Show how your proofs may be generalized to cover all cases of conformable matrices **A**, **B** up to order 3×3 (or $m\times n$ if you can manage it).

Theorem 28.1. *Let T be a linear transformation of the plane into itself, with matrix* **A**. *Then T is distance preserving* (*that is,* $|\mathbf{OP}| = |\mathbf{OP}_T|$ *for all positions of the point P*) *if and only if* **A** *is orthogonal.*

First observe that, if $\mathbf{OP} = \mathbf{r} = x\mathbf{i}+y\mathbf{j}$, then

$$\mathbf{r}'\mathbf{r} = (xy)\begin{pmatrix}x\\ y\end{pmatrix} = (x^2+y^2);$$

that is, the 1×1 matrix $\mathbf{r}'\mathbf{r}$ represents OP^2.

Now suppose that P is any point of the plane, that $\mathbf{OP} = \mathbf{r}$, and that

$$\mathbf{OP}_T = \mathbf{r}_T.$$

Then $$\mathbf{r}_T'\mathbf{r}_T = (\mathbf{Ar})'(\mathbf{Ar}) = \mathbf{r}'\mathbf{A}'\mathbf{Ar} \quad \text{(by Ex. 8)}$$

and thus

$$\mathbf{r}_T'\mathbf{r}_T = \mathbf{r}'\mathbf{r}$$

$$\Leftrightarrow \quad \mathbf{r}'(\mathbf{A}'\mathbf{A})\,\mathbf{r} = \mathbf{r}'\mathbf{r}$$

$$\Leftrightarrow \quad \mathbf{r}'(\mathbf{A}'\mathbf{A})\,\mathbf{r} = \mathbf{rIr}$$

$$\Leftrightarrow \mathbf{r}'(\mathbf{A}'\mathbf{A}-\mathbf{I})\,\mathbf{r} = \mathbf{0}.$$

† We use the notation P_T rather than P' here to avoid possible confusion with the notation for the transposed matrix.

(i) *Suppose* A *is orthgonal.*

Then $\mathbf{A'A} = \mathbf{I}$ and thus $\mathbf{r}'_T\mathbf{r}_T = \mathbf{r'r}$ for all points P of the plane and T is distance preserving.

(ii) *Suppose T is distance preserving*

Then $\mathbf{r}'(\mathbf{A'A}-\mathbf{I})\,\mathbf{r} = \mathbf{0}$ for all position vectors $\mathbf{r}$ and, in particular, for $\mathbf{r}_1 = \mathbf{i}$, $\mathbf{r}_2 = \mathbf{j}$ and $\mathbf{r}_3 = \mathbf{i}+\mathbf{j}$. Writing

$$\mathbf{A'A}-\mathbf{I} = \begin{pmatrix} b_1 & b_2 \\ b_3 & b_4 \end{pmatrix} = \mathbf{B}$$

we thus have

$$(1 \quad 0)\begin{pmatrix} b_1 & b_2 \\ b_3 & b_4 \end{pmatrix}\begin{pmatrix} 1 \\ 0 \end{pmatrix} = \begin{pmatrix} 0 \\ 0 \end{pmatrix} \Rightarrow b_1 = 0;$$

$$(0 \quad 1)\begin{pmatrix} b_1 & b_2 \\ b_2 & b_4 \end{pmatrix}\begin{pmatrix} 0 \\ 1 \end{pmatrix} = \begin{pmatrix} 0 \\ 0 \end{pmatrix} \Rightarrow b_4 = 0;$$

$$(1 \quad 1)\begin{pmatrix} b_1 & b_2 \\ b_3 & b_4 \end{pmatrix}\begin{pmatrix} 1 \\ 1 \end{pmatrix} = \begin{pmatrix} 0 \\ 0 \end{pmatrix} \Rightarrow b_1+b_2+b_3+b_4 = 0.$$

Thus $b_1 = b_4 = 0$ and $b_2 = -b_3 = b$ say, and we have

$$\mathbf{B} = \begin{pmatrix} 0 & b \\ -b & 0 \end{pmatrix} \quad \text{and} \quad \mathbf{A'A} = \begin{pmatrix} 1 & b \\ -b & 1 \end{pmatrix}.$$

But $(\mathbf{A'A})' = \mathbf{A'A}$ and therefore

$$\begin{pmatrix} 1 & b \\ -b & 1 \end{pmatrix} = \begin{pmatrix} 1 & -b \\ b & 1 \end{pmatrix}$$

and $b = 0$. This shows that $\mathbf{A'A} = \mathbf{I}$ and thus $\mathbf{A}$ is orthogonal.

**Ex.* 9. Prove that an orthogonal matrix $\mathbf{P}$ preserves separations, in the sense that

$$|\mathbf{Px}-\mathbf{Py}| = |\mathbf{x}-\mathbf{y}|.$$

**Ex.* 10. Prove that, if $\mathbf{P}$ is a 2×2 orthogonal matrix, then either

$$\mathbf{P} = \begin{pmatrix} \cos\theta & -\sin\theta \\ \sin\theta & \cos\theta \end{pmatrix}$$

corresponding to a rotation through an angle θ, or

$$\mathbf{P} = \begin{pmatrix} \cos\theta & \sin\theta \\ \sin\theta & -\cos\theta \end{pmatrix}$$

corresponding to a rotation through $-\theta$ followed by a reflection in the x axis (that is, a reflection in the axis $y = x\tan\frac{1}{2}\theta$).

**Ex.* 11. Prove that the eigenvalues of a 2×2 orthogonal matrix $\mathbf{P}$ are
 (i) conjugate complex numbers of modulus 1 if $\det\mathbf{P} = +1$, or
 (ii) the numbers ± 1 if $\det\mathbf{P} = -1$.

The reduction of a matrix $\mathbf{A}$ to diagonal form $\mathbf{P}^{-1}\mathbf{AP}$ is of particular importance if $\mathbf{A}$ is a symmetric matrix ($\mathbf{A}' = \mathbf{A}$). We shall show (Theorem 28.5) that, in this case, the reducing matrix $\mathbf{P}$ may be taken to be orthogonal. Before proving this, it is necessary to obtain some simple preliminary results. In the following four theorems, $\mathbf{A}$ is a 2×2 matrix with eigenvalues λ_1, λ_2 and corresponding eigenvectors $\mathbf{s}_1$, $\mathbf{s}_2$ and $\mathbf{P}$ is the matrix with first column $\mathbf{s}_1$ and second column $\mathbf{s}_2$.

Theorem 28.2. *If the eigenvalues of the* 2×2 *matrix* $\mathbf{A}$ *are distinct then the matrix* $\mathbf{P}$ *is non-singular.*

Proof.

$$\begin{aligned} & \det \mathbf{P} = 0 & \\ \Rightarrow\quad & \mathbf{s}_1 = k\mathbf{s}_2, & \text{where } k \text{ is a non-zero number,} \\ \Rightarrow\quad & \mathbf{As}_1 - k\mathbf{As}_2 = \mathbf{0}, & \text{on multiplying by } \mathbf{A}, \\ \Rightarrow\quad & \lambda_1\mathbf{s}_1 - \lambda k\mathbf{s}_2 = \mathbf{0} & \\ \Rightarrow\quad & (\lambda_1-\lambda_2)\,\mathbf{s}_1 = \mathbf{0}, & \text{since } k\mathbf{s}_2 = \mathbf{s}_1, \\ \Rightarrow\quad & \lambda_1 = \lambda_2, & \text{since } \mathbf{s}_1 \neq \mathbf{0}. \end{aligned}$$

Thus, since $\det \mathbf{P} = 0 \Rightarrow \lambda_1 = \lambda_2$, we have $\lambda_1 \neq \lambda_2 \Rightarrow \det \mathbf{P} \neq 0$.

Theorem 28.3. *If* $\mathbf{s}_1$ *and* $\mathbf{s}_2$ *are unit eigenvectors of* $\mathbf{A}$ *then* $\mathbf{s}_1$ *and* $\mathbf{s}_2$ *are perpendicular if and only if* $\mathbf{P}$ *is orthogonal.*

Proof.

Write
$$\mathbf{s}_1 = \begin{pmatrix} x_1 \\ y_1 \end{pmatrix}, \quad \mathbf{s}_2 = \begin{pmatrix} x_2 \\ y_2 \end{pmatrix};$$
then
$$\mathbf{P} = \begin{pmatrix} x_1 & x_2 \\ y_1 & y_2 \end{pmatrix}$$
and
$$\mathbf{P}'\mathbf{P} = \begin{pmatrix} x_1 & y_1 \\ x_2 & y_2 \end{pmatrix}\begin{pmatrix} x_1 & x_2 \\ y_1 & y_2 \end{pmatrix} = \begin{pmatrix} x_1^2+y_1^2 & x_1x_2+y_1y_2 \\ x_1x_2+y_1y_2 & x_2^2+y_2^2 \end{pmatrix}.$$

(i) Suppose $\mathbf{P}$ is orthogonal then
$$\begin{pmatrix} x_1^2+y_1^2 & x_1x_2+y_1y_2 \\ x_1x_2+y_1y_2 & x_2^2+y_2^2 \end{pmatrix} = \begin{pmatrix} 1 & 0 \\ 0 & 1 \end{pmatrix}$$
giving $x_1^2+y_1^2 = x_2^2+y_2^2 = 1$, $x_1x_2+y_1y_2 = 0$ and thus $\mathbf{s}_1$ and $\mathbf{s}_2$ are perpendicular unit vectors.

(ii) If $\mathbf{s}_1$ and $\mathbf{s}_2$ are perpendicular unit vectors then
$$\begin{aligned} \mathbf{P}'\mathbf{P} &= \begin{pmatrix} x_1^2+y_1^2 & x_1x_2+y_1y_2 \\ x_1x_2+y_1y_2 & x_2^2+y_2^2 \end{pmatrix} \\ &= \begin{pmatrix} 1 & 0 \\ 0 & 1 \end{pmatrix} \end{aligned}$$
and $\mathbf{P}$ is orthogonal.

Theorem 28.4. *If* **A** *is a* 2×2 *real symmetric matrix* $\neq k\mathbf{I}$ *then its eigenvalues are real and unequal.*

Proof. Write

$$\mathbf{A} = \begin{pmatrix} a & b \\ b & c \end{pmatrix}.$$

Then

$$\det(\mathbf{A}-\lambda\mathbf{I}) = 0,$$

$$\Leftrightarrow \quad (a-\lambda)(c-\lambda)-b^2 = 0$$

$$\Leftrightarrow \lambda^2-(a+c)\lambda+(ac-b^2) = 0.$$

But the discriminant of this quadratic equation is

$$\begin{aligned} &(a+c)^2-4(ac-b^2) \\ &= (a-c)^2+4b^2 \\ &> 0, \quad \text{since, if } b = 0,\ a \neq c. \end{aligned}$$

Thus, λ_1 and λ_2 are real and distinct.

Theorem 28.5. *The reduction to diagonal form* $\mathbf{P}^{-1}\mathbf{AP}$ *for a symmetric matrix* $\mathbf{A} \neq k\mathbf{I}$ *may always be effected, and* **P** *may be taken to be orthogonal.*

Proof. Notice first that, since $\mathbf{s}_1$ and $\mathbf{s}_2$ are non-zero,

$$\mathbf{s}_1 \text{ perpendicular to } \mathbf{s}_2 \Leftrightarrow \mathbf{s}_1'\mathbf{s}_2 = \mathbf{0}.$$

Now suppose that $\mathbf{s}_1$ and $\mathbf{s}_2$ are unit eigenvectors of **A**.

$$\begin{aligned} & \mathbf{As}_1 = \lambda_1\mathbf{s}_1 \\ \Rightarrow \quad & \mathbf{s}_2'\mathbf{As}_1 = \lambda_1\mathbf{s}_2'\mathbf{s}_1, \quad \text{on premultiplying by } \mathbf{s}_2', \\ \Leftrightarrow \quad & (\mathbf{s}_2'\mathbf{As}_1)' = \lambda_1(\mathbf{s}_1'\mathbf{s}_2), \quad \text{on taking the transpose of each side,} \\ \Leftrightarrow \quad & \mathbf{s}_1'\mathbf{As}_2 = \lambda_1\mathbf{s}_1'\mathbf{s}_2, \quad \text{by Ex. 8 and noting that } \mathbf{A}' = \mathbf{A}, \\ \Leftrightarrow \quad & \lambda_2\mathbf{s}_1'\mathbf{s}_2 = \lambda_1\mathbf{s}_1'\mathbf{s}_2, \quad \text{since } \mathbf{As}_2 = \lambda_2\mathbf{s}_2, \\ \Leftrightarrow \quad & (\lambda_1-\lambda_1)\,\mathbf{s}_1'\mathbf{s}_2 = \mathbf{0} \\ \Leftrightarrow \quad & \mathbf{s}_1'\mathbf{s}_2 = \mathbf{0}, \quad \text{since } \lambda_1 \neq \lambda_2 \text{ by Theorem 28.4.} \end{aligned}$$

Thus, $\mathbf{s}_1$ and $\mathbf{s}_2$ are perpendicular and, by Theorem 28.3, **P** is orthogonal. Furthermore, by Theorem 28.2, $\mathbf{P}^{-1}$ exists and the theorem is complete. (Notice that orthogonal matrices are always non-singular.)

Ex. 12. If $\mathbf{A} = \begin{pmatrix} 1 & 3 \\ 3 & 1 \end{pmatrix}$, find an orthogonal matrix **P** such that

$$\mathbf{P}'\mathbf{AP} = \begin{pmatrix} 4 & 0 \\ 0 & -2 \end{pmatrix}$$

and interpret **P**, **P**′ as matrices of a rotation transformation.

Ex. 13. $\mathbf{A}$ is a symmetric matrix corresponding to the linear transformation T. By writing

$$\mathbf{A} = \mathbf{PDP}',$$

where $\mathbf{P}$ is orthogonal and $\mathbf{D}$ is diagonal, give a geometrical interpretation of T. Illustrate your answer with the particular matrix $\mathbf{A}$ of Ex. 12.

Exercise 28(a)

1. Find the eigenvalues of the matrix

$$\mathbf{A} = \begin{pmatrix} 5 & 2 \\ 9 & 8 \end{pmatrix}$$

and the corresponding unit eigenvectors. What lines in the plane map into themselves under the linear transformation T with matrix $\mathbf{A}$?

2. Find the eigenvalues of the matrix

$$\mathbf{A} = \begin{pmatrix} 2 & 3 \\ 8 & 0 \end{pmatrix}$$

and the corresponding unit eigenvectors. What lines in the plane map into themselves under the linear transformation T with matrix $\mathbf{A}$?

3. Find the eigenvalues of the symmetric matrix

$$\mathbf{A} = \begin{pmatrix} 23 & 36 \\ 36 & 2 \end{pmatrix}$$

and hence find an orthogonal matrix $\mathbf{P}$ such that $\mathbf{P}^{-1}\mathbf{AP}$ is diagonal.

4. Find an orthogonal matrix $\mathbf{P}$ such that the matrix $\mathbf{P}^{-1}\mathbf{AP}$ is diagonal, where $\mathbf{A}$ is the symmetric matrix

$$\begin{pmatrix} 5 & -1 \\ -1 & 5 \end{pmatrix}.$$

Give a geometrical interpretation of the linear transformation T with matrix $\mathbf{A}$.

5. Give a geometrical description of the linear transformation T with matrix $\mathbf{A}$, where

$$\mathbf{A} = \begin{pmatrix} 2 & 2 \\ 1 & 3 \end{pmatrix}.$$

6. Give a geometrical description of the linear transformation T with matrix $\mathbf{A}$, where

$$\mathbf{A} = \begin{pmatrix} 3 & 4 \\ 2 & -4 \end{pmatrix}.$$

7. The 2×2 matrix $\mathbf{A}$ has the property that

$$\mathbf{A}^2 + \mathbf{I} = \mathbf{0}.$$

Prove that the eigenvalues of $\mathbf{A}$ are $\pm \mathrm{j}$.

8. A matrix $\mathbf{A}$ which has the property

$$\mathbf{A}^2 = \mathbf{A}$$

is said to be *idempotent*.

If $\mathbf{A}$ is a 2×2 idempotent matrix, prove that its eigenvalues are either 0 or 1 and that, if $\mathbf{A} \neq \mathbf{I}$, then $\mathbf{A}$ is singular.

Show that the matrix

$$\mathbf{A} = \begin{pmatrix} 3 & -2 \\ 3 & -2 \end{pmatrix}$$

is idempotent and give a geometric interpretation of the linear transformation for which $\mathbf{A}$ is the matrix.

9. Find $\mathbf{A}^n$ in the following cases:

(i) $\mathbf{A} = \begin{pmatrix} -3 & 10 \\ -3 & 8 \end{pmatrix}$; (ii) $\mathbf{A} = \begin{pmatrix} -13 & 6 \\ -35 & 16 \end{pmatrix}$;

(iii) $\mathbf{A} = \begin{pmatrix} -10 & 18 \\ -6 & 11 \end{pmatrix}$; (iv) $\mathbf{A} = \begin{pmatrix} -1 & 2 \\ -10 & 8 \end{pmatrix}$.

10. $\mathbf{A}$ is a 2×2 matrix and the eigenvalues of the matrix $\mathbf{A}-\mathbf{I}$ are $\pm \mathrm{j}$. Prove that $\det \mathbf{A} = 2$.

Is the converse of this result true?

11. a, b, c, d are unequal non-negative real numbers such that

$$a+b = c+d = 1.$$

Prove that the eigenvalues of the matrix

$$\begin{pmatrix} a & b \\ c & d \end{pmatrix}$$

are 1 and λ, where $0 < |\lambda| < 1$.

Find λ for the matrix

$$\mathbf{A} = \begin{pmatrix} \frac{1}{4} & \frac{3}{4} \\ \frac{1}{3} & \frac{2}{3} \end{pmatrix}$$

and describe geometrically the linear transformation with matrix $\mathbf{A}^n$ where n is a large positive integer.

12. Find the eigenvalues of the matrix

$$\mathbf{A} = \begin{pmatrix} 23 & -33 \\ 14 & -20 \end{pmatrix}$$

and deduce that

$$\mathbf{A}^n = \begin{pmatrix} 11.2^{n+1}-21 & -33.2^n+33 \\ 7.2^{n+1}-14 & -21.2^n+22 \end{pmatrix}.$$

If

$$\mathbf{B} = \mathbf{A}^n+\mathbf{A}^{n-1}+\mathbf{A}^{n-2}+\ldots+\mathbf{A}+\mathbf{I}$$

find an explicit form for $\mathbf{B}$.

What are the eigenvalues of $\mathbf{B}$?

2. EIGENVALUES AND EIGENVECTORS FOR 3×3 MATRICES

The results we have proved for eigenvalues and eigenvectors of 2×2 matrices hold in large part for 3×3 matrices too. As before, we define the *eigenvalues* of the 3×3 matrix $\mathbf{A}$ to be the roots $\lambda_1, \lambda_2, \lambda_3$ of the characteristic equation

$$\det(\mathbf{A}-\lambda\mathbf{I}) = 0,$$

and, corresponding to each λ_i, any non-zero vector $\mathbf{s}_i$ such that

$$\mathbf{A}\mathbf{s}_i = \lambda_i \mathbf{s}_i$$

is called an *eigenvector* of $\mathbf{A}$.

Geometrically, each real λ_i gives rise to a real $\mathbf{s}_i$ which defines a line through the origin, the points of which all map into points of the same line under the linear transformation for which $\mathbf{A}$ is the matrix. In contrast to the two-dimensional case, since a real cubic must possess at least one real root, there must be at least one such line.

Ex. 14. Show that, if $T(S) = S$ where S is not the origin, then T maps each point of the line OS into itself.

Ex. 15. U, V, W are distinct non-collinear points in space with $T(U) = U$, $T(V) = V$, $T(W) = W$. Prove that, if the plane UVW does not contain the origin, then T is the identity transformation.

As in two dimensions, if the 3×3 matrix $\mathbf{A}$ has three distinct eigenvalues $\lambda_1, \lambda_2, \lambda_3$ with corresponding eigenvectors $\mathbf{s}_1, \mathbf{s}_2, \mathbf{s}_3$ and if the matrix $\mathbf{P}$, with first column $\mathbf{s}_1$, second column $\mathbf{s}_2$, and third column $\mathbf{s}_3$ is non-singular, then $\mathbf{P}^{-1}\mathbf{A}\mathbf{P}$ is a diagonal matrix. For

$$\mathbf{A}\mathbf{s}_1 = \lambda_1\mathbf{s}_1,\quad \mathbf{A}\mathbf{s}_2 = \lambda_2\mathbf{s}_2,\quad \mathbf{A}\mathbf{s}_3 = \lambda_3\mathbf{s}_3$$

$$\Rightarrow\quad \mathbf{A}\mathbf{P} = \mathbf{P}\begin{pmatrix}\lambda_1 & 0 & 0\\ 0 & \lambda_2 & 0\\ 0 & 0 & \lambda_3\end{pmatrix}$$

$$\Rightarrow\quad \mathbf{P}^{-1}\mathbf{A}\mathbf{P} = \begin{pmatrix}\lambda_1 & 0 & 0\\ 0 & \lambda_2 & 0\\ 0 & 0 & \lambda_3\end{pmatrix},\quad \text{since } \mathbf{P} \text{ is non-singular.}$$

If such a diagonalizing procedure exists, $\mathbf{A}$ is said to be *reducible to a diagonal matrix.* Not all 3×3 matrices are reducible to a diagonal matrix; the discussion of necessary conditions upon $\mathbf{A}$ for it to be so and the possibility of reducing a matrix $\mathbf{A}$ which does not satisfy these conditions to a form (the Jordan canonical form) which approximates to a diagonal matrix is beyond the scope of this book: the interested reader should consult one of the more advanced algebra texts mentioned in the bibliography in Volume 1. (See, however, Exercise 28(*b*), Question 25.)

**Ex.* 16. Given two 3×3 matrices $\mathbf{A}$ and $\mathbf{B}$, if a non-singular matrix $\mathbf{P}$ exists such that

$$\mathbf{B} = \mathbf{P}^{-1}\mathbf{A}\mathbf{P},$$

$\mathbf{B}$ is said to be *similar* to $\mathbf{A}$.

Prove that:
(i) **A** is similar to **A**;
(ii) if **B** is similar to **A**, then **A** is similar to **B**;
(iii) if **B** is similar to **A** and **C** is similar to **B**, then **C** is similar to **A**.

Example 2. *Find the eigenvalues and corresponding unit eigenvectors of the matrix*

$$\mathbf{A} = \begin{pmatrix} 1 & 0 & 0 \\ 10 & -7 & 10 \\ 7 & -5 & 8 \end{pmatrix}$$

and interpret the results geometrically.

Find a matrix **P** *such that* $\mathbf{P}^{-1}\mathbf{AP}$ *is diagonal.*

The characteristic equation of **A** is

$$\begin{vmatrix} 1-\lambda & 0 & 0 \\ 10 & -7-\lambda & 10 \\ 7 & -5 & 8-\lambda \end{vmatrix} = 0;$$

that is
$$(1-\lambda)(\lambda^2-\lambda-6) = 0,$$

giving the three eigenvalues $\lambda = 1$, $\lambda = -2$, $\lambda = 3$.

(i) $\lambda = 1$

The equations
$$10x-8y+10z = 0,$$
$$7x-5y+7z = 0,$$

have solutions $x = 1$, $y = 0$, $z = -1$ and the corresponding unit eigenvector is

$$\mathbf{e}_1 = \begin{pmatrix} 1/\sqrt{2} \\ 0 \\ -1/\sqrt{2} \end{pmatrix}.$$

All points of the line
$$\frac{x}{1} = \frac{y}{0} = \frac{z}{-1}$$

map into points of the same line (in fact, in this case, into themselves, since $\lambda = 1$).

(ii) $\lambda = -2$

The equations
$$3x \qquad = 0,$$
$$10x-5y+10z = 0,$$
$$7x-5y+10z = 0,$$

have solution $x = 0, y = 2, z = 1$ and the corresponding unit eigenvector is

$$\mathbf{e}_2 = \begin{pmatrix} 0 \\ 2/\sqrt{5} \\ 1/\sqrt{5} \end{pmatrix}.$$

All points of the line

$$\frac{x}{0} = \frac{y}{2} = \frac{z}{1}$$

map into points of the same line (but not, apart from the origin, into themselves).

(iii) $\lambda = 3$

The equations

$$\begin{aligned} -2x \qquad\qquad &= 0, \\ 10x - 10y + 10z &= 0, \\ 7x - \ 5y + \ 5z &= 0, \end{aligned}$$

have solution $x = 0, y = 1, z = 1$ and the corresponding unit eigenvector is

$$\mathbf{e}_3 = \begin{pmatrix} 0 \\ 1/\sqrt{2} \\ 1/\sqrt{2} \end{pmatrix}.$$

All points of the line

$$\frac{x}{0} = \frac{y}{1} = \frac{z}{1}$$

map into points of the same line.

The matrix

$$\mathbf{P} = \begin{pmatrix} 1 & 0 & 0 \\ 0 & 2 & 1 \\ -1 & 1 & 1 \end{pmatrix}$$

has the required property, that

$$\mathbf{P}^{-1}\mathbf{A}\mathbf{P} = \begin{pmatrix} 1 & 0 & 0 \\ 0 & -2 & 0 \\ 0 & 0 & 3 \end{pmatrix}.$$

Exactly as in the two-dimensional case (see Example 1) the geometrical interpretation of the linear transformation T, which has matrix $\mathbf{A}$, relative to $\mathbf{i}, \mathbf{j}, \mathbf{k}$ as base vectors, is facilitated by expressing the position vector of a point in terms of $\mathbf{s}_1$, $\mathbf{s}_2$ and $\mathbf{s}_3$. For the point $\mathbf{Q}$, with position vector

$$\mathbf{q} = \alpha\mathbf{s}_1 + \beta\mathbf{s}_2 + \gamma\mathbf{s}_3,$$

maps into the point $\mathbf{Q}'$ with position vector

$$\mathbf{q}' = \alpha\mathbf{s}_1 - 2\beta\mathbf{s}_2 + 3\gamma\mathbf{s}_3.$$

Since the coefficient of $\mathbf{s}_1$ is unchanged, the displacement $\mathbf{QQ}'$ is seen to be parallel to the plane determined by $\mathbf{s}_2$ and $\mathbf{s}_3$.

Ex. 17. Verify that, in the notation of Example 2, $\mathbf{P}^{-1}\mathbf{AP}$ is a diagonal matrix.

Pursuing the analogy with 2×2 matrices, we define a 3×3 *orthogonal matrix* $\mathbf{P}$ as a matrix which has the property that $\mathbf{PP}' = \mathbf{I}$.

**Ex.* 18. If $\mathbf{P}$ is a 3×3 orthogonal matrix, prove that $\mathbf{P}$ is non-singular, $\det \mathbf{P} = \pm 1$ and $\mathbf{P}^{-1} = \mathbf{P}'$.

**Ex.* 19. Follow through the steps of the proof of Theorem 28.1 to show that a 3×3 matrix is distance preserving if and only if it is orthogonal.

Theorem 28.6. *If* $\mathbf{P}$ *is a* 3×3 *orthogonal matrix with det* $\mathbf{P} = +1$, *then* 1 *is an eigenvalue of* $\mathbf{P}$.

Proof.

Since
$$(\mathbf{P}-\mathbf{I})\,\mathbf{P}' = \mathbf{I}-\mathbf{P}' = -(\mathbf{P}-\mathbf{I})',$$
we have
$$\det(\mathbf{P}-\mathbf{I})\det\mathbf{P}' = \det(-(\mathbf{P}-\mathbf{I})').$$
But $\det\mathbf{P}' = \det\mathbf{P} = 1$, and also
$$\det(-(\mathbf{P}-\mathbf{I})') = \det(-(\mathbf{P}-\mathbf{I})) = -\det(\mathbf{P}-\mathbf{I})$$
(since $\mathbf{P}-\mathbf{I}$ is a 3×3 matrix)

and thus
$$\det(\mathbf{P}-\mathbf{I}) = -\det(\mathbf{P}-\mathbf{I})$$
$$\Rightarrow \det(\mathbf{P}-\mathbf{I}) = 0$$
$$\Rightarrow 1 \text{ is an eigenvalue of } \mathbf{P}.$$

Corollary. If $\mathbf{P}$ *is a* 3×3 *orthogonal matrix with* $\det\mathbf{P} = -1$, *then* -1 *is an eigenvalue of* $\mathbf{P}$.

This follows immediately on writing $-\mathbf{P}$ for $\mathbf{P}$, since $\det(-\mathbf{P}) = +1$.

**Ex.* 20. In the case $\det\mathbf{P} = +1$ ($\mathbf{P}$ orthogonal), let $\mathbf{s}$ be an eigenvector corresponding to the eigenvalue $\lambda = 1$. The linear transformation T of which P is the matrix:

(i) preserves distances between all pairs of points;
(ii) leaves all points of the line OS fixed.

Show that T may be interpreted geometrically as a rotation about the axis OS.

**Ex.* 21. Show that, if $\det\mathbf{P} = -1$ ($\mathbf{P}$ orthogonal), then the corresponding linear transformation may be interpreted as a rotation about the axis defined by the eigenvalue $\lambda = -1$ followed by a reflection in the origin.

Ex. 22. Show that the matrix

$$\mathbf{P} = \begin{pmatrix} \cos\theta & -\sin\theta & 0 \\ \sin\theta & \cos\theta & 0 \\ 0 & 0 & 1 \end{pmatrix}$$

is orthogonal and represents a rotation about the z axis through a positive angle θ.

Ex. 23. Show that the columns of a 3×3 orthogonal matrix $\mathbf{P}$ represent three mutually perpendicular unit vectors and interpret this result geometrically.

In summary, if $\mathbf{P}$ is a 3×3 orthogonal matrix representing the linear transformation T, then the columns of $\mathbf{P}$ are the components of three mutually perpendicular unit vectors and T represents either a pure rotation, or a rotation followed by a reflection in the origin.

Ex. 24. If

$$\mathbf{P} = \tfrac{1}{3}\begin{pmatrix} 1 & -2 & -2 \\ -2 & 1 & -2 \\ -2 & -2 & 1 \end{pmatrix},$$

prove that $\mathbf{P}$ is orthogonal and interpret $\mathbf{P}$ as the matrix of a linear transformation.

**Ex.* 25. If $\mathbf{u}$ is a unit vector, show that it is possible to construct an orthogonal matrix with $\mathbf{u}$ as its first column.

**Ex.* 26. If $\mathbf{P}$, $\mathbf{Q}$ are two 3×3 orthogonal matrices, prove that:
(i) $\mathbf{P}^{-1}$ is orthogonal; (ii) $\mathbf{PQ}$ is orthogonal.
Interpret both these results geometrically.

Ex. 27. Find a result corresponding to Theorem 28.6 for 4×4 orthogonal matrices. (Be careful when you take determinants!)

We saw earlier (Theorems 28.4 and 28.5) that the eigenvalues of a 2×2 symmetric matrix $\mathbf{A} \neq k\mathbf{I}$ were always real and distinct and this enabled us to construct an orthogonal matrix $\mathbf{P}$ such that $\mathbf{P}^{-1}\mathbf{AP}$ was diagonal. 3×3 symmetric matrices are not quite so amenable: although it can be shown that their eigenvalues are always real, they are not necessarily distinct. Nevertheless, the diagonalization process is still always possible as we shall now show.

Theorem 28.7. *Given any* 3×3 *symmetric matrix* $\mathbf{A}$ *it is always possible to find an orthogonal matrix* $\mathbf{P}$ *such that* $\mathbf{P}^{-1}\mathbf{AP}$ *is diagonal.*

Proof. We reduce the problem to the 2×2 case. First, if λ_1 is a real eigenvalue (such an eigenvalue must exist, since the characteristic equation is a cubic) with $\mathbf{s}_1$ the corresponding *unit* eigenvector, then $\mathbf{As}_1 = \lambda_1\mathbf{s}_1$ and thus

$$\mathbf{AP} = \mathbf{P}\begin{pmatrix} \lambda_1 & \alpha_1 & \alpha_2 \\ 0 & \beta_1 & \beta_2 \\ 0 & \gamma_1 & \gamma_2 \end{pmatrix},$$

where $\mathbf{P}$ is an orthogonal matrix with $\mathbf{s}_1$ as its first column and $\alpha_1 \dots \gamma_2$ are constants (see Ex. 25). Premultiplying by $\mathbf{P}^{-1}$ we then have

$$\mathbf{P}^{-1}\mathbf{AP} = \begin{pmatrix} \lambda_1 & \alpha_1 & \alpha_2 \\ 0 & \beta_1 & \beta_2 \\ 0 & \gamma_1 & \gamma_2 \end{pmatrix}.$$

But $(\mathbf{P}^{-1}\mathbf{AP})' = \mathbf{P}'\mathbf{A}'(\mathbf{P}^{-1})' = \mathbf{P}'\mathbf{A}(\mathbf{P}')' = \mathbf{P}^{-1}\mathbf{AP}$ (see Ex. 8), since $\mathbf{A}' = \mathbf{A}$ and $\mathbf{P}' = \mathbf{P}^{-1}$. It follows that $\mathbf{P}^{-1}\mathbf{AP}$ is a symmetric matrix:

$$\mathbf{P}^{-1}\mathbf{AP} = \begin{pmatrix} \lambda_1 & 0 & 0 \\ 0 & \beta_1 & \beta_2 \\ 0 & \beta_2 & \gamma_2 \end{pmatrix}.$$

Now, by Theorem 28.5, there exists an orthogonal matrix

$$\begin{pmatrix} q_1 & q_2 \\ r_1 & r_2 \end{pmatrix}$$

which reduces the symmetric matrix

$$\begin{pmatrix} \beta_1 & \beta_2 \\ \beta_2 & \gamma_2 \end{pmatrix}$$

to diagonal form. Thus the matrix

$$\mathbf{Q} = \begin{pmatrix} 1 & 0 & 0 \\ 0 & q_1 & q_2 \\ 0 & r_1 & r_2 \end{pmatrix}$$

reduces $\mathbf{P}^{-1}\mathbf{AP}$ to diagonal form:

$$\mathbf{Q}^{-1}\mathbf{P}^{-1}\mathbf{APQ} = \begin{pmatrix} \lambda_1 & 0 & 0 \\ 0 & \mu_2 & 0 \\ 0 & 0 & \mu_3 \end{pmatrix}.$$

Finally, we may write $\mathbf{PQ} = \mathbf{R}$ where, by Ex. 26(ii), $\mathbf{R}$ is orthogonal.

Ex. 28. Is the converse result true that, if $\mathbf{P}$ is orthogonal and $\mathbf{P}^{-1}\mathbf{AP}$ is diagonal, then $\mathbf{A}$ is a symmetric matrix?

The determination of an inverse matrix is generally a burdensome operation, but, if $\mathbf{P}$ is orthogonal, we have

$$\mathbf{P}^{-1} = \mathbf{P}'$$

and thus the diagonalization process for a symmetric matrix is particularly simple to effect.

3. THE CAYLEY–HAMILTON THEOREM AND POWERS OF MATRICES

A matrix $\mathbf{A}$ is said to *satisfy the equation*

$$a_0\lambda^n + a_1\lambda^{n-1} + a_2\lambda^{n-2} + \ldots + a_{n-1}\lambda + a_n = 0,$$

if

$$a_0\mathbf{A}^n + a_1\mathbf{A}^{n-1} + a_2\mathbf{A}^{n-2} + \ldots + a_{n-1}\mathbf{A} + a_n\mathbf{I} = \mathbf{0}.$$

Theorem 28.8. (*The Cayley–Hamilton Theorem.*) *Every square matrix* $\mathbf{A}$ *satisfies its own characteristic equation.*

Proof. (We prove the result for a 3×3 matrix $\mathbf{A}$, but the proof clearly generalizes very readily to the $n\times n$ case.)

If $\det(\mathbf{A}-\lambda\mathbf{I}) = a_0 + a_1\lambda + a_2\lambda^2 - \lambda^3$, we have to show that

$$a_0\mathbf{I} + a_1\mathbf{A} + a_2\mathbf{A}^2 - \mathbf{A}^3 = \mathbf{0}.$$

Now each element of the matrix $\operatorname{adj}(\mathbf{A}-\lambda\mathbf{I})$, being a cofactor of $(\mathbf{A}-\lambda\mathbf{I})$, is a polynomial of degree at most two in λ. Thus

$$\operatorname{adj}(\mathbf{A}-\lambda\mathbf{I}) = \mathbf{C}_0 + \lambda\mathbf{C}_1 + \lambda^2\mathbf{C}_2,$$

where $\mathbf{C}_0$, $\mathbf{C}_1$, $\mathbf{C}_2$ are 3×3 matrices whose elements do not contain λ. But

$$(\mathbf{A}-\lambda\mathbf{I})\operatorname{adj}(\mathbf{A}-\lambda\mathbf{I}) = \mathbf{I}\det(\mathbf{A}-\lambda\mathbf{I}) \quad \text{(Theorem 14.2)}$$

and thus

$$(\mathbf{A}-\lambda\mathbf{I})(\mathbf{C}_0 + \lambda\mathbf{C}_1 + \lambda^2\mathbf{C}_2) \equiv \mathbf{I}(a_0 + a_1\lambda + a_2\lambda^2 - \lambda^3).$$

Comparing coefficients of powers of λ in this identity

$$\mathbf{AC}_0 = a_0\mathbf{I}, \quad (1)$$

$$\mathbf{AC}_1 - \mathbf{C}_0 = a_1\mathbf{I}, \quad (2)$$

$$\mathbf{AC}_2 - \mathbf{C}_1 = a_2\mathbf{I}, \quad (3)$$

$$-\mathbf{C}_2 = -\mathbf{I}. \quad (4)$$

Multiplying (2) by $\mathbf{A}$, (3) by $\mathbf{A}^2$, and (4) by $\mathbf{A}^3$ and adding this gives

$$\mathbf{0} = a_0\mathbf{I} + a_1\mathbf{A} + a_2\mathbf{A}^2 - \mathbf{A}^3$$

and the proof is complete.

**Ex.* 29. Verify the Cayley–Hamilton theorem for 2×2 matrices by direct substitution.

Ex. 30. If

$$\mathbf{A} = \begin{pmatrix} 1 & 0 & -4 \\ -1 & 2 & 1 \\ 0 & 0 & -3 \end{pmatrix},$$

find $\mathbf{A}^3$ by applying the Cayley–Hamilton theorem.

Example 3. *If*

$$\mathbf{A} = \begin{pmatrix} 1 & -1 \\ 2 & 3 \end{pmatrix},$$

find $\mathbf{A}^8$.

The characteristic equation of **A** is

$$\lambda^2 - 4\lambda + 5 = 0$$

and thus by the Cayley–Hamilton theorem

$$\mathbf{A}^2 - 4\mathbf{A} + 5\mathbf{I} = \mathbf{0}.$$

But
$$\lambda^8 \equiv (\lambda^2 - 4\lambda + 5) f(\lambda) + \alpha\lambda + \beta,$$

where α, β are integers. Since $\lambda = 2+\mathrm{j} \Rightarrow \lambda^2 - 4\lambda + 5 = 0$ we have

$$(2+\mathrm{j})^8 = \alpha(2+\mathrm{j}) + \beta.$$

Using the Binomial Theorem and equating real and imaginary parts,

$$\begin{cases} 2^8 - 28.2^6 + 70.2^4 - 28.2^2 + 1 = 2\alpha + \beta \\ 8.2^7 - 56.2^5 + 56.2^3 - 8.2 \quad = \alpha \end{cases}$$

$$\Rightarrow \begin{cases} 16[16 - 112 + 70 - 7] + 1 \quad = 2\alpha + \beta \\ 16[64 - 112 + 28 - 1] \quad = \alpha \end{cases}$$

$$\Rightarrow \qquad \begin{cases} \alpha = -336 \\ \beta = 145. \end{cases}$$

Thus
$$\lambda^8 \equiv (\lambda^2 - 4\lambda + 5) f(\lambda) - 336\lambda + 145$$

and
$$\mathbf{A}^8 = -336\mathbf{A} + 145\mathbf{I}, \quad \text{since} \quad \mathbf{A}^2 - 4\mathbf{A} + 5\mathbf{I} = \mathbf{0},$$

$$= \begin{pmatrix} -336 & 336 \\ -672 & -1008 \end{pmatrix} + \begin{pmatrix} 145 & 0 \\ 0 & 145 \end{pmatrix}$$

$$= \begin{pmatrix} -191 & 336 \\ -672 & -863 \end{pmatrix}.$$

Alternatively, one can 'build up' $\mathbf{A}^8$:

$$\begin{aligned} \mathbf{A}^2 &= 4\mathbf{A} - 5\mathbf{I}, \quad \text{as before;} \\ \mathbf{A}^4 &= 16\mathbf{A}^2 - 40\mathbf{A} + 25\mathbf{I}, \quad \text{using the commutative rule,} \\ &= 16(4\mathbf{A} - 5\mathbf{I}) - 40\mathbf{A} + 25\mathbf{I} \\ &= 24\mathbf{A} - 55\mathbf{I}; \\ \mathbf{A}^8 &= 24^2\mathbf{A}^2 - 48.55\mathbf{A} + 55^2\mathbf{I} \\ &= 24^2(4\mathbf{A} - 5\mathbf{I}) - 48.55\mathbf{A} + 55^2\mathbf{I} \\ &= -48.7\mathbf{A} - 5(576 - 605)\,\mathbf{I} \\ &= -336\mathbf{A} + 145\mathbf{I}, \end{aligned}$$

and proceed as before.

Notice that, by multiplying the Cayley–Hamilton result by $\mathbf{A}^r$ we establish a recurrence relation between successive powers of $\mathbf{A}$:

$$\mathbf{A}^{r+2} = 4\mathbf{A}^{r+1} - 5\mathbf{A}^r.$$

Exercise 28(b)

1. Find the eigenvalues of the matrix

$$\mathbf{A} = \begin{pmatrix} 3 & 4 & 1 \\ 2 & 5 & 1 \\ 2 & 3 & 3 \end{pmatrix}$$

and determine corresponding unit eigenvectors.

2. Find the eigenvalues of the matrix

$$\mathbf{A} = \begin{pmatrix} 1 & -2 & 1 \\ 3 & -4 & 1 \\ 3 & -7 & 4 \end{pmatrix}.$$

What does the fact that zero is one of the eigenvalues tell us about $\mathbf{A}$?

Find a matrix $\mathbf{P}$ such that $\mathbf{P}^{-1}\mathbf{A}\mathbf{P} = \mathbf{D}$, where $\mathbf{D}$ is a diagonal matrix.

3. Find the eigenvalues of the matrix

$$\mathbf{A} = \begin{pmatrix} 3 & 1 & -2 \\ 4 & 0 & -2 \\ 4 & -1 & -1 \end{pmatrix}$$

and hence write down the value of det $\mathbf{A}$.

Find $\mathbf{A}^{-1}$ and determine the eigenvalues of $\mathbf{A}^{-1}$.

Suggest and prove a general result about the eigenvalues of an inverse matrix.

4. Find the eigenvalues of the matrix

$$\mathbf{A} = \begin{pmatrix} 1 & 2 & -2 \\ 6 & 4 & -6 \\ 6 & 5 & -7 \end{pmatrix}$$

and corresponding unit eigenvectors. Find a matrix $\mathbf{P}$ such that $\mathbf{P}^{-1}\mathbf{A}\mathbf{P} = \mathbf{D}$ where $\mathbf{D}$ is a diagonal matrix.

Show that, if $\mathbf{A}$ is the matrix of the linear transformation T, then T maps all points of the plane $2x+y-2z = 0$ into points of the same plane. Find the equations of the other two planes through the origin which have this property.

5. Find the eigenvalues of the symmetric matrix

$$\mathbf{A} = \tfrac{1}{3}\begin{pmatrix} -1 & -6 & -4 \\ -6 & -2 & -2 \\ -4 & -2 & 3 \end{pmatrix}$$

and hence find an orthogonal matrix $\mathbf{P}$ such that $\mathbf{P}^{-1}\mathbf{A}\mathbf{P} = \mathbf{D}$, where $\mathbf{D}$ is diagonal.

6. Find the equations of the planes through the origin, which map into themselves under the linear transformation T with matrix

$$\mathbf{A} = \begin{pmatrix} 1 & 3 & -2 \\ 3 & 1 & -2 \\ 3 & 4 & -5 \end{pmatrix}.$$

7. $\mathbf{P}$ is a 2×2 orthogonal matrix with $\det \mathbf{P} = 1$. Show that the eigenvalues of $\mathbf{P}$ are $e^{j\theta}$ and $e^{-j\theta}$, and interpret θ.

Show that, although the geometrical property of a 2×2 orthogonal matrix $\mathbf{P}$ is characterized by its eigenvalues if $\det \mathbf{P} = +1$, this is no longer true if $\det \mathbf{P} = -1$.

8. The *trace* of the matrix $\mathbf{A}$, written $\operatorname{tr}\mathbf{A}$, is defined to be the sum of elements of $\mathbf{A}$ in the leading diagonal (top left to bottom right). If $\mathbf{A}$ is a 3×3 matrix with eigenvalues $\lambda_1, \lambda_2, \lambda_3$, prove that

$$\operatorname{tr}\mathbf{A} = \lambda_1+\lambda_2+\lambda_3.$$

Prove also that $\operatorname{tr}(\mathbf{A}+\mathbf{B}) = \operatorname{tr}\mathbf{A}+\operatorname{tr}\mathbf{B}$ and that $\operatorname{tr}(\mu\mathbf{A}) = \mu \operatorname{tr}(\mathbf{A})$.

9. $\mathbf{A}$ and $\mathbf{B}$ are both 2×2 matrices. Prove that $\mathbf{A}$ and $\mathbf{B}$ have the same eigenvalues if and only if both $\det \mathbf{A} = \det \mathbf{B}$ and $\operatorname{tr}\mathbf{A} = \operatorname{tr}\mathbf{B}$. (See Question 8 for definition of trace.)

10. Prove that, if $\mathbf{A}$ is a non-singular 3×3 matrix, then

$$\det(\mathbf{A}-\lambda\mathbf{I}) = \det\mathbf{A}-\lambda\det\mathbf{A}\operatorname{tr}(\mathbf{A}^{-1})+\lambda^2\operatorname{tr}\mathbf{A}-\lambda^3.$$

11. If $\mathbf{A}$, $\mathbf{B}$ have a common eigenvector $\mathbf{s}$, with corresponding eigenvalues λ, μ, prove that $\mathbf{s}$ is an eigenvector of (i) $\mathbf{A}+\mathbf{B}$; (ii) $\mathbf{AB}$. What are the corresponding eigenvalues?

12. $\mathbf{A}$ is a 3×3 skew-symmetric matrix ($\mathbf{A} = -\mathbf{A}'$). Prove that

$$\det(\mathbf{A}-\lambda\mathbf{I}) = -\det(\mathbf{A}+\lambda\mathbf{I})$$

and deduce that if $\mathbf{A}$ possesses a non-zero eigenvalue α, then $-\alpha$ is also an eigenvalue.

If $f(\lambda) = \det(\mathbf{A}^2-\lambda\mathbf{I})$, prove that $-f(\lambda)/\lambda$ is the square of a linear polynomial in λ.

13. If $\mathbf{B} = \mathbf{P}^{-1}\mathbf{AP}$ and $\mathbf{s}$ is an eigenvector of $\mathbf{A}$ prove that $\mathbf{P}^{-1}\mathbf{s}$ is an eigenvector of $\mathbf{B}$.

14. If λ is an eigenvalue of the 3×3 matrix $\mathbf{A}$ prove that λ^n is an eigenvalue of $\mathbf{A}^n$. Deduce that, if

$$\mathbf{A}^r = a_2\mathbf{A}^2+a_1\mathbf{A}+a_0\mathbf{I},$$

then

$$\lambda^r = a_2\lambda^2+a_1\lambda+a_0.$$

15. A matrix $\mathbf{A}$ is said to be *nilpotent* if there exists a positive integer n such that $\mathbf{A}^n = \mathbf{0}$. Prove that, if $\mathbf{A}$ is nilpotent, then all the eigenvalues are zero.

16. Find $\mathbf{A}^4+\mathbf{A}^2+\mathbf{I}$ if

$$\mathbf{A} = \begin{pmatrix} 3 & 5 \\ -1 & -2 \end{pmatrix}.$$

17. If

$$A = \begin{pmatrix} 5 & -2 \\ 12 & -5 \end{pmatrix},$$

find A^n, where n is a positive integer.

18. If

$$A = \begin{pmatrix} 3 & -7 \\ -4 & 1 \end{pmatrix},$$

find $A^3 + 3A^2 + 12A$.

19. Show that the characteristic equation of the matrix

$$A = \begin{pmatrix} 5 & 7 & 3 \\ 1 & 5 & 2 \\ 3 & 2 & 1 \end{pmatrix}$$

is

$$\lambda^3 - 11\lambda^2 + 15\lambda - 1 = 0.$$

Deduce that A is non-singular and that

$$A^{-1} = A^2 - 11A + 15I.$$

20. If

$$A = \begin{pmatrix} 1 & 1 & 2 \\ 0 & 2 & 1 \\ 1 & 0 & 2 \end{pmatrix},$$

show that

$$A^3 = (5A - I)(A - I)$$

and

$$A^{-1} = (A - 3I)(A - 2I).$$

Deduce explicit forms for A^3 and A^{-1}.

21. The matrix A is defined by

$$A = \begin{pmatrix} 1 & 2 & 3 \\ 3 & 1 & 2 \\ 2 & 3 & 1 \end{pmatrix}.$$

Show that

$$A^3 - 3A^2 - 16A - 16I = 2I - A$$

and express $(2I - A)^{-1}$ as a quadratic polynomial in A.

22. Show that 1, 2, 3 are the eigenvalues of the matrix

$$A = \begin{pmatrix} 17 & -16 & 8 \\ 10 & -8 & 6 \\ -10 & 11 & -3 \end{pmatrix}.$$

Hence, or otherwise, find an explicit form for A^n.

Write down the inverse matrix of A.

23. The 2×2 matrix A has equal eigenvalues λ. Prove that, for $n \geqslant 2$

$$A^n = n\lambda^{n-1}A - (n-1)\lambda^n I.$$

Prove further that, if $\lambda \neq 0$, this formula is true for all integral n.

24. If **S** is a 3×3 skew-symmetric matrix, prove that its eigenvalues are 0 and $\pm j\alpha$, α real.

If **P** is a 3×3 orthogonal matrix and $\mathbf{P}+\mathbf{I}$ is non-singular, prove that

$$\mathbf{S} = (\mathbf{P}-\mathbf{I})\,(\mathbf{P}+\mathbf{I})^{-1}$$

is skew-symmetric, and that, if λ is an eigenvalue of **S**, then

$$\mathbf{A} = \mathbf{P}-\mathbf{I}-\lambda(\mathbf{P}+\mathbf{I})$$

is singular.

The eigenvalues of **P** are 1, $e^{j\theta}$, $e^{-j\theta}$; what are the corresponding eigenvalues of **S**?

25. The 3×3 matrix **A** has characteristic equation

$$(\lambda-\lambda_1)^2\,(\lambda-\lambda_2) = 0,$$

where λ_1, λ_2 are real and $\lambda_1 \neq \lambda_2$.

Show geometrically that, if the equations

$$\mathbf{A}\mathbf{x} = \lambda_1\mathbf{x}$$

have a *plane* of solutions, then it is possible to form a matrix **P** such that

$$\mathbf{P}^{-1}\mathbf{A}\mathbf{P} = \begin{pmatrix} \lambda_1 & 0 & 0 \\ 0 & \lambda_1 & 0 \\ 0 & 0 & \lambda_2 \end{pmatrix}.$$

If **A** is the matrix of the linear transformation *T*, and if *l* is the invariant line through the origin corresponding to the eigenvalue λ_2, show that every plane containing *l* is mapped into itself by *T*.

If

$$\mathbf{A} = \begin{pmatrix} -4 & -10 & 30 \\ -3 & -5 & 18 \\ -2 & -4 & 13 \end{pmatrix}$$

show that *T* leaves all the points of a certain plane containing the origin fixed and find the equation of this plane. Show also that every plane containing the line

$$\frac{x}{5} = \frac{y}{3} = \frac{z}{2}$$

is mapped into itself by *T*.

Find a diagonalizing matrix **P** for this particular matrix **A**.

26. The linear transformation *T* has matrix **A** where **A** is non-singular with an eigenvector $\mathbf{d} = \mathbf{OD}$. *P* is a given point not on *OD* and $T(P) = Q$. Show that *T* maps *PQ* into itself if *PQ* is parallel to *OD*. Show further that, if this is the case, then the linear transformation *S* with matrix $\mathbf{A}-\mathbf{I}$ maps the line *PQ* into the line *OD*.

27. **P** is a non-singular 3×3 matrix, ω is a cube root of unity and r is an integer. Prove that, if

$$\mathbf{\Omega} = \mathbf{P}^{-1}\begin{pmatrix} \omega^r & 0 & 0 \\ 0 & \omega^{r+1} & 0 \\ 0 & 0 & \omega^{r+2} \end{pmatrix}\mathbf{P},$$

then $$\mathbf{\Omega}^3 = \mathbf{I}.$$

Prove conversely, that if $\mathbf{\Omega}$ is a 3×3 matrix, with real determinant, having the property

$$\mathbf{\Omega}^3 = \mathbf{I}$$

then

$$\mathbf{\Omega} = \mathbf{P}^{-1}\begin{pmatrix} \omega^r & 0 & 0 \\ 0 & \omega^{r+1} & 0 \\ 0 & 0 & \omega^{r+2} \end{pmatrix}\mathbf{P},$$

for some matrix **P** and integer r.

Prove further that, if $\mathbf{\Omega} \neq \mathbf{I}$, then the matrix

$$\mathbf{I} + \mathbf{\Omega} + \mathbf{\Omega}^2$$

is singular.

28. Show that the linear transformation T with matrix

$$\mathbf{A} = \begin{pmatrix} 1 & 3 & -2 \\ -1 & 5 & -2 \\ -1 & 4 & -1 \end{pmatrix}$$

leaves just two lines through the origin invariant and find their equations.

Explain geometrically why it is impossible to find a matrix **P** such that the matrix

$$\mathbf{P}^{-1}\mathbf{A}\mathbf{P}$$

is diagonal.

29. *Further coordinate geometry*

1. PARAMETRIC FORMS FOR PLANE CURVES

We have seen in Chapters 22 and 27 that the coordinates of points on certain plane curves may be given in terms of parameters. Indeed, we could *define*, for example, the parabola as the set of all points P with position vectors

$$\mathbf{r} = at^2\mathbf{i}+2at\mathbf{j}$$

relative to some origin O, where t is a scalar parameter. More generally a plane curve is completely specified if we are given a parametric representation

$$\mathbf{r} = f(t)\,\mathbf{i}+g(t)\,\mathbf{j},$$

where f, g are two (continuous) functions. We shall also demand that there is a 1–1 correspondence between points of the curve and values of the parameter – with the possible exception of a limited number of multiple points, for which several values of the parameter may yield the same point (see Ex. 5). If $f(t)$, $g(t)$ happen to be algebraic expressions, the resulting analysis is much simplified and we are often able to employ the theory of polynomial equations in problem solving.

Ex. 1. Show that an algebraic parametric form for the ellipse

$$b^2x^2+a^2y^2 = a^2b^2$$

is

$$\mathbf{r} = \frac{a(1-t^2)}{1+t^2}\mathbf{i}+\frac{2bt}{1+t^2}\mathbf{j}.$$

What point of the ellipse is excluded by this parametric representation?

Find a similar algebraic parametric form for the hyperbola.

$$b^2x^2-a^2y^2 = a^2b^2.$$

In the following two examples we illustrate some of the methods available for the solution of geometrical problems employing algebraic parameters. In the first example, we define a curve parametrically and deduce a geometric property; in the second, we obtain a geometrical property of the parabola using the standard parametric form.

Example 1. *The semi-cubical parabola is defined parametrically by*

$$\mathbf{r} = at^2\mathbf{i}+at^3\mathbf{j}.$$

Prove that a straight line not parallel to the y axis cuts the curve in either one or three real points and that, if a straight line not through the origin cuts the curve at P, Q, R and the tangents to the curve at these three points cut the curve again at P′, Q′, R′, then P′Q′R′ is a straight line.

Any line not parallel to the y axis has an equation of the form

$$y = mx+c.$$

This cuts the curve in points with parameters given by the roots of the equation

$$at^3 = mat^2+c,$$

that is, by
$$at^3-mat^2-c = 0.$$

Since this is a cubic equation in t, it has either one or three real roots, due regard being paid to their multiplicity. This proves the first result.

Now suppose that the parameters of the points P, Q, R are respectively p, q, r. Then, if the equation of the line PQR is $y = mx+c$ we see that the roots of the cubic equation

$$at^3-mat^2-c = 0$$

are p, q, r; thus $qr+rp+pq = 0$.

But the line does not contain the origin; thus $p, q, r \neq 0$ and we have

$$1/p+1/q+1/r = 0.$$

Again, since the tangent at P cuts the curve at P', parameter p', we have, as above,

$$(1/p)+(1/p)+(1/p') = 0,$$

giving, $1/p' = -2/p$, similarly $1/q' = -2/q$, and $1/r' = -2/r$ and we have

$$(1/p')+(1/q')+(1/r') = -2[(1/p)+(1/q)+(1/r)] = 0.$$

It follows that p', q', r' are the roots of a cubic equation of the form

$$at^3-m'at^2-c = 0$$

(where $m' = p'+q'+r'$ and $c'/a = p'q'r'$) and thus the points P', Q', R' lie on the line

$$y = m'x+c'.$$

The shape of the semi-cubical parabola is shown in Figure 29.1. O is called the *pole* and Ox is the *axis* of the curve. The Cartesian equation is $x^3 = ay^2$. A connection between the parabola and the semi-cubical parabola is obtained in Example 2.

Example 2. *Given a parabola and a point C, prove that, in general, either one or three normals may be drawn to the parabola to pass through C but that, if just two such normals may be drawn, then C lies on a certain semi-cubical parabola.*

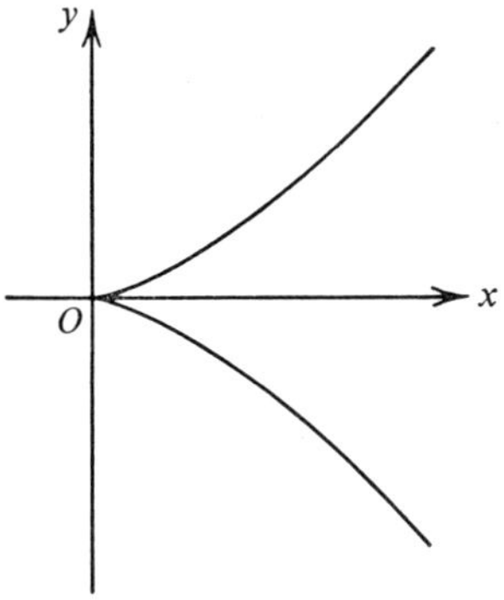

Fig. 29.1

If the normals at three points, P, Q, R on a parabola are concurrent, show that the circumcircle of the triangle PQR passes through the vertex of the parabola.

Take the parabola in the standard form $y^2 = 4ax$ and let C be the point (h, k). The equation of the normal at the point $P(at^2, 2at)$ is

$$tx+y = at^3+2at$$

and this passes through C if

$$at^3+(2a-h)\,t-k = 0.$$

This is a cubic equation in t, with, in general, either one or three real roots, and the first part of the question is proved.

If just *two* normals pass through C, then the cubic equation above has a repeated root and the condition for this (see Chapter 23) is

$$\frac{4}{27}\left(2-\frac{h}{a}\right)^3+\frac{k^2}{a^2} = 0,$$

showing that C in this case lies on the curve

$$4(x-2a)^3 = 27ay^2.$$

By Example 1, this is a semi-cubical parabola with pole at $(2a, 0)$ and axis coinciding with the axis of the parabola.

Now suppose that the normals at $P(ap^2, 2ap)$, $Q(aq^2, 2aq)$, $R(ar^2, 2ar)$ meet at C; then p, q, r are the roots of the equation

$$at^3+(2a-h)\,t-k = 0$$

and thus $p+q+r = 0$.

Suppose that the circumcircle of triangle PQR has equation

$$(x-\alpha)^2+(y-\beta)^2 = \gamma^2.$$

This cuts the parabola at points with parameters given by the quartic equation

$$(at^2-\alpha)^2+(2at-\beta)^2 = \gamma^2.$$

But the coefficient of t^3 in this equation is zero and thus the sum of the roots is zero. Since three of the roots are, by definition p, q and r, and since $p+q+r = 0$, the fourth root must be zero and we have proved the final result.

Ex. 2. Prove that the normals to the parabola $y^2 = 4ax$ touch the semi-cubical parabola of Example 2. Draw in the same sketch the parabola $y^2 = 4ax$ and the semi-cubical parabola $4(x-2a)^3 = 27ay^2$. (It is worth drawing the parabola accurately and then constructing a large number of normals—which may be done very rapidly using the geometrical property that, in the notation of Exercise 22(*b*), $NG = 2a$.)

Ex. 3. Draw the parabola $y^2 = 4ax$ and shade in the set of all points from which three real normals may be drawn to the parabola.

Ex. 4. The Folium of Descartes has Cartesian equation

$$x^3+y^3 = 3axy.$$

By considering the intersection of the line $y = tx$ with the folium, show that the curve may be represented parametrically by

$$x = \frac{3at}{1+t^3}, \quad y = \frac{3at^2}{1+t^3}.$$

Taking $t = 0$ it is seen that the origin lies on the curve and that the x axis is a tangent there; by writing $u = t^{-1}$, show that the y axis is also a tangent at the origin (which is thus a *double point* of the curve, that is, a point through which pass two separate branches of the curve).

Show that the line $x-y = 0$ is an axis of symmetry and that the line $x+y = a$ is an asymptote. Sketch the curve.

Ex. 5. Using the method of Ex. 4, find a parametric form for the curve

$$y^2 = x^2(1+x),$$

and deduce that no point of the curve has an x coordinate less than -1. Prove also that the x axis is an axis of symmetry for the curve.

Show that the origin lies on the curve and that this point arises from two distinct values of the parameter. Deduce that the origin is a double point of the curve and determine the equations of the two tangents there.

Exercise 29(a)

1. Show that, if $\mathbf{OP} = \mathbf{r} = at\,\mathbf{i} + at^3\mathbf{j}$, then P lies on the curve $\Gamma: x^3 = a^2y$. Prove that any line in the plane cuts Γ in either one or three real points.

The tangents at three points P, Q, R of the curve meet Γ again at P', Q', R' respectively. Prove that:

(i) if the centroid of the triangle PQR lies on the y axis, so also does the centroid of the triangle $P'Q'R'$;

(ii) if P, Q, R are collinear, so also are P', Q', R'.

2. The *cissoid* is defined parametrically by

$$\mathbf{r} = \frac{at^2}{1+t^2}\mathbf{i}+\frac{at^3}{1+t^2}\mathbf{j}.$$

Find the Cartesian equation of the cissoid and give a rough sketch of the curve.

Prove that, if the chord PQ subtends a right angle at O, then the mid-point M of PQ lies on a fixed straight line.

3. A curve Γ is given parametrically by

$$\mathbf{r} = \frac{t^3+1}{t}\mathbf{i}+\frac{t^2-1}{t}\mathbf{j} \quad (t \neq 0).$$

Prove that, if the points with parameters t_1, t_2, t_3 are collinear, then

$$1+t_1+t_2+t_3+t_1t_2t_3 = 0,$$

and conversely.

Show that, given a point P lying on Γ, two lines may be drawn through P to touch the curve at Q and R. If QR cuts Γ again at S find the parameter of the point S in terms of the parameter of P.

Show that the roots of the equation $t^2+t-1 = 0$ give the double point of the curve.

4. A curve Γ is given parametrically by

$$\mathbf{r} = (t^2+1/t)\,\mathbf{i}+(t^2-1/t)\,\mathbf{j}.$$

Obtain a necessary and sufficient condition for the three points P, Q, R of the curve to be collinear.

The tangent at P_1 meets Γ again at P_2, the tangent at P_2 meets Γ again at P_3 and the tangent at $\mathbf{P}_3$ meets Γ again at P_4. P_1P_3 meets the curve at Q while P_2P_4 meets the curve at R. Prove that QR is a tangent to Γ at Q.

5. A curve is given parametrically by the equations

$$x = a(1-t^2), \quad y = a(t-t^3).$$

Prove:

(i) that an arbitrary line $lx+my+na = 0$ meets the curve in three points;

(ii) that if three points $t = t_1$, $t = t_2$, $t = t_3$ are collinear then

$$t_2t_3+t_3t_1+t_1t_2 = -1; \qquad \text{(over)}$$

(iii) that if t_1, t_2 and t_3 satisfy the above equation then the points

$$t = t_1, \quad t = t_2, \quad t = t_3$$

are collinear.

A chord through the point $(a, 0)$ meets the curve again at the points P and Q. Prove that the locus of the middle point of PQ is a curve with parametric equations

$$x = -\tfrac{1}{2}aT^2, \quad y = -\tfrac{1}{2}T(T^2+2). \qquad \text{(O \& C)}$$

6. Prove that, in general, three tangents may be drawn from a point C to the cubic curve $x^3 = a^2y$ and that, if these tangents cut the curve again at P, Q, R then the tangents at P, Q, R are concurrent.

7. A rectangular hyperbola is given parametrically by the equations

$$x = ct, \quad y = c/t.$$

If the four points of the curve with parameters t_1, t_2, t_3, t_4 lie on a circle, show that $t_1t_2t_3t_4 = 1$. Show conversely that, if $t_1t_2t_3t_4 = 1$, then the four points lie on a circle.

A variable circle passes through the fixed points A, B of a rectangular hyperbola, and meets the hyperbola again at P, Q. Show that the direction of PQ is fixed.

8. P, Q are variable points on the parabola $y^2 = 4ax$ such that PQ is parallel to the fixed line $x+ky = 0$. The normals to the parabola at P and Q meet at R. Prove that the locus of R is the normal to the parabola at a fixed point on the parabola, and find the coordinates of this point. (O & C)

9. Find the equation of the normal to the rectangular hyperbola $xy = c^2$ at the point $(ct, c/t)$.

The normals to the rectangular hyperbola at the points P, Q, R, S are concurrent; prove that each of these points is the orthocentre of the triangle formed by the other three.

10. The normals at three points P, Q, R of the parabola $y^2 = 4ax$ meet at a point N. Prove that the centroid of the triangle PQR lies on the axis of the parabola.

If N coincides with P, prove that QR passes through a fixed point (that is, a point whose position is independent of P, Q, R). (O & C)

11. Find the coordinates of the point P, other than the origin in which the line $y = tx$ meets the curve $x^3+y^3 = axy$. (t is called the *parameter* of P.)

If a line meets the curve in three points whose parameters are t_1, t_2, t_3, prove that $t_1\,t_2\,t_3 = -1$.

If Q is any point on the curve with negative parameter, not equal to -1, prove that there are two points P_1, P_2 of the curve (other than Q) such that the tangents at P_1 and P_2 pass through Q. Prove that OP_1 and OP_2 make equal angles (apart from the sense) with either of the coordinate axes. (O & C)

12. Prove that the equation of the normal to the curve $y = x^3$ at the point (t, t^3) is

$$x+3t^2y = t+3t^5.$$

By considering the maximum and minimum values of a certain function of t and drawing a rough graph of the function, or otherwise, prove that three normals can be drawn to curve $y = x^3$ from a point $(0, b)$ when $b > 4/3\sqrt{3}$.

How many real normals can be drawn to the curve from a point $(0, c)$ when $0 < c < 4/3\sqrt{3}$? (O & C)

13. A circle has a diameter OA of length a, and the tangent at A is l. A variable line through O meets the circle again at Q, and l at R; P is the point on OR such that $OP = QR$. If O is the origin and OA is the x axis, show that P has coordinates $(at^2/(1+t^2), at^3/(1+t^3))$, where t is a suitable parameter.

Prove that, if a line meets the locus of P in three points with parameters t_1, t_2, t_3 then $t_2t_3+t_3t_1+t_1t_2 = 0$.

Hence, or otherwise, prove that the tangent to the locus at P meets the locus again at the point with parameter $-\frac{1}{2}t$. (O & C)

14. A point on the curve $ay^2 = x^3$ is given parametrically in the form (at^2, at^3). If the points on the curve with parameters t_1, t_2, t_3 are collinear, prove that

$$(1/t_1)+(1/t_2)+(1/t_3) = 0.$$

Hence show that, if the tangent at the point with parameter t_1 meets the curve again at the point with parameter t_4, then $t_1+2t_4 = 0$.

Perpendicular lines through O meet the curve at P, Q; PQ meets the curve again at R and S is the point of the curve such that the tangent at S passes through R. Prove that OP, OQ are the bisectors of the angles between OS and Ox.

(O & C)

15. The rectangular hyperbola $xy = k^2$ is met by a circle, passing through its centre O, in four points A_1, A_2, B_1, B_2. The lengths of the perpendiculars from O to A_1A_2 and B_1B_2 are a and b. Prove that $ab = k^2$. (C. S.)

16. Show that there are three values of t, not necessarily real, for which the point (t^2, t^3) lies on a given straight line.

P, Q and R are distinct points (p^2, p^3), (q^2, q^3) and (r^2, r^3) on the curve $y^2 = x^3$. Show that:

(i) if these points are collinear then $\Sigma pq = 0$;

(ii) if the tangents at these points are concurrent, then $\Sigma p = 0$;

(iii) there are no real points on the curve for which these two conditions co-exist.

The tangent at the point P on this curve meets the curve again at P'. Find the ratio in which PP' is divided by the x axis. (London)

2. SURFACES AND CURVES IN IN THREE DIMENSIONS

The simplest surface in three-dimensional space is the *plane*. Given three points A, B, C, the vector equation of the plane through ABC is

$$\mathbf{r} = \mathbf{a}+\lambda(\mathbf{b}-\mathbf{a})+\mu(\mathbf{c}-\mathbf{a}),$$

where λ, μ are scalar parameters. Notice that we need *two* parameters to define the surface.

Two surfaces intersect in a curve (which need not necessarily be a plane curve). Thus, two planes intersect in a *line*; the vector equation of the line through A, B is

$$\mathbf{r} = \mathbf{a}+\lambda(\mathbf{b}-\mathbf{a}),$$

where we have the *single* scalar parameter λ.

Another familiar surface is the *sphere*, which is defined as the set of points in three dimensions lying at a fixed distance from a fixed point. If the fixed distance (*radius*) is c and the fixed point (*centre*) is A, then the vector equation of the sphere is

$$(\mathbf{r}-\mathbf{a}).(\mathbf{r}-\mathbf{a}) = c^2.$$

Ex. 6. Prove that

$$\mathbf{r} = \mathbf{a}+\cos\theta\cos\phi\mathbf{i}+\sin\theta\cos\phi\mathbf{j}+\sin\phi\mathbf{k}$$

gives a parametric representation of the sphere. (Notice now that we have *two* parameters; suggest names for them.)

**Ex.* 7. Prove that the vector equation $(\mathbf{r}-\mathbf{a}).(\mathbf{r}-\mathbf{b}) = 0$ represents a sphere, and locate the points A and B as points on the sphere.

The Cartesian equation of a sphere, centre $A(a_1, a_2, a_3)$ is thus

$$(x-a_1)^2+(y-a_2)^2+(z-a_3)^2 = c^2,$$

which may be rewritten

$$x^2+y^2+z^2+2ux+2vy+2wz+d = 0.$$

Conversely, an equation of the form

$$x^2+y^2+z^2+2ux+2vy+2wz+d = 0$$

may be rewritten as

$$(x+u)^2+(y+v)^2+(z+w)^2 = u^2+v^2+w^2-d,$$

and thus represents a sphere, centre $(-u, -v, -w)$, provided

$$u^2+v^2+w^2 > d.$$

Ex. 8. Show that the equation

$$x^2+y^2+z^2-2x-4y+2z-10 = 0$$

represents a sphere, and find its centre and radius.

**Ex.* 9. Show that four points in space in general define a unique sphere. What exceptional cases may arise?

The intersection of the sphere

$$(x-a_1)^2+(y-a_2)^2+(z-a_3)^2 = c^2$$

with the plane $$z = 0$$

is the curve in the Oxy plane with equation

$$(x-a_1)^2+(y-a_2)^2 = c^2-a_3^2,$$

which is seen to be a circle provided $c^2 > a_3^2$. If $c^2 = a_3^2$, the equation represents a single point and, if $c^2 < a_3^2$, the sphere and plane do not intersect.

Thus, a sphere and a plane, if they intersect at all, intersect in a circle, which may be of zero radius if the plane is tangential to the sphere. The centre of the circle lies on the perpendicular from the centre of the sphere on to the plane. If the sphere and the plane have vector equations

$$(\mathbf{r}-\mathbf{a}).(\mathbf{r}-\mathbf{a}) = c^2, \quad \mathbf{r}.\mathbf{n} = p \quad (\text{where } |\mathbf{n}| = 1),$$

this perpendicular will be $$\mathbf{r} = \mathbf{a}+\lambda\mathbf{n}.$$

To find the value of λ which gives the centre of the circle determined by the sphere and the plane we solve

$$\mathbf{r} = \mathbf{a}+\lambda\mathbf{n} \quad \text{and} \quad \mathbf{r}.\mathbf{n} = p,$$

giving $$(\mathbf{a}+\lambda\mathbf{n}).\mathbf{n} = p.$$

Thus, $\lambda = p-\mathbf{a}.\mathbf{n}$, since $\mathbf{n}.\mathbf{n} = 1$ and the position vector of the centre of the circle is

$$\mathbf{r} = \mathbf{a}+(p-\mathbf{a}.\mathbf{n})\,\mathbf{n}.$$

Ex. 10. Find the centre and radius of the circle determined by the sphere

$$(x-1)^2+(y-3)^2+(z-2)^2 = 20$$

and the plane $$x-y-3z = 3.$$

Now consider two spheres, with equations

$$(\mathbf{r}-\mathbf{a}).(\mathbf{r}-\mathbf{a}) = c^2, \quad (\mathbf{r}-\mathbf{b}).(\mathbf{r}-\mathbf{b}) = d^2.$$

These equations may be rewritten

$$\mathbf{r}.\mathbf{r}-2\mathbf{r}.\mathbf{a} = c^2-\mathbf{a}.\mathbf{a} \quad \text{and} \quad \mathbf{r}.\mathbf{r}-2\mathbf{r}.\mathbf{b} = d^2-\mathbf{b}.\mathbf{b}$$

which on subtraction, give

$$2\mathbf{r}.(\mathbf{b}-\mathbf{a}) = c^2-d^2-\mathbf{a}.\mathbf{a}+\mathbf{b}.\mathbf{b}.$$

But this is the equation of a plane, perpendicular to the vector $\mathbf{b}-\mathbf{a}$, that is, to the line of centres AB and, since a plane and a sphere determine a circle, we see that two spheres, provided they intersect at all, intersect in a circle.

The line $$\mathbf{r} = \mathbf{a}+\lambda\mathbf{n} \quad (|\mathbf{n}| = 1)$$
cuts the sphere $$(\mathbf{r}-\mathbf{b}).(\mathbf{r}-\mathbf{b}) = c^2$$
in points with parameters given by
$$(\lambda\mathbf{n}+\mathbf{a}-\mathbf{b}).(\lambda\mathbf{n}+\mathbf{a}-\mathbf{b}) = c^2;$$
that is, by the quadratic equation
$$\lambda^2+2\lambda\mathbf{n}.(\mathbf{a}-\mathbf{b})+(\mathbf{a}-\mathbf{b}).(\mathbf{a}-\mathbf{b})-c^2 = 0.$$
Thus, a line cuts a sphere in two points (which may be coincident or imaginary).

If $\mathbf{n}$ is given and we take A to be the mid-point of any chord in the direction defined by $\mathbf{n}$, the two roots λ_1 and λ_2 of the above quadratic equation have the same magnitude but opposite sign; that is, $\lambda_1+\lambda_2 = 0$. It follows that $\mathbf{n}.(\mathbf{a}-\mathbf{b}) = 0$ and A lies on the plane $\mathbf{n}.(\mathbf{r}-\mathbf{b}) = 0$ which passes through the centre, B, of the sphere and is perpendicular to $\mathbf{n}$; thus the locus of the mid-points of parallel chords of a sphere is a plane through the centre of the sphere (a *diametral plane*).

Example 3. *Find the equation of the tangent plane to the sphere*
$$x^2+y^2+z^2-8x-6z-2 = 0$$
at the point (1, 3, 0).

Prove that the plane
$$x-5y+z+20 = 0$$
is a tangent to the sphere and find its point of contact.

Rewriting the equation of the sphere in the form
$$(x-4)^2+y^2+(z-3)^2 = 27$$
we see that the centre of the sphere is the point (4, 0, 3).

(i) The vector $(4\mathbf{i}+3\mathbf{k})-(\mathbf{i}+3\mathbf{j}) = 3\mathbf{i}-3\mathbf{j}+3\mathbf{k}$ is normal to the required tangent plane whose equation is thus
$$1(x-1)-1(y-3)+1(z-0) = 0,$$
that is $$x-y+z+2 = 0.$$

(ii) The radius of the sphere is $\sqrt{27}$; but the perpendicular distance from (4, 0, 3) to the plane $x-5y+z+20 = 0$ is
$$\frac{4+3+20}{\sqrt{(1^2+5^2+1^2)}} = \sqrt{27}.$$
Thus the plane $x-5y+z+20 = 0$ is tangential to the sphere.

A normal to the given plane is $\mathbf{i}-5\mathbf{j}+\mathbf{k}$ and thus any point on the radius to the point of contact has coordinates $(4+\lambda, -5\lambda, 3+\lambda)$. This lies on the given plane if

$$(4+\lambda)-5(-5\lambda)+(3+\lambda)+20 = 0,$$

that is, if $\lambda = -1$. This gives the coordinates of the point of contact as (3, 5, 2).

The vector equation of the tangent plane at a point T of the sphere, centre A and radius c, is easily derived by the same method as that used in the last example. For a unit vector in the direction AT is $(\mathbf{t}-\mathbf{a})/c$ and thus the equation of the tangent plane is

$$(\mathbf{r}-\mathbf{a}).(\mathbf{t}-\mathbf{a}) = c^2.$$

Notice that the equation of the tangent plane is obtained from the equation of the sphere by writing **t** for **r** in one of the brackets; this simple rule enables us to write down the equation of the tangent plane at any point of the sphere. For example, the equation of the tangent to the sphere

$$x^2+y^2+z^2-2x+4y+12z+4 = 0$$

at the point $(1, -3, 0)$ is

$$x.1+y.(-3)+z.0-(x+1)+2(y-3)+6(z+0)+4 = 0$$

or
$$y-6z+3 = 0.$$

Example 4. *Determine whether or not the circles*

$$S_1\colon x^2+y^2+z^2-18x-6y-4z+14 = 0, \quad x+3y+5z = 0,$$

$$S_2\colon x^2+y^2+z^2-6x+14y-12z+22 = 0, \quad x+y+z = 2,$$

are linked (*as in the links of a chain*).

The planes of the two circles meet in the line, l, with equations

$$x+3y+5z = 0, \quad x+y+z = 2$$

which reduce to
$$\frac{x-3}{1} = \frac{y+1}{-2} = \frac{z}{1}$$

and any point on this line has coordinates $(3+\lambda, -1-2\lambda, \lambda)$. Thus l meets the given sphere through S_1 at points given by

$$(3+\lambda)^2+(-1-2\lambda)^2+\lambda^2-18(3+\lambda)+6(1+2\lambda)-4\lambda+14 = 0,$$

or
$$6\lambda^2-24 = 0,$$

giving $\lambda = +2$ or -2.

Again, l meets the given sphere through S_2 at points given by

$$(3+\lambda)^2+(-1-2\lambda)^2+\lambda^2-6(3+\lambda)-14(1+2\lambda)-12\lambda+22 = 0,$$

or
$$6\lambda^2-36\lambda = 0$$

giving $\lambda = 0$ or 6.

Thus, if S_1 meets l at A_1 ($\lambda = -2$) and B_1 ($\lambda = +2$) and S_2 meets l at A_2 ($\lambda = 0$) and B_2 ($\lambda = 6$), the order of the points on the line is $A_1A_2B_1B_2$ and the circles must therefore be linked.

Another familiar surface in three dimensions is the *circular cylinder*. If the axis of the cylinder passes through the point $A(a_1, a_2, a_3)$ and is in the direction of the unit vector $\mathbf{u} = l\mathbf{i}+m\mathbf{j}+n\mathbf{k}$, any point $P(x, y, z)$ lies on the cylinder if its perpendicular distance from the axis is a constant, b. Thus, if $\mathbf{OA} = \mathbf{a}$, $\mathbf{OP} = \mathbf{r}$, we have (see Figure 29.2) $|(\mathbf{r}-\mathbf{a}) \wedge \mathbf{u}| = b$, or in Cartesian form,

$$\Sigma[n(y-a_2)-m(z-a_3)]^2 = b^2.$$

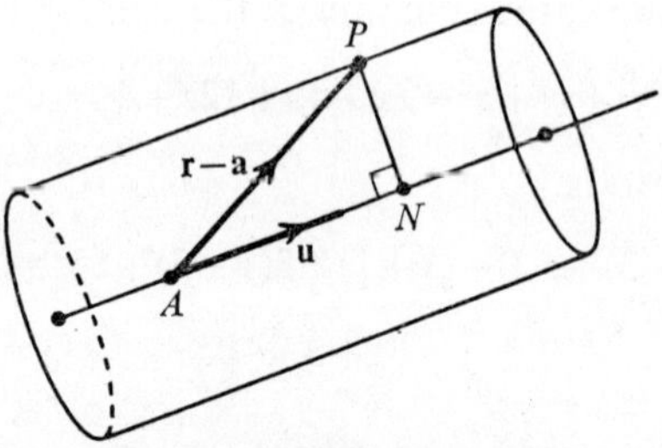

Fig. 29.2

If we take the x axis as the axis of the cylinder

(and thus $l = 1$, $m = n = 0$ and $a_1 = a_2 = a_3 = 0$)

this equation reduces to the much simpler form

$$y^2+z^2 = b^2.$$

**Ex*. 11. Show that a parametric form for a cylinder, radius b, with its axis along the z axis, is

$$\mathbf{r} = b \cos \theta\mathbf{i}+b \sin \theta\mathbf{j}+\lambda\mathbf{k}.$$

Notice once again that we have *two* parameters, θ and λ for this surface.

Ex. 12. What does the surface

$$\frac{x^2}{a^2}+\frac{y^2}{b^2} = 1$$

represent in three dimensions?

A cylindrical spring is a three-dimensional curve lying on the surface of a circular cylinder. The curve is called a *helix* and is defined parametrically by

$$\mathbf{r} = b\cos\theta\mathbf{i}+b\sin\theta\mathbf{j}+\frac{p\theta}{2\pi}\mathbf{k}$$

where p (the *pitch* of the helix) and b (the radius of the cylinder on which the helix lies) are constants. Note that the parametric representation of the helix involves only *one* parameter.

**Ex.* 13. Interpret geometrically the constant p, the pitch of the helix.

Ex. 14. Show that the helix defined above is one of the two curves of intersection of the cylinder $x^2+y^2 = b^2$ and the corrugated surface $x = a\cos\dfrac{2\pi z}{p}$. Show also that the other curve formed is an oppositely twined helix.

The *right-circular cone* is the surface traced out by a variable line through a fixed point O (the *vertex* of the cone) and making a constant angle with a fixed line though O (the *axis* of the cone). If we take O as the origin and the unit vector $\mathbf{u} = l\mathbf{i}+m\mathbf{j}+n\mathbf{k}$ as defining the direction of the axis we have, for any point P of the cone,

$$\mathbf{r}.\mathbf{u} = |\mathbf{r}|\cos\alpha,$$

where $\mathbf{r} = \mathbf{OP}$ and α is the angle of the cone (Figure 29.3). In Cartesian form, this gives

$$(lx+my+nz)^2 = (x^2+y^2+z^2)\cos^2\alpha$$

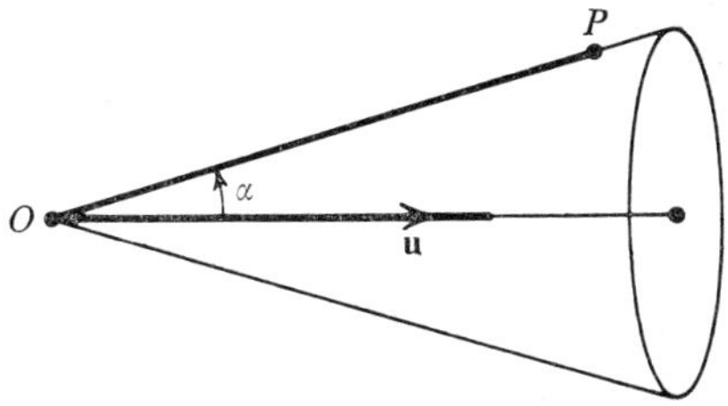

Fig. 29.3

and we see that the equation of a cone with its vertex at the origin is homogeneous in x, y, z. We saw in Chapter 26 that the section of a plane with a cone is a conic.

Example 5. *Obtain the equation of the cone with vertex* V (2, -1, -2) *which touches the sphere* $x^2+y^2+z^2 = 1$ *at the points of a circle on the surface of the sphere.*

Since the centre of the sphere is the origin, O, and the cone touches the sphere, VO is the axis of the cone. Let VT be a line on the surface of the cone (a *generator of the cone*) (see Figure 29.4). Then, if $\angle OVT = \alpha$, since $OV = 3$ and the radius of the sphere is 1,

$$\cos \alpha = \frac{2\sqrt{2}}{3}.$$

Now consider any point $P(x, y, z)$ on the cone. A unit vector along OV is $\mathbf{u} = \frac{2}{3}\mathbf{i} - \frac{1}{3}\mathbf{j} - \frac{2}{3}\mathbf{k}$ and thus, since

$$\mathbf{VP}.\mathbf{u} = |\mathbf{VP}| \cos \alpha$$

we obtain

$$[\tfrac{2}{3}(x-2) - \tfrac{1}{3}(y+1) - \tfrac{2}{3}(z+2)]^2 = \tfrac{8}{9}[(x-2)^2 + (y+1)^2 + (z+2)^2],$$

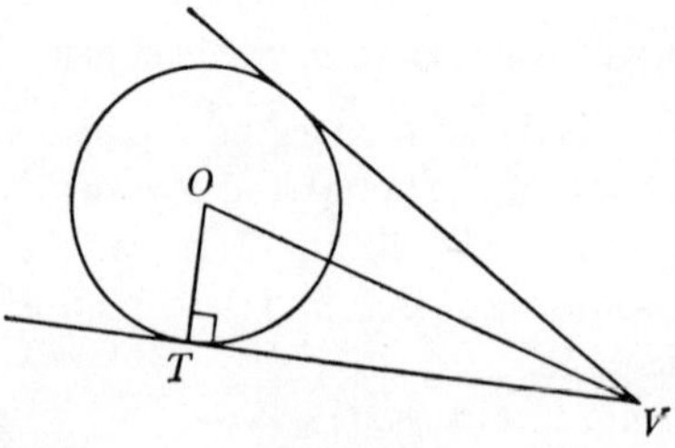

Fig. 29.4

which may be rewritten as

$$4(x-2)^2 + 7(y+1)^2 + 4(z+2)^2 - 4(y+1)(z+2) + 8(z+2)(x-2)$$
$$+4(x-2)(y+1) = 0.$$

3. CHANGING THE COORDINATE SYSTEM IN A PLANE: ROTATION OF AXES

In Chapter 13, a linear transformation of the plane into itself of the form

$$x' = ax + by, \quad y' = cx + dy$$

was regarded as a mapping of the plane into itself, in which the point $P(h, k)$ is mapped into the point $P'(ah + bk, ch + dk)$, coordinates *being referred to the same axes.* In particular, the transformation T whose matrix $\mathbf{P}$, is orthogonal,

$$\mathbf{P} = \begin{pmatrix} \cos \theta & -\sin \theta \\ \sin \theta & \cos \theta \end{pmatrix}$$

has the effect of rotating any line OA through the angle θ, measured in a positive sense, into the position OA', where $OA = OA'$.

There is, however, an alternative way of looking at an orthogonal linear transformation. Suppose we leave each point of the plane where it is but *re-name the coordinates* by taking a new pair of perpendicular axes, with the origin still at O, but with the new axes Ox', Oy' making an angle $-\theta$ with the old axes Ox, Oy (Figure 29.5).

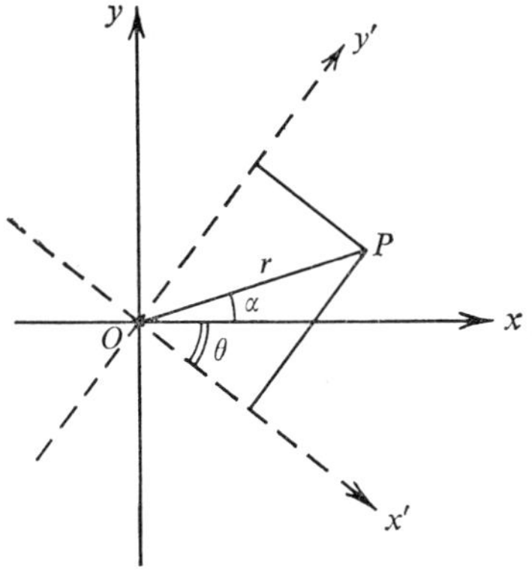

Fig. 29.5

Let $OP = r$, and suppose OP makes an angle α with Ox. Then

$$x = r\cos\alpha, \quad y = r\sin\alpha$$

$$x' = r\cos(\alpha+\theta), \quad y' = r\sin(\alpha+\theta).$$

Thus,

$$\begin{aligned} x' &= r\cos\alpha\cos\theta - r\sin\alpha\sin\theta \\ &= x\cos\theta - y\sin\theta, \\ y' &= r\cos\alpha\sin\theta + r\sin\alpha\cos\theta \\ &= x\sin\theta + y\cos\theta. \end{aligned}$$

In matrix notation, this means

$$\begin{pmatrix} x' \\ y' \end{pmatrix} = \begin{pmatrix} \cos\theta & -\sin\theta \\ \sin\theta & \cos\theta \end{pmatrix} \begin{pmatrix} x \\ y \end{pmatrix}$$

and we have precisely the orthogonal linear transformation discussed in the previous paragraph.

In summary, the linear transformation T, with orthogonal matrix

$$\mathbf{P} = \begin{pmatrix} \cos\theta & -\sin\theta \\ \sin\theta & \cos\theta \end{pmatrix}$$

may be regarded either (i) as mapping each point A into another point A', where $OA = OA'$ and $A\hat{O}A' = +\theta$ or (ii) leaving each point in its original position but rotating the axes of coordinates through an angle $-\theta$ (i.e. through θ clockwise).

We shall now show how this new interpretation of an orthogonal matrix may be employed to find the major and minor axes of a central conic. Consider first the equation

$$ax^2+by^2 = 1 \quad (a, b \text{ not both negative}),$$

which represents a conic with its centre at the origin and its major and minor axes along the axes of coordinates. If we rotate the axes through an angle $-\theta$, that is, if we rename each point by the rule

$$x' = x\cos\theta - y\sin\theta,$$

$$y' = x\sin\theta + y\cos\theta,$$

then

$$x = x'\cos\theta + y'\sin\theta,$$

$$y = -x'\sin\theta + y'\cos\theta,$$

and the new equation of the conic takes the form

$$a(x'\cos\theta + y'\sin\theta)^2 + b(-x'\sin\theta + y'\cos\theta)^2 = 1,$$

or

$$(a\cos^2\theta + b\sin^2\theta)\,x'^2 + 2(a-b)\,x'y'\sin\theta\cos\theta + (a\sin^2\theta + b\cos^2\theta)\,y'^2 = 1.$$

We may rewrite this as

$$a'x'^2 + 2h'x'y' + b'y'^2 = 1,$$

which is the typical form of equation of a central conic with its centre at the origin.

Now consider the converse problem: given the equation of a central conic in the form

$$ax^2+2hxy+by^2 = 1,$$

how can we rotate the axes to obtain the equation in the form

$$a'x'^2+b'y'^2 = 1?$$

The equation $ax^2+2hxy+by^2 = 1$ may be rewritten in matrix form as

$$(x \quad y)\begin{pmatrix} a & h \\ h & b \end{pmatrix}\begin{pmatrix} x \\ y \end{pmatrix} = 1,$$

or

$$\mathbf{u}'\mathbf{A}\mathbf{u} = 1,$$

where

$$\mathbf{u} = \begin{pmatrix} x \\ y \end{pmatrix} \quad \text{and} \quad \mathbf{A} = \begin{pmatrix} a & h \\ h & b \end{pmatrix}.$$

The problem thus reduces to that of diagonalizing the matrix **A** for then the equation would become, in the new coordinate system,

$$(x' \quad y')\begin{pmatrix} \lambda_1 & 0 \\ 0 & \lambda_2 \end{pmatrix}\begin{pmatrix} x' \\ y' \end{pmatrix} = 1$$

or
$$\lambda_1 x'^2 + \lambda_2 y'^2 = 1.$$

But **A** is a symmetric matrix and, by Theorems 28.3 and 28.4, we can find an orthogonal matrix **P** such that $\mathbf{PAP}' = \mathbf{D}$ where

$$\mathbf{D} = \begin{pmatrix} \lambda_1 & 0 \\ 0 & \lambda_2 \end{pmatrix},$$

λ_1 and λ_2 being the eigenvalues of **A**. (Recall that, for an orthogonal matrix, $\mathbf{P}^{-1} = \mathbf{P}'$. Notice too that, to obtain $\mathbf{PAP}'$ in diagonal form (rather than $\mathbf{P}'\mathbf{AP}$), the matrix **P** is obtained by *transposing the matrix with columns which are the eigenvectors of the matrix* **A**; that is, the eigenvectors appear as the *rows*.) The fact that **P** is orthogonal tells us that the new axes, Ox', Oy' will be perpendicular. The details of the transformation are given below.

Consider the linear transformation defined by

$$\mathbf{v} = \mathbf{Pu},$$

where $\mathbf{v} = \begin{pmatrix} x' \\ y' \end{pmatrix}$, the new position vector of the point $\mathbf{u} = \begin{pmatrix} x \\ y \end{pmatrix}$. Then

$$\mathbf{P}'\mathbf{v} = \mathbf{P}^{-1}\mathbf{v} = \mathbf{u}$$

and the equation
$$\mathbf{u}'\mathbf{Au} = 1$$
transforms into the equation

$$(\mathbf{P}'\mathbf{v})'\,\mathbf{A}(\mathbf{P}'\mathbf{v}) = 1$$

or, since $(\mathbf{P}')' = \mathbf{P}$, and using the results of Ex. 8, Chapter 28,

$$\mathbf{v}'\mathbf{PAP}'\mathbf{v} = 1.$$

But $\mathbf{PAP}' = \mathbf{D}$, a diagonal matrix and the equation has been transformed into

$$(x' \quad y') \begin{pmatrix} \lambda_1 & 0 \\ 0 & \lambda_2 \end{pmatrix} \begin{pmatrix} x' \\ y' \end{pmatrix} = 1,$$

or
$$\lambda_1 x'^2 + \lambda y^{2\prime} = 1.$$

If we define the trace of the matrix **A**, tr **A**, as

$$\text{tr}\,\mathbf{A} = a + b$$

(see Exercise 28(*b*), Questions 8–10), the characteristic equation of **A**,

$$\begin{vmatrix} a-\lambda & h \\ h & b-\lambda \end{vmatrix} = 0,$$

may be written in the form

$$\lambda^2 - \lambda\,\text{tr}\,\mathbf{A} + \det\mathbf{A} = 0.$$

Thus we have
$$\operatorname{tr} \mathbf{A} = \lambda_1+\lambda_2,$$
$$\det \mathbf{A} = \lambda_1\lambda_2;$$
but
$$\operatorname{tr} \mathbf{D} = \lambda_1+\lambda_2,$$
$$\det \mathbf{D} = \lambda_1\lambda_2,$$
and thus both the trace and the determinant of the matrix defining the conic remain invariant under the given linear transformation.

**Ex.* 15. Show how the invariance of the trace and determinant enables us to deduce immediately the nature (ellipse, hyperbola) of a central conic
$$ax^2+2hxy+by^2 = 1.$$

Ex. 16. Prove that, if $\mathbf{Q}$ is any non-singular 2×2 matrix, then
$$\operatorname{tr}(\mathbf{Q}^{-1}\mathbf{AQ}) = \operatorname{tr}\mathbf{A} \quad \text{and} \quad \det(\mathbf{Q}^{-1}\mathbf{AQ}) = \det \mathbf{A}.$$

Can these results be generalized for higher-order matrices (the trace still being defined as the sum of the elements in the leading diagonal)?

Example 6. *Determine the nature of the central conic*
$$5x^2+8xy+11y^2 = 1$$
and find the lengths of its axes.

The matrix $\mathbf{A}$ of the conic is given by
$$\mathbf{A} = \begin{pmatrix} 5 & 4 \\ 4 & 11 \end{pmatrix}.$$

Let the eigenvalues of $\mathbf{A}$ be λ_1, λ_2; then
$$\lambda_1+\lambda_2 = 5+11 = 16,$$
$$\lambda_1\lambda_2 = 55-16 = 39$$
giving
$$\lambda_1 = 3, \quad \lambda_2 = 13.$$

The equation of the conic may thus be reduced to the form
$$3x'^2+13y'^2 = 1.$$

This is seen to be an ellipse, with major semi-axis $1/\sqrt{3}$ and minor semi-axis $1/\sqrt{13}$.

To draw a sketch of the ellipse, showing it in relation to the original axes, we need to find the inclination of the axes of the conic to the axes Ox, Oy. This may be done by obtaining an explicit form for the reducing matrix $\mathbf{P}$.

Corresponding to the eigenvalue $\lambda_1 = 3$ we have the unit eigenvector $\begin{pmatrix} 2/\sqrt{5} \\ -1/\sqrt{5} \end{pmatrix}$, while corresponding to $\lambda_2 = 13$ we have $\begin{pmatrix} 1/\sqrt{5} \\ 2/\sqrt{5} \end{pmatrix}$.

Thus
$$\mathbf{P}' = \sqrt{\tfrac{1}{5}}\begin{pmatrix} 2 & 1 \\ -1 & 2 \end{pmatrix}$$
and
$$\mathbf{P} = \sqrt{\tfrac{1}{5}}\begin{pmatrix} 2 & -1 \\ 1 & 2 \end{pmatrix}.$$
The transformation
$$\mathbf{v} = \mathbf{Pu}$$
may be interpreted as a rotation of axes clockwise through an angle
$$\theta = \arctan \tfrac{1}{2} \quad \text{(Figure 29.6)}.$$

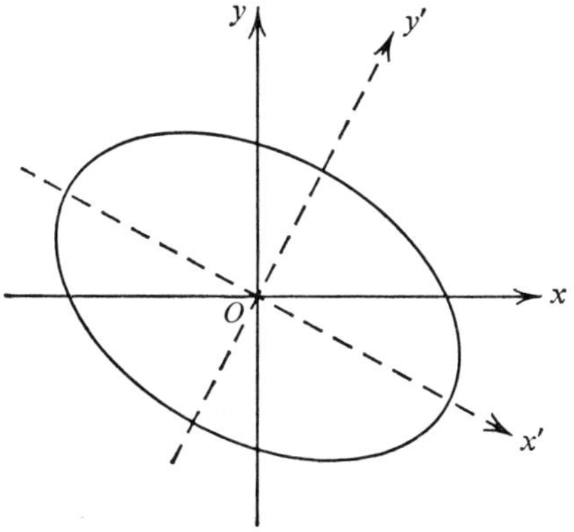

Fig. 29.6

In other words, in terms of the original coordinate system, the major and minor axes of the conic lie respectively along
$$x+2y = 0 \quad \text{and} \quad 2x-y = 0.$$

**Ex.* 17. Show that the equation
$$ax^2+2hxy+by^2 = 1$$
(i) represents an ellipse if $h^2-ab < 0$, and $a > 0$;
(ii) represents a hyperbola if $h^2-ab > 0$;
(iii) represents a rectangular hyperbola of $a+b = 0$;
(iv) represents a pair of parallel straight lines if $h^2 = ab$ $(h \neq 0)$.

Ex. 18. Let us write the conic of Ex. 17 in the form
$$\mathbf{x}'\mathbf{A}\mathbf{x} = 1,$$
where
$$\mathbf{A} = \begin{pmatrix} a & h \\ h & b \end{pmatrix}, \quad \mathbf{x} = \begin{pmatrix} x \\ y \end{pmatrix};$$
$\mathbf{u}$ is a unit direction vector and P is a point with position vector $\mathbf{p}$. The point with position vector $\mathbf{p}+\lambda\mathbf{u}$ lies on the conic if
$$(\mathbf{p}+\lambda\mathbf{u})'\,\mathbf{A}(\mathbf{p}+\lambda\mathbf{u}) = 1;$$

show that this reduces to

$$\lambda^2\mathbf{u}'\mathbf{A}\mathbf{u}+2\lambda\mathbf{u}'\mathbf{A}\mathbf{p}+\mathbf{p}'\mathbf{A}\mathbf{p} = 1.$$

If P is the mid-point of a chord parallel to **u**, prove that

$$\mathbf{u}'\mathbf{A}\mathbf{p} = 0;$$

what does this tell us about the vector **Au**?

The axes of a central conic (other than a circle) may be defined as the (unique) pair of *perpendicular conjugate diameters*: use the results proved above to deduce that **u** is in the direction of an axis if there exists a number k such that

$$\mathbf{A}\mathbf{u} = k\mathbf{u}.$$

Interpret your results in terms of eigenvectors.

Exercise 29(b)

1. Obtain the equation of the sphere:
(i) with centre $(0, -1, 0)$ and radius 1;
(ii) with centre $(2, -3, -1)$ and radius 5;
(iii) with centre (a, b, c) and radius $\sqrt{(a^2+b^2+c^2)}$.

2. Find the centre and radius of the sphere:
(i) with equation $x^2+y^2+z^2-6x-2y+2z-5 = 0$;
(ii) with equation $x^2+y^2+z^2+4x-8y-2z-60 = 0$;
(iii) with equation $x^2+y^2+z^2-2ax-2by+2cz+2bc-2ca+2ab = 0$.

3. Find the equation of the sphere:
(i) through the points $(4, 4, 4)$, $(5, 6, 1)$, $(0, -4, 2)$, $(7, 3, 2)$;
(ii) through the points $(2, 4, 4)$, $(5, 1, 4)$, $(3, 4, 3)$, $(-3, -1, 0)$;
(iii) through the points $(0, 0, 0)$, $(a, 0, 0)$, $(0, b, 0)$, $(0, 0, c)$.

4. Find the centre and radius of the circle:
(i) with equations $x^2+y^2+z^2-12x+2y-10z+44 = 0$, $x-y+z = 6$;
(ii) with equations $x^2+y^2+z^2-18x-10y+12z-55 = 0$, $2x+3y-4z = -1$.

5. The position vectors of points A and B relative to some origin O are respectively **a** and **b**. Show that the locus of a point P which moves in space in such a way that $PA^2+PB^2 = d^2$, where d is a constant, is a sphere with its centre at the mid-point of AB. Find the radius of the sphere in terms of **a**, **b** and d. What can you say about the sphere if $d = |\mathbf{a}-\mathbf{b}|$?

6. With the notation of Question 5, show that the locus of a point P which moves in space in such a way that $PA^2+PB^2+PC^2 = d^2$, where C is a point with position vector **c**, is a sphere and find its centre.

Can you generalize this result in any way?

7. Show that the line $\mathbf{r} = \lambda\mathbf{u}$, where **u** is a unit vector, cuts the sphere

$$(\mathbf{r}-3\mathbf{i}-2\mathbf{j}-2\mathbf{k}).(\mathbf{r}-3\mathbf{i}-2\mathbf{j}-2\mathbf{k}) = 6$$

at two points with parameters given by the quadratic equation

$$\lambda^2-2\lambda\mathbf{u}.(3\mathbf{i}+2\mathbf{j}+2\mathbf{k})+11 = 0.$$

Deduce that the line through the origin in the direction of the vector $\mathbf{i}+\mathbf{j}+3\mathbf{k}$ touches the given sphere.

Prove more generally that the line through the origin in the direction of the vector $l\mathbf{i}+m\mathbf{j}+n\mathbf{k}$ is tangential to the sphere if and only if

$$(3l+2m+2n)^2 = 11(l^2+m^2+n^2),$$

and hence show that the equation of the cone with vertex the origin which circumscribes the sphere is

$$(3x+2y+2z)^2 = 11(x^2+y^2+z^2).$$

8. Find the equation of the cone obtained by rotating the line

$$x/p = y/q = z/r$$

about the line $$x/a = y/b = z/c.$$

9. A right-circular cone of semi-angle α has its vertex at the origin and contains lines in the directions $\mathbf{i}$, $\mathbf{i}+\mathbf{j}$ and $\mathbf{i}+\mathbf{j}+\mathbf{k}$. Show that

$$\sec^2\alpha = 9-2\sqrt{2}-2\sqrt{6}.$$

10. Show that the equation $$yx+zx+xy = 0$$

represents a cone, vertex the origin and semi-angle arctan $\sqrt{2}$. What is the curve of intersection of this cone:

(i) with the plane $x+y+z = 1$;
(ii) with the plane $x-2y+z = 0$;
(iii) with the plane $x-2y+z = 1$?

11. A circle, S, is defined by the vector equation

$$(\mathbf{r}-\mathbf{a}).(\mathbf{r}-\mathbf{a})-c^2 = 0, \quad \mathbf{r}.\mathbf{n} = p.$$

Show that any sphere passing through the circle S has vector equation

$$(\mathbf{r}-\mathbf{a}).(\mathbf{r}-\mathbf{a})-c^2+\lambda[\mathbf{r}.\mathbf{n}-p] = 0$$

and deduce that, if B is a point not on the plane $\mathbf{r}.\mathbf{n} = p$ and if $\mathbf{OB} = \mathbf{b}$, then the vector equation of the sphere through S and containing the point B is

$$(\mathbf{b}.\mathbf{n}-p)[(\mathbf{r}-\mathbf{a}).(\mathbf{r}-\mathbf{a})-c^2]-[(\mathbf{b}-\mathbf{a}).(\mathbf{b}-\mathbf{a})-c^2](\mathbf{r}.\mathbf{n}-p) = 0.$$

Find the centre and radius of the sphere which passes through the point $(1, -1, 1)$ and contains the circle.

$$x^2+y^2+z^2 = 4, \quad x+2y+z = 1.$$

12. A is the fixed point with position vector $a\mathbf{k}$, and L, M are variable points with position vectors $\lambda\mathbf{i}$, $\mu\mathbf{j}$ respectively, where λ, μ are scalar parameters. If P is a point such that the angles LPM, MPA, APL are all right angles show that, whatever the values of λ and μ, P lies on a sphere, centre A and radius a.

13. Find the equation of the tangent to the sphere

$$x^2+y^2+z^2-10x-2y-12z+35 = 0$$

at the point (4, 2, 1).

Prove that the plane $x+y+z = 3$ is also a tangent to the sphere and find the point of contact.

14. Find the values of k for which the plane

$$4x+2y+3z = k$$

touches the sphere

$$x^2+y^2+z^2-10x-6y-10z+30 = 0.$$

15. The point B lies outside the sphere $(\mathbf{r}-\mathbf{a}).(\mathbf{r}-\mathbf{a})-c^2 = 0$. If $\mathbf{OB} = \mathbf{b}$, prove that the length of the tangent from B to the sphere is $(\mathbf{b}-\mathbf{a}).(\mathbf{b}-\mathbf{a})-c^2$.

Prove that the set of points from which the tangents to two non-intersecting spheres are of equal length is a plane which is perpendicular to the line of centres of the two spheres.

Deduce that the set of such points for three non-intersecting spheres is a line and that, given four non-intersecting spheres with non-coplanar centres, there is a point B from which the tangents to the four spheres are of equal lengths.

Explain how the last result can be used to show that a sphere may be circumscribed about any tetrahedron.

16. Describe the curve of intersection of the surface with equation

$$x^2+y^2-z^2 = 0;$$

(i) with the plane $z = 1$;
(ii) with the plane $x = 1$.
What surface does the equation represent?

17. Show that the parametric equation of the helix of pitch 8 which lies on the surface of the cylinder $x^2+z^2 = 1$ is

$$\mathbf{r} = \cos\theta\mathbf{i}+\frac{4\theta}{\pi}\mathbf{j}+\sin\theta\mathbf{k}.$$

Find the coordinates of the point in which the helix cuts the plane

$$\sqrt{2}x-y+2\sqrt{2}z = 0.$$

Show also that the helix cuts the plane $x+z = 0$ at an infinite number of points, and explain this geometrically.

18. Prove that the mid-points P, Q, R, S, T, U of the edges BC, CA, AB, AD, BD, CD of a tetrahedron $ABCD$, in which the opposite edges (BC, AD), (CA, BD), (AB, CD) are perpendicular pairs, lie on a sphere. Is the converse result, that if the points P, Q, R, S, T, U lie on a sphere then the opposite pairs of edges are perpendicular, true or false?

If the sphere cuts BC again at P', prove that BC is perpendicular to the plane $P'AD$.

19. Two spheres, centres A_1 and A_2 and radii c_1 and c_2 intersect in a circle of radius d. If $A_1A_2 = l$, where $l^2 = c_1^2+c_2^2$, prove that $d = c_1c_2/l$ and find the ratio in which the centre of the circle divides the line A_1A_2.

20. Prove that there is one and only one sphere which contains a given circle and passes through a given point not in the plane of the circle. Two circles, not in the same plane, meet at points A and B. A plane meets one of these circles at C and D and the other circle at E and F. Prove that:

(i) the points C, D, E, F lie on a circle;
(ii) the lines AB, CD, EF are either concurrent or parallel. (O & C)

21. Show that, if the ellipse $(x^2/a^2)+(y^2/b^2) = 1$ lying on the plane $z = 0$ is rotated through four right angles about the x axis, the equation of the surface obtained is

$$\frac{x^2}{a^2}+\frac{y^2}{b^2}+\frac{z^2}{b^2} = 1.$$

The surface thus obtained is called an *oblate spheroid* if $a < b$ and *a prolate spheroid* if $a > b$. Both surfaces are particular cases of the *ellipsoid.*

$$\frac{x^2}{a^2}+\frac{y^2}{b^2}+\frac{z^2}{c^2} = 1.$$

By considering sections of this ellipsoid with planes $x = p$, $y = q$, $z = r$, describe the shape of the general ellipsoid and in particular, show that it is a closed surface.

22. Describe the nature of the *hyperboloid of one sheet*

$$\frac{x^2}{a^2}+\frac{y^2}{b^2}-\frac{z^2}{c^2} = 1$$

and of the *hyperboloid of two sheets*

$$\frac{x^2}{a^2}-\frac{y^2}{b^2}-\frac{z^2}{c^2} = 1.$$

23. Describe the nature of the *elliptic paraboloid*

$$\frac{x^2}{a^2}+\frac{y^2}{b^2} = \frac{2z}{c}$$

and of the *hyperbolic paraboloid*

$$\frac{x^2}{a^2}-\frac{y^2}{b^2} = \frac{2z}{c}.$$

24. Prove that the hyperboloid of one sheet (see Question 22)

$$\frac{x^2}{a^2}+\frac{y^2}{b^2}-\frac{z^2}{c^2} = 1$$

contains the line determined by the two planes

$$\frac{x}{a}+\frac{z}{c} = \lambda\left(1+\frac{y}{b}\right) \quad \text{and} \quad \lambda\left(\frac{x}{a}-\frac{z}{c}\right) = 1-\frac{y}{b}.$$

Call this line a λ *generator*. Prove also that the surface contains μ *generators* of the form

$$\frac{x}{a}-\frac{z}{c} = \mu\left(1+\frac{y}{b}\right), \quad \mu\left(\frac{x}{a}+\frac{z}{c}\right) = 1-\frac{y}{b}.$$

Prove also that any pair of λ generators and any pair of μ generators are necessarily skew lines but that each λ generator meets every μ generator at a point with coordinates of the form

$$\left(\frac{a(\lambda+\mu)}{1+\lambda\mu}, \frac{b(1-\lambda\mu)}{1+\lambda\mu}, \frac{c(\lambda-\mu)}{1+\lambda\mu}\right).$$

(Notice that we have obtained a parametric representation of the hyperboloid of one sheet in terms of the two parameters λ, μ.)

25. (See Questions 23, 24.) Prove that the hyperbolic paraboloid

$$\frac{x^2}{a^2}-\frac{y^2}{b^2}=\frac{2z}{c}$$

contains a system of λ generators and a system of μ generators with the properties that any pair of λ generators are skew, any pair of μ generators are skew, but each λ generator meets every μ generator in a point of the form

$$(a(\lambda+\mu),\quad b(\lambda-\mu),\quad 2c\lambda\mu).$$

26. Determine the nature of the central conic

$$7x^2-8xy+13y^2 = 150$$

and find its eccentricity and the equations of its axes.

27. Determine the nature of the central conic

$$6x^2+5xy-6y^2 = 169$$

and find its foci and the equations of its axes.

28. Show that the equation

$$x^2+6xy+y^2+6 = 0$$

represents a hyperbola and find its eccentricity and the equations of its asymptotes.

29. Two lines, l_1 and l_2, intersect at an angle α. Find the locus of the centre of a sphere of given radius which touches both lines.

Revision exercise C

1. The polynomials $g(x)$, $h(x)$ both leave the same remainder on division by $(x-a)$. Prove that the polynomial

$$xg(x)-ah(x)$$

is divisible by $(x-a)$.

The polynomial $f(x)$ (degree $\geqslant 4$) leaves a remainder of $rx+s$ on division by $(x-b)^3$. Prove that $f''(x)$ is divisible by $(x-b)$.

2. If $z \neq 1$ and

$$w = \frac{1+z}{1-z}$$

show that $|z| = 1$ if and only if $\text{Re}\,(w) = 0$.

3. If $a < b \Rightarrow a^n < b^n$ for all $a, b \in R$ where $n \in Z$, what further conditions does n satisfy? Are these further conditions still necessary if a, b are restricted to positive real numbers? (S.M.P.)

4. The function $f: R \to R$ is defined by

$$f(x) = x^2 - 2|x|.$$

Sketch the graph of f and find what values of x are invariant under f.

5. Solve the equation $\quad 3\sin\theta + 2\sin(60° - \theta) = 2$

for values of θ lying between 0° and 360°.

6. A random number table consists of a succession of digits each chosen at random independently from the set 0, 1, 2, ..., 9. Two successive digits are taken from such a table; show that the probability that their sum is 9 is 1/10.

Four successive digits are taken from the table. Calculate the probability, p, that the sum of the first two equals the sum of the third and fourth. (M.E.I.)

7. Prove that

$$a + ar + ar^2 + \ldots + ar^{n-1} = \frac{a(1-r^n)}{1-r},$$

where $r \neq 1$.

Prove that in general

$$(1+x+x^2+\ldots+x^{2n})(1-x+x^2-\ldots+x^{2n}) = 1+x^2+x^4+\ldots+x^{4n}.$$

State any values of x for which your proof does not apply and obtain the appropriate results in each of these cases. (O & C)

8. Prove that, if the points O, A, C are non-collinear, then the position vector with respect to O of any point in the plane OAC may be expressed uniquely in terms of $\mathbf{a} = \mathbf{OA}$ and $\mathbf{c} = \mathbf{OC}$. (over)

$OABC$ is a parallelogram, L is the mid-point of AB and T is the point which divides OB in the ratio 3:2. CT meets OA at X and LT meets OC at Y. XY meets OB at Z. Find the ratio $OZ:ZB$ in which Z divides OB.

9. If

$$\mathbf{A} = \begin{pmatrix} 5 & 1 \\ 3 & 3 \end{pmatrix},$$

prove by induction that

$$\mathbf{A}^n = \tfrac{1}{4}(6^n - 2^n)(\mathbf{A} - 2\mathbf{I}) + 2^n\mathbf{I}.$$

10. If

$$\mathbf{M} = \begin{pmatrix} 2 & 3 \\ \frac{1}{2} & 1 \end{pmatrix}, \quad \mathbf{N} = \begin{pmatrix} 4 & -1 \\ 5 & -8 \end{pmatrix}, \quad \mathbf{Q} = \begin{pmatrix} 2 & 4 \\ 1 & 2 \end{pmatrix},$$

form the products **MQ** and **NQ**. What law, which is true for products in the algebra of real numbers but which is not true for products of matrices, is exemplified by this? (S.M.P.)

11. Express

$$p = \frac{5+\mathrm{j}}{2+3\mathrm{j}} \quad \text{and} \quad q = \frac{2\sqrt{2}}{1-\mathrm{j}\sqrt{3}}$$

in the form $x+\mathrm{j}y$, where x and y are real.

Find also the modulus and the argument of p and of q. Hence write down the argument of $p+q$. (M.E.I.)

12. Express in partial fractions:

(i) $\dfrac{6-3x}{(x-1)(x+2)}$; (ii) $\dfrac{2x-3}{x(1+x^2)}$; (iii) $\dfrac{3x^2-7x+3}{(1-x)^2(1-2x)}$.

13. Two concentric circles have radii 10 and 15 cm. If a point is taken at random in the annulus formed by them (all points of the annulus being equally likely) find the expected distance of the point from the centre of the circles. Explain why you would expect your answer to be greater than 12·5 cm.

14. P is the point $(at^2, 2at)$ on the parabola $y^2 = 4ax$ and OQ is the chord passing through the origin O and parallel to the tangent at P. Find the coordinates of the point of intersection, R, of the tangents at P and Q. Give a reason to show that the locus of R is another parabola.

If S is the mid-point of OQ, prove that PSR is a right-angled triangle, and that the area of triangle PQR is $\frac{1}{2}a^2t^3$. (O & C)

15. The sets A, B, C, D are defined as follows:

$$A = \{x \in R: -2 < x \leqslant 1\}, \quad B = \{x \in R: -1 \leqslant x < 2\},$$
$$C = \{x \in R: x \geqslant 1\}, \quad D = \{x \in R: x \leqslant -2\}.$$

Express in the form {:}, the sets:
(i) $A \cap B$; (ii) $A' \cap C$; (iii) $A \cap C$; (iv) $B' \cap C' \cap D'$;
(v) $(A \cap B) \cup (B \cap C')$.
Express the sets

$$E = \{x \in R: x < -1\} \quad \text{and} \quad F = \{x \in R: x < 2\}$$

in terms of unions and intersections of the sets A, B, C, D and their complements.

16. OAB is a triangle, right-angled at O; $OA = a$, $OB = b$. Points X, Y are taken on OA, OB (or on these lines produced) in such a way that

$$OX+OY = a+b.$$

AY and BX meet at P. Find the locus of P.

17. Assuming a population in which equal numbers of births occur in each of the twelve months of the year calculate the probabilities that, of four persons taken at random:

(i) no two will be found to have birthdays in the same month;

(ii) exactly two will have birthdays in the same month and the others in different months. (M.E.I.)

18. The polynomial ax^2+bx+c can be represented by the vector

$$\begin{pmatrix} a \\ b \\ c \end{pmatrix}.$$

Find a 3×3 matrix $\mathbf{M}$ which premultiplies this vector to give the vector representing the derivative of the polynomial. Check that $\mathbf{M}^3 = \mathbf{0}$ and comment on the significance of this result. (S.M.P).

19. Prove that

$$1^4-2^4+3^4-4^4+\ldots+(-1)^{n-1}\,n^4 = (-1)^{n-1}\,\tfrac{1}{2}n(n^3+2n^2-1).$$

20. Find all angles in the interval $0 \leqslant \theta \leqslant 2\pi$ satisfying the equation

$$\sin 2\theta+\cos(\theta-\tfrac{1}{4}\pi) = 0.$$

21. If k is very small, show that $x = 1-\frac{1}{2}k$ is an approximation to a root of the equation

$$kx^3+x^2-1 = 0,$$

and find approximate values for the other two roots.

22. Solve the inequality

$$\frac{(x-1)(x+1)}{(x-7)(x+3)} \leqslant 1.$$

23. Express

$$y = \frac{4x-1}{(2x-1)(3x-1)}$$

in partial fractions and deduce a quadratic approximation for y in the neighbourhood of the point $(0, -1)$.

Deduce that the line

$$x+y = -1$$

is the tangent to the graph of y at the point $(0, -1)$ and that, near this point, the curve lies in the region

$$\{(x, y): x+y \geqslant -1\}.$$

24. a, p, q, r are four non-zero complex numbers such that

$$p+q+r = a, \quad p^2+q^2+r^2 = a^2, \quad pqr = a^*.$$

Find, in terms of a,

(i) $|(a-p)(a-q)(a-r)|$; (ii) $1/p+1/q+1/r$.

25. Three points P, Q, R vary in position on the surface of a fixed sphere in such a way that

$$PA^2+QA^2+RA^2$$

is constant, where A is a fixed point. Find the locus of the centroid of the triangle PQR.

26. If θ is real and $z = \cos\theta + \mathrm{j}\sin\theta$, find

$$z^n+\frac{1}{z^n} \quad \text{and} \quad z^n-\frac{1}{z^n}.$$

Determine a, b, c, d such that

$$2^7\cos^3\theta\sin^3\theta = a\sin 8\theta + b\sin 6\theta + c\sin 4\theta + d\sin 2\theta. \qquad \text{(M.E.I.)}$$

27. Show that the cubic equation

$$2x^3+3x^2-12x+k = 0 \quad (k \text{ real})$$

has three distinct real roots if and only if

$$-20 < k < 7.$$

Solve the equation completely if $k = -13$.

28. Two real numbers are taken at random from the set of real numbers

$$\{x \in R: 0 \leqslant x \leqslant 1\}.$$

Find the probability:
(i) that the square of their sum is greater than 1;
(ii) that the sum of their squares is greater than 1.

29. Show that the point

$$\left(a(1-t^2)/(1+t^2),\ \ 2bt/(1+t^2)\right)$$

lies on the ellipse

$$\frac{x^2}{a^2}+\frac{y^2}{b^2} = 1$$

for all values of t.

Deduce that, if the line

$$lx+my = 1$$

is a tangent to the ellipse, then

$$l^2a^2+m^2b^2 = 1.$$

30. Prove that

$$1^2+2^2+\ldots+n^2 = \tfrac{1}{6}n(n+1)(2n+1).$$

If

$$s_1 = 1^2+4^2+7^2+\ldots+(3n-2)^2,$$

and

$$s_2 = 2^2+5^2+8^2+\ldots+(3n-1)^2$$

show that

$$s_2-s_1 = 3n^2.$$

and

$$s_1+s_2 = 6n^3-n.$$

Hence, or otherwise, find s_1 and s_2. (O & C)

31. Prove that the medians of a triangle ABC are concurrent at a point G.

O is an arbitrary point in the plane of triangle ABC and L, M, N are the mid-points of BC, CA, AB. Prove that the lines through L, M, N drawn parallel to OA, OB, OC respectively are concurrent at a point H.

Prove also that the line GH passes through O and determine the ratio $OG:GH$.

32. It is observed that

$$\begin{aligned} 1+2 &= 3 \\ 4+5+6 &= 7+8 \\ 9+10+11+12 &= 13+14+15 \\ 16+17+18+19+20 &= 21+22+23+24. \end{aligned}$$

Enunciate a general result of which the above are particular cases and prove your conjecture.

33. For what values of θ in the interval $0° < \theta < 360°$ is the expression

$$6\cos^2\theta - \cos\theta$$

less than one? Give your answers to the nearest tenth of a degree.

Answer the same question for the expression

$$6\sin^2\theta - \sin\theta.$$

34. Express

$$\frac{x-4}{(x-3)(x-2)}$$

in partial fractions.

If x is so large that $(1/x)^3$ and higher powers of $1/x$ may be neglected, show that the given expression is approximately equal to $(x^2+x-1)/x^3$.

35. Show that there is just one real value of the constant λ for which the equations $\lambda x+y+z = 2x+\lambda y+3z = 4x-2y+\lambda z = 0$ have a common solution other than $x = y = z = 0$. Find, for this value of λ, the solution of these equations which satisfies also the equation $x+y+z = 3$. (M.E.I.)

36. A and B play a match of five games, each of which must be won or lost. In each of the first three games the probability that A will win is 2/3 and in the remaining two games the probability is 3/4. Find:

(i) the generating function for the probability that A wins r games;
(ii) the probability that A will win the match;
(iii) the average number of games that A will win. (O and C modified)

37. Prove that, if $\mathbf{A}$, $\mathbf{B}$ are two 3×3 matrices, then

$$(\mathbf{BA})' = \mathbf{A}'\mathbf{B}'.$$

What is the transpose of a 1×1 matrix?

Let T, T' be two linear transformations of three-dimensional space into itself, with matrices $\mathbf{A}$, $\mathbf{A}'$, where $\mathbf{A}'$ is the transpose of $\mathbf{A}$. Prove that, if

$$T(P) = Q \quad\text{and}\quad T'(P) = R,$$

then OP is perpendicular to QR.

38. Show that the points $O(0, 0, 0)$, $A(0, 2, 2)$, $B(2, 0, 2)$, $C(2, 2, 0)$ are the vertices of a regular tetrahedron. (over)

L, M are the mid-points of the edges OA, OB respectively. The tetrahedron $OABC$ is now cut through by a plane CML. Find the cosines of the acute angles between the plane CML and
(i) the plane OML; (ii) the plane OCM.

39. Prove that, for any complex number z, $zz^* = |z|^2$.
The reflection of the point z_1 in the line $\theta = \alpha$ through the origin in the Argand diagram is z_2 and the reflection of z_2 in the line $\theta = \beta$ through the origin is z_3. Prove that

$$z_2 = z_1(\cos 2\alpha + \mathrm{j} \sin 2\alpha)$$

and that

$$z_3 = z_1[\cos(2\beta - 2\alpha) + \mathrm{j} \sin(2\beta - 2\alpha)].$$

40. Prove, by induction, or otherwise, that

$$\sum_{r=1}^{n} r(r+1)\left(\tfrac{1}{2}\right)^{r-1} = 16 - (n^2+5n+8)\left(\tfrac{1}{2}\right)^{n-1}.$$ (M.E.I.)

41. If $\mathbf{OA} = \mathbf{a}$, $\mathbf{OB} = \mathbf{b}$, prove that the position vector, relative to O, of the mid-point of AB is $\frac{1}{2}(\mathbf{a}+\mathbf{b})$. Deduce that the lines joining the mid-points of opposite edges of any tetrahedron are concurrent.
If $ABCD$, $A'B'C'D'$ are any two parallelograms, not necessarily coplanar, and if the mid-points of AA', BB', CC', DD' are respectively W, X, Y, Z, prove that $WXYZ$ is a parallelogram.

42. Give rough sketches of the curves:
(i) $y = \sin x$; (ii) $y = \sin(x+\frac{1}{3}\pi)$; (iii) $y = 1+2\sin(x+\frac{1}{3}\pi)$.
How many real roots has the equation

$$x = 1+2\sin(x+\tfrac{1}{3}\pi)?$$

43. Sketch the graph of

$$y = \frac{3x+14}{x^2-5x-14}.$$

For what values of x is $y < -1$?

44. Find the coordinates of the point of intersection, P, of the plane

$$x+3y+z = 0,$$

and the line joining $A(0, 1, 15)$ and $B(3, -5, 3)$.
Find the ratio $AP : PB$.

45. The line l is a tangent to a parabola with focus F and N is the foot of the perpendicular from F to l. Prove that N lies on the tangent at the vertex and that FN produced meets the directrix in a point M such that $FN = NM$.
The triangle ABC is right-angled at B. It is reflected in a line m in such a way that the image of C lies in the original line of AB. Prove that m is a tangent to a parabola with focus at C and directrix AB. (M.E.I.)

46. Interpret geometrically the transformations performed by premultiplying the position vectors of points in a plane by each of the matrices

$$R = \begin{pmatrix} \cos\theta & -\sin\theta \\ \sin\theta & \cos\theta \end{pmatrix}, \quad \mathbf{S} = \begin{pmatrix} 1 & c \\ 0 & 1 \end{pmatrix}, \quad \mathbf{T} = \begin{pmatrix} 1/k & 0 \\ 0 & k \end{pmatrix}.$$

State in the form of a single matrix each of the expressions $\mathbf{R}^n$, $\mathbf{S}^n$, $\mathbf{T}^n$, where n is a positive integer, and prove one of these results by induction. (M.E.I.)

47. Show that there are two complex numbers z such that

$$|z-2-\mathrm{j}| = 1$$

and $$\arg z = \tfrac{1}{4}\pi$$

and find their moduli.

48. The roots of the equation

$$ax^2+bx+c = 0,$$

where a, b and c are non-zero constants, are α, β and the roots of the equation

$$ax^2+2bx+c = 0$$

are γ, δ. Without solving either equation, show that the equation whose roots are $\alpha\gamma+\beta\delta$ and $\alpha\delta+\beta\gamma$ is

$$a^3x^2-2ab^2x+c(5b^2-4ac) = 0.$$

Show also that if this last equation has equal roots then one and only one of the original equations has equal roots. (M.E.I.)

49. Show that:

(i) $k^{\frac{1}{3}}-1 = \dfrac{k-1}{k^{\frac{2}{3}}+k^{\frac{1}{3}}+1}$,

(ii) $(1+x)^{\frac{1}{3}} < 1+x/3 \quad (x > 0)$,

and deduce that, when $x > 0$,

$$(1+x)^{\frac{1}{3}}-1 > \frac{9x}{27+9x+x^2}.$$

Hence show that the error in taking $(1+x)^{\frac{1}{3}}$ as $1+x/3$ is less than

$$\frac{9x^2+x^3}{81+27x+3x^2}.$$

(M.E.I.)

50. A machine produces components to any required length specification with a standard deviation of 1·40 mm. At a certain setting it produces to a mean length of 102·30 mm. Assuming the distribution of lengths to be normal, calculate:

(i) what percentage would be rejected as less than 100 mm long;

(ii) to what value, to the nearest 0·01 mm, the mean should be adjusted if this rejection rate is to be 1%;

(iii) whether at the new setting more than 1% of components would exceed 107 mm in length. (M.E.I.)

51. Prove that, if z, w are complex numbers such that

$$|z-2| = |w+2| = 1,$$

then $$-\tfrac{1}{6}\pi \leqslant \arg(z-w) \leqslant \tfrac{1}{6}\pi.$$

Find the limits between which $|z-w|$ must lie.

52. If $A = \begin{pmatrix} 1 & 3 \\ 1 & 1 \end{pmatrix}$, $\mathbf{u} = \begin{pmatrix} 1 \\ 1 \end{pmatrix}$, $A^n\mathbf{u} = \begin{pmatrix} x_n \\ y_n \end{pmatrix}$ prove that, for large n, $x_n \approx y_n\sqrt{3}$ and interpret your result geometrically.

53. Prove that, for any positive integer r,

$$\tan(2^{r-1}\theta) \equiv \cot(2^{r-1}\theta) - 2\cot(2^r\theta).$$

Hence evaluate

$$\sum_{r=0}^{n} 2^{r-1}\tan(2^{r-1}\theta).$$

54. Two players stake £1 each in a game in which eight counters are tossed on the ground. The counters are painted black on one side and white on the other. If the number of blacks showing is odd the thrower wins the other stake, if all counters are of the same colour he wins a double stake, and otherwise he loses his stake. Calculate the expected gain of the thrower and state what assumptions you make. (M.E.I.)

55. Prove that

$$\sum_{r=1}^{n} r^2 = \tfrac{1}{6}n(n+1)(2n+1).$$

The x axis from $x = 1$ to $x = 2$ is divided into n equal intervals by the points

$$(1, 0), \quad \left(1+\frac{1}{n}, 0\right), \quad \left(1+\frac{2}{n}, 0\right), \quad \ldots, \quad \left(1+\frac{n-1}{n}, 0\right), \quad (2, 0).$$

Rectangles with sides parallel to the axes are drawn on these intervals with one vertex on the curve $y = x^2$ and the rest of each rectangle between the curve and the x axis. Show that the sum of the areas of these rectangles is

$$1+\left(1-\frac{1}{n}\right)\left(\frac{4}{3}-\frac{1}{6n}\right).$$

Show further that as n increases this sum increases but remains less than $\frac{7}{3}$.

56. Show that, if the roots α and β of the quadratic

$$ax^2+(b+c)x+d = 0$$

are unequal, the equation

$$au_n u_{n+1}+bu_n+cu_{n+1}+d = 0$$

can be put in the form

$$\frac{u_{n+1}-\alpha}{u_{n+1}-\beta} = \lambda\frac{u_n-\alpha}{u_n-\beta},$$

where λ is given by the equation

$$\frac{\lambda+1}{\lambda-1}(\alpha-\beta) = \frac{b-c}{a}.$$

57. If $a/b - 1$ is a small positive quantity not greater than 10^{-N}, show that

$$\left(\frac{a}{b}\right)^{\frac{1}{2}} - \frac{3a+b}{a+3b}$$

is not greater than 10^{-3N-1}.

58. The linear transformation, T, has matrix (relative to **i**, **j**, **k**)

$$\mathbf{A} = \begin{pmatrix} -1 & -3 & -1 \\ 2 & 4 & 1 \\ 0 & 0 & 1 \end{pmatrix}.$$

Show that T leaves every point of the plane

$$2x+3y+z = 0$$

fixed and that, if Π is any plane containing the line of intersection of the two planes

$$x+y = 0, \quad z = 0$$

then Π is mapped into itself by T.

59. OAB is a fixed triangle and points X, Y are taken on OA, OB (or these sides produced) in such a way that AB and XY are parallel. AY and BX meet at P. Show that, for varying positions of X, Y, the locus of P is OC, where C is a vertex of the parallelogram $OACB$.

60. Explain what is meant by the set $R \times R$.

The function f: $R \times R \rightarrow R$ is defined by the equation

$$f(x, y) = x+y.$$

Find the image, under f, of:

(i) $\{(x, y) \in R \times R: 2x+y = 1\}$; (ii) $\{(x, y) \in R \times R: x^2+y^2 = 1\}$.

61. Find the modulus and argument of $1+\cos\theta+\mathrm{j}\sin\theta$, where $-\pi < \theta < \pi$, and hence or otherwise express $(1+\cos\theta+\mathrm{j}\sin\theta)^n$ in the form $a+\mathrm{j}b$, where n is a positive integer.

By writing $\cos\theta+\mathrm{j}\sin\theta = z$, and using the binomial expansion, prove that

$$1+\binom{n}{1}\cos\theta+\binom{n}{2}\cos 2\theta+\ldots+\cos n\theta \equiv (2\cos\tfrac{1}{2}\theta)^n \cos\tfrac{1}{2}n\theta,$$

and obtain a similar identity with sines on the left-hand side. (O & C)

62. If the equation $$x^3+ax^2+bx+c = 0$$

has the property that one root is equal to the product of the other two, prove that

$$(b-c)^2+c(a-1)^2 = 0.$$

If $a = \frac{5}{2}$ and $b = \frac{1}{2}$, find the two values of c which satisfy this condition, and solve the resulting cubic equations. (O & C)

63. Prove that the ellipse

$$\frac{x^2}{a^2}+\frac{y^2}{b^2} = 1$$

and the hyperbola $$xy = c^2$$

touch each other if $2c^2 = ab$. Is the converse true?

Let D be one of the points of contact in the case $2c^2 = ab$. Show that if the common normal at D meets the hyperbola again at A and the coordinate axes at P, Q, then the mid-point of PQ coincides with the mid-point of AD. (M.E.I.)

64. A probability density $p(x)$ is given by

$$p(x) = \begin{cases} Cx(4-x) & (0 \leqslant x \leqslant 4), \\ 0 & (x < 0, 4 < x), \end{cases}$$

where C is a positive constant. Find the value of C and sketch the distribution.

Explain why the standard deviation of this distribution is the same as the standard deviation of the distribution with probability density

$$p(x) = \begin{cases} C(4-x^2) & (|x| \leqslant 2), \\ 0 & (|x| > 2). \end{cases}$$

For the first distribution, calculate the mean, the standard deviation and the mode. Calculate also the probability that a value of the variable x taken at random will lie further than one standard deviation away from the mean. (M.E.I.)

65. If a circle passes through the origin, show that its equation, written in complex form, is $zz^*+g^*z+gz^* = 0$, where $z = x+jy$, $z^* = x-jy$, and g is a complex constant.

If a variable point Z on the circle is transformed into a point W by means of the equation $w = a^2/z$, where a is a real constant, find the equation satisfied by w. What geometrical locus does this equation represent? (O & C)

66. By solving the set of equations

$$\begin{aligned} x_1 \qquad\qquad &= a, \\ -x_1+x_2 \qquad &= b, \\ 3x_1+2x_2+x_3 &= c \end{aligned}$$

or otherwise, find the inverse of the matrix

$$\mathbf{A} = \begin{pmatrix} 1 & 0 & 0 \\ -1 & 1 & 0 \\ 3 & 2 & 1 \end{pmatrix}.$$

Given
$$\mathbf{B} = \begin{pmatrix} 1 & 4 & -2 \\ 0 & 1 & 3 \\ 0 & 0 & 1 \end{pmatrix},$$

find $\mathbf{B}^{-1}$ and also, by any other method, the inverse of $\mathbf{AB}$. State a set of equations which this inverse $(\mathbf{AB})^{-1}$ would help you to solve. (M.E.I.)

67. An unbiased six-sided die is thrown. If a six is thrown the die is thrown a second time and the scores for the two throws are added; find

(i) the chance of a score of more than eight;

(ii) the average score.

If the die is thrown again whenever a six has been thrown, find the average number of throws and the average total score. (O & C)

68. The position vectors relative to the origin O of the points A, B, P are respectively **a**, **b**, **p**. X is the point of trisection of AP nearer A, and Y is the mid-point of BP. Z is a vertex of the parallelogram $PXZY$. Write down, in terms of **a**, **b**, **p**, the position vectors of X, Y, Z.

If A, B are fixed points on a unit sphere, centre O, and P is a variable point on the sphere, deduce that the locus of Z is a sphere and find its radius and the position vector of its centre.

69. Prove that the lines

$$\tfrac{1}{2}(x-3) = y+4 \quad = z-3,$$
$$x+2 = \tfrac{1}{2}(y-3) = z+3,$$
$$\tfrac{1}{2}(x+2) = -\tfrac{1}{5}y \quad = \tfrac{1}{2}z$$

are skew and find the equations of the line through the origin which intersects all three lines.

70. UV is a diameter of an ellipse and P, Q are points on the ellipse. PU, QV meet at X and PV, QU meet at Y; prove that the chord defined by XY is bisected by UV.

71. Prove that, if ξ is any root of the equation

$$z^n = (z-2\mathrm{j})^n,$$

then $\mathrm{Im}\,(\xi) = 1$.

Draw a sketch to show the position of the roots in the case $n = 3$ and show how they are related to the equations

(i) $z^3 = (z+2\mathrm{j})^3$; (ii) $z^3 = (z+2)^3$.

72. Express the vector $\mathbf{i}-4\mathbf{j}-5\mathbf{k}$ as

(i) the sum of three vectors in the directions $\mathbf{i}$, $\mathbf{i}+\mathbf{j}$, $\mathbf{i}+\mathbf{j}+\mathbf{k}$;

(ii) the sum of two vectors, one in the plane $2x+4y+3z = 1$ and one perpendicular to this plane.

73. A single die is thrown six times. Find the probability that:

(i) exactly one six is thrown;

(ii) at least two sixes are thrown;

(iii) each face of the die turns up once.

Find the probability that, when a die is thrown repeatedly, a six appears for the first time at the rth throw. Find the least value of r such that there is more than a 50 per cent chance of obtaining a six on the rth throw or earlier.

Write down expressions, in terms of $p = \frac{1}{6}$ and $q = \frac{5}{6}$, for the expectations $\mathscr{E}(r)$, $\mathscr{E}(r^2)$ and calculate the numerical values of these expressions. (M.E.I.)

74. A transformation is defined by

$$\begin{pmatrix} x' \\ y' \\ z' \end{pmatrix} = \begin{pmatrix} -2 & 1 & 4 \\ 1 & 0 & -2 \\ 3 & 4 & -6 \end{pmatrix} \begin{pmatrix} x \\ y \\ z \end{pmatrix}.$$

Show that the set of planes perpendicular to the y axis maps into a set of parallel lines. What is the direction of these lines? Which of these planes is mapped onto a line through the origin? (S.M.P.)

75. In the binomial expansion of $(1-x)^{\frac{1}{3}}$, where x is positive and less than 1, show that the ratio of the term in x^{n+1} to that in x^n is

$$\frac{3n-1}{3n+3}x. \qquad \text{(over)}$$

By comparing the sum of all the terms after the first three with a suitable infinite geometric series, prove that the magnitude of the error involved in neglecting these terms is less than $(5x^3/81)(1-x)^{-1}$.

Show that, if $x \leqslant 0{\cdot}1$, this error is, in magnitude, less than $0{\cdot}00007$ and hence, using the binomial expansion, evaluate $(0{\cdot}9)^{\frac{1}{3}}$ correct to three decimal places.

76. Prove that the points with coordinates $(0, -2, -\frac{1}{2})$, $(1, 1, 1)$ $(5, 3, 4)$ and $(8, 2, 5\frac{1}{2})$ are the vertices of a trapezium in which the non-parallel sides are equal in length. (S.M.P.)

77. The fixed point C and a variable point P have position vectors $\mathbf{c}$, $\mathbf{r}$ respectively in three-dimensional Euclidean space. Describe in geometrical terms the locus Γ defined by

$$\mathbf{r}.\mathbf{c} = |\mathbf{r}||\mathbf{c}| \cos \alpha,$$

where α is a given acute angle. If C is the point (a, b, c) and P the point (x, y, z) express the equation of Γ in cartesian form and show that, for all real numbers λ,

$$(x, y, z) \in \Gamma \Rightarrow (\lambda x, \lambda y, \lambda z) \in \Gamma.$$

Taking the special case with C the point $(4, 0, 3)$, $\cos \alpha = 4/5$, show that Γ meets the plane $z = 6$ in a parabola and that the sphere $|\mathbf{r}-\mathbf{c}| = 3$ touches the plane at the focus of this parabola. (M.E.I.)

78. Obtain a quadratic approximation for the expression

$$\sqrt{(1+x)/(1-2x)}$$

in the neighbourhood of $x = 0$.

Write down the equation of the tangent to the curve

$$y = \sqrt{(1+x)/(1-2x)}$$

at the point $(0, 1)$ and draw a sketch showing the relative position of the curve and the tangent at this point.

79. $ABCDA'B'C'D'$ is a parallelepiped, with faces $ABCD$, $A'B'C'D'$ and parallel edges AA', BB', CC', DD', etc. The mid-points of CC', AB', $B'C'$, $C'D'$ are respectively W, X, Y, Z. Prove that the plane XYZ cuts AW externally in the ratio 1:3.

80. Find the five roots of the equation

$$z^5+1 = 0,$$

and plot their positions on an Argand diagram.

A regular pentagon $ABCDE$ is inscribed in a circle of unit radius. Prove that, if P is any point of the circle, then $PA.PB.PC.PD.PE \leqslant \sqrt{2}$.

81. If $\mathbf{A}$ is a 2×2 singular matrix, prove that there is a number a such that, for any integer $k > 0$,

$$\mathbf{A}^k = a^{k-1}\mathbf{A}.$$

Find a if $\mathbf{A}$ is the matrix $\begin{pmatrix} 4 & 6 \\ 2 & 3 \end{pmatrix}$.

$\mathbf{B}$ is the matrix $\begin{pmatrix} 5 & 6 \\ 2 & 4 \end{pmatrix}$;

prove that $\mathbf{B}^n = \frac{1}{7}(8^n-1)\,\mathbf{A}+\mathbf{I}$

for any positive integer n.

82. Prove that $$\binom{n}{r} = \binom{n}{n-r} \qquad (0 \leqslant r \leqslant n).$$

By considering the product $(1-x)^n\,(1+x)^n$,

prove that $$\sum_{r=0}^{n} (-1)^r \left[\binom{n}{r}\right]^2 = \binom{2n}{n}.$$

83. A boy and girl play a game of 'spotting' approaching red cars when travelling along a road on which 25% of the cars are red. If a car is red the first one to 'claim' it gains one point (a claim is always made by either the boy or girl for a red car, but never by both simultaneously). If a car is not red it is found that on half the occasions either the boy or the girl (but not both) makes a claim, for which there is a penalty of two points. The first to be five or more points ahead is the winner.

The events 'a claim by the boy' and 'the car is red' are independent and the probability that he makes a claim on a car is 2/5.

(i) Draw a tree diagram in which the first two branches lead to the events 'red' and 'not red', and mark on the secondary branches the probabilities of claims by the children.

(ii) If it is known that a claim has been made for a car which is not red, what is the probability it was made by the girl?

(iii) If the score is 8–5 in favour of the girl, show that the probability that she will have won by the time two more cars have passed is 0·48. (S.M.P.)

84. Find the solution set of the equations

$$\begin{pmatrix} 1 & 3 & a \\ 2 & -1 & -5 \\ 1 & 1 & 2 \end{pmatrix}\begin{pmatrix} x \\ y \\ z \end{pmatrix} = \begin{pmatrix} 4 \\ b \\ 1 \end{pmatrix}$$

in the following cases:

(i) $a = 9, b = -1$; (ii) $a = 8, b = -1$; (iii) $a = 8, b = -2\frac{1}{2}$.

For what value of a does the equation

(i) $$\begin{pmatrix} a & -5 & 2 \\ 3 & -1 & 1 \\ 1 & 2 & 1 \end{pmatrix}\begin{pmatrix} x \\ y \\ z \end{pmatrix} = \begin{pmatrix} 4 \\ 7 \\ 8 \end{pmatrix}$$ not have a unique solution;

(ii) $$\begin{pmatrix} a & -10 & 4 \\ 3 & -1 & 1 \\ 1 & 2 & 1 \end{pmatrix}\begin{pmatrix} x \\ y \\ z \end{pmatrix} = \begin{pmatrix} 4 \\ 7 \\ 9 \end{pmatrix}$$ not have a unique solution? (S.M.P.)

85. Use de Moivres's Theorem to prove that

$$\cos n\theta = \Sigma(-1)^r \binom{n}{2r} \cos^{n-2r}\theta \sin^{2r}\theta,$$

and deduce that $\cos n\theta$ may be expressed as a polynomial in $\cos\theta$.

Write $$\cos n\theta = \sum_{r=0}^{n} a_r \cos^r\theta.$$

By taking $\theta = 0$ prove that $a_n = 2^{n-1}$. What is a_0 if n is even?

86. Find and illustrate graphically the set of points (x, y) that satisfy simultaneously the inequalities

$$(x-1)^3 \leqslant x-1,$$
$$y^2 \leqslant |x|. \qquad \text{(M.E.I.)}$$

87.
$$\mathbf{M} = \begin{pmatrix} 1 & a & b \\ 1 & a^2 & b^2 \end{pmatrix}.$$

By considering the matrix product $\mathbf{M}'\mathbf{M}$, or otherwise, prove that

$$\begin{vmatrix} 2 & a+a^2 & b+b^2 \\ a+a^2 & a^2+a^4 & ab+a^2b^2 \\ b+b^2 & ab+a^2b^2 & b^2+b^4 \end{vmatrix} = 0.$$

Evaluate the determinant

$$\begin{vmatrix} 2 & x+y & x+y \\ x+y & x^2+y^2 & 2xy \\ x+y & 2xy & x^2+y^2 \end{vmatrix}.$$

88. Describe geometrically the sets of points in the Argand diagram defined by
(i) $|z-1| = |z+\mathrm{j}|$;
(ii) $|z-1| = 2|z+\mathrm{j}|$;
(iii) $|z-1|+|z+\mathrm{j}| = 2$;
(iv) $\mathrm{Re}\,(z+1) = |z+\mathrm{j}|$.

89. The points $P(c\alpha, c/\alpha)$ and $Q\ (c\beta, c/\beta)$ lie on the rectangular hyperbola $xy = c^2$, and (X, Y) is the mid-point of PQ. Find X and Y in terms of c, α and β. Write down a quadratic equation in t whose coefficients involve only c, X and Y and whose roots are α and β.

Determine what points (X, Y) can arise as mid-points of real chords of the hyperbola. On a rough sketch shade in the regions they occupy. (O & C)

90. Show how the matrix
$$\mathbf{A} = \begin{pmatrix} \cos 2\theta & \sin 2\theta \\ \sin 2\theta & -\cos 2\theta \end{pmatrix}$$
can be associated with reflection in the line $y = x \tan \theta$.

If $\mathbf{B}$ is the matrix associated similarly with reflection in the line $y = x\sqrt{3}$, find the values of θ for which $\mathbf{AB} = \mathbf{BA}$.

Describe how these values could be predicted by geometrical argument
(M.E.I.)

91. A rectangular box has faces $OABC$, $O'A'B'C'$ and OO', etc. vertical. $OA = 2$ units, $OC = 1$ unit, $OO' = 1$ unit. Taking O as origin and axes Ox, Oy, Oz along OA, OC, OO', find the equations of the planes $OA'C'$, ACM, where M is the mid-point of $B'C'$. Hence show that
(i) the line of intersection of these planes is parallel to the face $OABC$;
(ii) the cosine of the acute angle between planes is $7\sqrt{6}/18$.

92. A, B are two fixed points with position vectors $\mathbf{a}$, $\mathbf{b}$ relative to some origin O. X, Y are two variable points with position vectors $\mathbf{x}$, $\mathbf{y}$. Show that, if

$$\mathbf{x}+\mathbf{y} = \mathbf{a} \quad \text{and} \quad \mathbf{x} \wedge \mathbf{y} = \mathbf{b},$$

then O, X, Y, A are coplanar and OB is perpendicular to this plane. Hence find the most general solution, in terms of **a**, **b** and a scalar parameter, of the vector equations

$$\mathbf{x}+\mathbf{y} = \mathbf{a}, \quad \mathbf{x} \wedge \mathbf{y} = \mathbf{b}.$$

93. Prove that the matrix

$$\mathbf{P} = \begin{pmatrix} \cos\theta & -\sin\theta & 0 \\ \sin\theta & \cos\theta & 0 \\ 0 & 0 & 1 \end{pmatrix}$$

is orthogonal and interpret **P** as the matrix of a linear transformation. Show that, if x is very small, the matrix

$$\mathbf{Q} = \begin{pmatrix} 1 & -x & 0 \\ x & 1 & 0 \\ 0 & 0 & 1 \end{pmatrix}$$

represents a small rotation, if we ignore powers of x higher than the first. If $\delta\mathbf{S}$ is a matrix all of whose elements are very small, and if

$$\mathbf{I}+\delta\mathbf{S}$$

is orthogonal, prove that the matrix $\delta\mathbf{S}$ is skew-symmetric.

94. Two men, A and B, play a match consisting of separate games, the probability of A's winning a game being p and the probability of B's winning being q. ($p+q$ is not necessarily equal to one.) They start the match with n counters each and the winner of each game receives a counter from the loser; the first player to win all $2n$ counters wins the match. If the probability that, when A has k counters he will eventually win the match, is denoted by u_k, prove that

$$(p+q)\, u_k = pu_{k+1}+qu_{k-1}.$$

What is the initial probability that A will win the match?

95. Prove that, if n is a positive integer, then

$$0 < (2-\sqrt{2})^n < 1.$$

Use the Binomial Theorem to deduce the existence of an integer m such that

$$(2m-1)\,(2-\sqrt{2})^n < 2^n < 2m(2-\sqrt{2})^n.$$

96. In the complex equation

$$z^n+a_{n-1}z^{n-1}+a_{n-2}z^{n-2}+\ldots+a_1z+1 = 0$$

it is given that the complex numbers $a_{n-1}, a_{n-2}, \ldots, a_1$ satisfy

$$|a_{n-1}| \leqslant 1, \quad |a_{n-2}| \leqslant 1, \quad \ldots, \quad |a_1| \leqslant 1.$$

Show that any root of the equation in the complex plane must lie in the annular region

$$\tfrac{1}{2} < |z| < 2. \qquad \text{(C.S.)}$$

97. The expressions E_1, E_2, E_3 are defined as follows:

$$E_1 \equiv x+y-2z-3,$$

$$E_2 \equiv 2x-3y-4z+4,$$

$$E_3 \equiv 3x+5y+3z-5. \qquad \text{(over)}$$

If λ, μ are two numbers, explain the geometrical significance of the equation

$$E_1+\lambda E_2+\mu E_3 = 0.$$

Find the equation of the plane which contains the line

$$(x+1) = -7(y-1) = -7(z-1)$$

and passes through the point of intersection of the planes $E_1 = 0$, $E_2 = 0$, $E_3 = 0$.

98. Show that the equation

$$x^4-3x^2-3x+1 = 0$$

has real roots between 0 and 1 and between 2 and 3. Find the larger of these roots correct to three significant figures and check that three figure accuracy has been achieved. (O & C)

99. The plane Π is said to be invariant under the linear transformation T of three-dimensional space into itself if every point of Π maps into another (possibly the same) point of Π.

Find the equations of the planes through the origin which are invariant under the linear transformation T with matrix $\mathbf{A}$ (relative to the usual base vectors) given by

$$\mathbf{A} = \begin{pmatrix} 4 & 5 & 2 \\ 5 & 4 & 2 \\ 5 & -1 & 7 \end{pmatrix}.$$

100. The complex numbers z, z' are represented in the complex plane by the points $P(x, y)$ and $P'(x', y')$. State in geometrical terms the transformation of the plane $P \to P'$ given by $z' = z(\cos\theta+\mathrm{j}\sin\theta)$. Verify your statement by deriving from $z' = z(\cos\theta+\mathrm{j}\sin\theta)$ the matrix form

$$\begin{pmatrix} x' \\ y' \end{pmatrix} = \mathbf{M}\begin{pmatrix} x \\ y \end{pmatrix}.$$

What geometrical transformation is given by

$$z'-c = (z-c)(\cos\theta+\mathrm{j}\sin\theta),$$

where c is a complex constant? If the transformation $z \to z'$ corresponds to a turn of 120° about the point (1, 0) and the transformation $z' \to z''$ corresponds to a turn of 60° about the point (−3, 0), show that $z'' = -z+2\sqrt{3}\mathrm{j}$, and hence express in geometrical terms the single transformation equivalent to the two transformations in the order given. (S.M.P.)

101. The points F_1, F_2 are the foci of hyperbola and B is a point of the branch defined by $F_1B-F_2B = k > 0$. Indicate on a sketch the position of points P, Q such that

$$F_1P-F_2P > k, \quad F_1Q-F_2Q < k,$$

and justify your answer.

By reflecting the figure in the line l which bisects the angle F_2BF_1 internally, or otherwise, show that B is the unique point of the hyperbola which lies on l. Deduce a theorem relating the tangent at a point of the curve and the lines joining that point to the foci. (M.E.I.)

102. If α is one of the complex fifth roots of unity, and $x = \alpha - \alpha^4$, show that

$$x^4 + 5x^2 + 5 = 0.$$

Express the other roots of this equation in terms of α. (O & C)

103. The tangent to the cubic curve $y = ax^3$ at $P_1(x_1, y_1)$ meets the curve again at $P_2(x_2, y_2)$. (P_1 is not at the origin.) The tangent at P_2 meets the curve again at $P_3(x_3, y_3)$ and so on. Show that $x_1, x_2, x_3, \ldots$ are in geometric progression.

The tangent at P_1 meets the perpendicular from P_3 to the x axis at P. Show that P_1 is the mid-point of P_2P and that the locus of P is another cubic curve. (M.E.I.)

104. S stands for three-dimensional space in which points are specified in the usual way by coordinates (x, y, z), and P stands for the plane $z = 0$ in which points are specified by just two coordinates (x, y). A transformation

$$T_1\colon \begin{pmatrix} x \\ y \\ z \end{pmatrix} \to \begin{pmatrix} 1 & 0 & -3 \\ 0 & 1 & -2 \end{pmatrix} \begin{pmatrix} x \\ y \\ z \end{pmatrix}$$

maps points of S onto points of P; and a transformation

$$T_2\colon \begin{pmatrix} x \\ y \end{pmatrix} \to \begin{pmatrix} -2 & -3 \\ -2 & -1 \\ -1 & -1 \end{pmatrix} \begin{pmatrix} x \\ y \end{pmatrix}$$

maps points of P onto points of S. Show that for T_1 any given image point in P is associated with a set of points of S, and describe this set; and that for T_2 the range is a proper subset of S_1, to be specified.

Show that for both transformations the line joining a point and its image is in a fixed direction. Hence describe in geometrical terms the transformations T_1T_2 and T_2T_1. (S.M.P.)

105. Show that the number of solutions of the equation

$$x + 2y + 3z = n,$$

where x, y, z, n are positive integers, is equal to the coefficient of t^n in

$$1/\{(1-t)(1-t^2)(1-t^3)\}.$$

106. Show that the roots of the equation

$$\frac{A}{x-a} + \frac{B}{x-b} + \frac{C}{x-c} = 0,$$

where A, B, C are positive real numbers and $a > b > c$ lie one between a and b and one between b and c.

If A is small compared with B and C show that one of the roots is approximately

$$a + \frac{A(b-a)(c-a)}{B(c-a) + C(b-a)}.$$

107. Under what circumstances is it true that $\mathbf{a}.\mathbf{b} = 0$?
Prove that, for any angle θ and any positive integer n,

$$\sum_{r=0}^{n-1} \cos\left(\theta+\frac{2r\pi}{n}\right) = 0.$$

Prove also that
$$\sum_{r=0}^{n-1} \cos^2\left(\theta+\frac{r\pi}{n}\right) = \frac{n}{2}$$

and find
$$\sum_{r=0}^{n-1} \sin^2\left(\theta+\frac{r\pi}{n}\right).$$

108. A square matrix is called 'magic' if there exists a number k such that the sum of the elements of each row is k and the sum of the elements of each column is k. Prove that, if $\mathbf{A}$ is a non-singular magic matrix, so also is adj $\mathbf{A}$.

109. X, Y and $(n-2)$ other players play a game in which each player has an equal chance of winning each round. The game is won by winning 8 consecutive rounds of the game. Let

$$q_k = \Pr\,(X \textit{ wins}|\, Y \textit{ has won the last } k \textit{ times}),$$

$$r_k = \Pr\,(X \textit{ wins}|\, X \textit{ has won the last } k \textit{ times})$$

for $k = 1, 2, \ldots, 7$.

Show that

$$(n-1)\,q_k + r_k = 1,$$

$$r_k = \frac{1}{n} r_{k+1} + \left(1-\frac{1}{n}\right) q_1, \quad \text{for} \quad k = 1, 2, \ldots, 6,$$

and find the corresponding result for $k = 7$.

Hence show that the chance that X will win the game is $(n^7-1)/(n^8-1)$ if Y has just won one round but did not win the round before that. (M.E.I.)

110. The remainder when a polynomial P in x is divided by $(x-\alpha)^2$ is the linear polynomial $rx+s$. Prove that

$$r = P'(\alpha) \quad \text{and} \quad s = P(\alpha)-\alpha P'(a).$$

The expression $-s/r$ is of significance in the solution of the equation $P(x) = 0$ by a certain approximate method. Explain this, and show how the division process for polynomials could be used to approximate to the roots of this equation.

Illustrate your answer by finding in this way closer approximation than -1, 3, 5, to the roots of

$$x^3-7x^2+7x+14 = 0.$$

Find the sum and product of your three approximations and explain how these serve to check the answers. (S.M.P.)

111. Let H be the set of matrices of the form

$$\mathbf{X} = \begin{pmatrix} a+b\mathrm{j} & -c+d\mathrm{j} \\ c+d\mathrm{j} & a-b\mathrm{j} \end{pmatrix},$$

where $j = \sqrt{(-1)}$ and a, b, c, d are real numbers not all 0. Show that H contains a subset representing the complex numbers and prove that each element of H has an inverse in H.

The matrix $\mathbf{X}$ is expressed in the form

$$\mathbf{X} = a\mathbf{I} + c\mathbf{J} + b\mathbf{K} + d\mathbf{L},$$

where

$$\mathbf{I} = \begin{pmatrix} 1 & 0 \\ 0 & 1 \end{pmatrix}, \quad \mathbf{J} = \begin{pmatrix} 0 & -1 \\ 1 & 0 \end{pmatrix}, \quad \mathbf{K} = \begin{pmatrix} j & 0 \\ 0 & -j \end{pmatrix}, \quad \mathbf{L} = \begin{pmatrix} 0 & j \\ j & 0 \end{pmatrix}.$$

Prove that $\mathbf{J}^2 = \mathbf{K}^2 = \mathbf{L}^2 = -\mathbf{I}$ and that $\mathbf{JK} = -\mathbf{KJ} = \mathbf{L}$.

Given that $\mathbf{KL} = -\mathbf{LK} = \mathbf{J}$ and $\mathbf{LJ} = -\mathbf{JL} = \mathbf{K}$, evaluate $\mathbf{XY}$, where

$$\mathbf{Y} = \begin{pmatrix} a' + b'j & -c' + d'j \\ c' + d'j & a' - b'j \end{pmatrix},$$

by multiplying the expressions for $\mathbf{X}$ and $\mathbf{Y}$ in terms of $\mathbf{I}$, $\mathbf{J}$, $\mathbf{K}$, $\mathbf{L}$. (M.E.I.)

112. X and Y are independent Poisson variates from distributions with parameters μ and λ respectively. Find

$$\Pr(X = r \mid X + Y = n)$$

for $r = 0, 1, \ldots, n$.

The telephone exchange in a large firm handles both internal and external calls. It is desired to investigate the distribution of internal calls and equipment is available for recording the number of internal calls in each set of 10 calls. Calculate the probabilities of 0, 1, 2 and 3 internal calls in sets of 10 if the average number of calls per set is 7. What is the most likely number of internal calls? (M.E.I.)

113. Show that if tabular values are given for $x = 0, 1, 2, 3$ of the function

$$f(x) \equiv a + bx + cx(x-1) + dx(x-1)(x-2),$$

then the values of b, c, d can be obtained from the differences derived from the table, e.g. b from $f(1) - f(0)$.

Given the following table for a cubic polynomial $g(x)$, find and check the values of a, b, c, d required to express it in the form given above; and hence or otherwise evaluate $g(5)$, $g(6)$, $g(7)$ and $g(8)$.

x	0	1	2	3	4
$g(x)$	87·25	89·37	91·49	96·85	108·69

(M.E.I.)

114. Three bags each contain N tickets numbered 1 to N. Three tickets are drawn, one from each bag; find the probability that the sum of the numbers obtained is $2N$.

115. Two circles, centres A and B and radii one unit, are drawn respectively in the planes $\mathbf{r}.\mathbf{n} = p_1$, $\mathbf{r}.\mathbf{n}_2 = p_2$. Prove that they lie on the surface of a sphere if and only if

$$(\mathbf{a}-\mathbf{b}).(\mathbf{n}_1 \wedge \mathbf{n}_2) = 0 \quad \text{and} \quad (p_1 - \mathbf{n}_1.\mathbf{b})^2 = (p_2 - \mathbf{n}_2.\mathbf{a})^2.$$

116. *ABCDEF* is a hexagon inscribed in a circle centre *O*. If

$$\angle AOB = \angle COD = \angle EOF = \pi/3,$$

prove that the mid-points of *BC*, *DE*, *FA* are the vertices of an equilateral triangle.

117. The region E_0 is defined by

$$ax^2+by^2 \leqslant 1,$$

where $a > 0$ and $b > 0$. *Without assuming any properties of conics*, prove that, if $\mathbf{p}_1$, $\mathbf{p}_2$ are the position vectors of points of E_0, then

(i) $-\mathbf{p}_1 \in E_0$, (ii) $\frac{1}{2}(\mathbf{p}_1+\mathbf{p}_2) \in E_0$,

(iii) $2\mathbf{p}_1 \in E$, where E is the region $ax^2+by^2 \leqslant 4$.

A system of elliptical discs congruent to E_0 is obtained by translating the centre O of E_0 to all points $\mathbf{g}$ whose coordinates are integers. Prove that, if the discs with centres at $\mathbf{g}_1$, $\mathbf{g}_2$ overlap, then

(iv) for a common $\mathbf{t}$, $\mathbf{g}_1-\mathbf{t} \in E_0$ and $\mathbf{t}-\mathbf{g}_2 \in E_0$,

(v) $\frac{1}{2}(\mathbf{g}_1-\mathbf{g}_2) \in E_0$, (vi) $\mathbf{g}_1-\mathbf{g}_2 \in E$.

Deduce that if E contains no point, other than the origin, with integral coefficients, then no members of the system S overlap. (M.E.I.)

118. The log-normal distribution is a probability distribution given by

$$f(x) = \frac{1}{x\sqrt{(2\pi\alpha)}} \exp\left[-\{\ln(x/\rho)\}^2/2\alpha\right]$$

for $0 < x < \infty$. By using the transformation $x = \rho e^u$ and comparing with the normal distribution show that:

(i) $\displaystyle\int_0^\infty f(x)\,dx = 1$, (ii) $\displaystyle\int_0^\infty xf(x)\,dx = \rho e^{\frac{1}{2}\alpha}$, (iii) $\displaystyle\int_0^\infty x^2 f(x)\,dx = \rho^2 e^{2\alpha}$.

Deduce the values of the mean and variance of the distribution. (O & C)

119. If x is a close rational approximation to the cube root of 2, and

$$x' = (x^2+x+2)/(x^2+x+1),$$

show, by expressing $x'-x$ in the form $(2^{\frac{1}{3}}-x)f(x)$ and verifying that $1 < f(x) < 2$, or otherwise, that x' is a better approximation than x, and that if one is below the correct value the other is above. Use this method to find the cube root of 2 correct to two places of decimals. (O & C)

120. The surface of a sphere is divided into regions by r circles; prove that, if a further circle is drawn on the surface, the number of regions is not increased by more than $2r$.

Prove that n circles on the surface of a sphere divide the surface into at most n^2-n+2 regions.

Hence, or otherwise, prove that n spheres divide space into at most

$$\tfrac{1}{3}(n^3-3n^2+8n)$$

regions. (O & C)

121. Prove that $$\sum_{r=1}^{n} \frac{1}{r(r+1)(r+2)} = \frac{1}{4} - \frac{1}{2(n+1)(n+2)}.$$

It is desired to estimate $$\sum_{r=1}^{1000} \frac{1}{r(r+1\cdot01)(r+2\cdot001)}$$

to within 0·001 by summing only the first N terms. Find a suitable value for N.
(M.E.I.)

122. Six integers a_i ($i = 1$ to 6) are independently and uniformly distributed in the interval $-N \leqslant a_i \leqslant N$. Find the probability that the simultaneous equations

$$a_1x + a_2y = a_3, \quad a_4x + a_5y = a_6$$

have a unique solution.

Check your answer by enumerating the various possibilities for $N = 1$.

Show that the probability of obtaining a unique solution tends to 1 as N tends to ∞.

123. Show that, if a, b, c, d, p and q are positive integers such that

$$\frac{a}{b} < \frac{p}{q} < \frac{c}{d},$$

then $$p \geqslant (a+c)/\Delta \quad \text{and} \quad q \geqslant (b+d)/\Delta,$$
where $\Delta = bc - ad$.

Construct the rational numbers of denominator 20 or less between $\frac{10}{13}$ and $\frac{4}{5}$.
(C.S.)

124. If in the expression

$$g(x) \equiv a_0x^n + a_1x^{n-1} + \ldots + a_n \quad (a_i \text{ all real}),$$

$a_i = a_{i+1} = 0$, for some i in the interval $1 \leqslant i \leqslant n-2$, prove that the equation $g(x) = 0$ has at least two complex roots.

The polynomial $f(x)$ is given by

$$f(x) = b_0x^n + b_1x^{n-1} + \ldots + b_n \quad (b_i \text{ all real}).$$

Show that, by a suitable choice of α, β, the equation

$$\alpha x(b_s x - b_{s+1}) f(x) + \beta(b_{s-1}x - b_s) f(x) = 0 \quad (1 < s \leqslant n-1)$$

has at least two complex roots.

Deduce that, if $b_s^2 < b_{s-1}b_{s+1}$, then the equation $f(x) = 0$ has at least two complex roots.

125. If ζ, ζ^* are conjugate complex numbers, give a geometric description of those numbers z for which $$|z-\zeta| < |z-\zeta^*|.$$

Let $z_1, \ldots z_n$ be n complex numbers, the imaginary parts of which are strictly positive, and put

$$\prod_{i=1}^{n} (z-z_i) = z^n + (a_1 + jb_1) z^{n-1} + \ldots + (a_n + jb_n),$$

where the $a_1, \ldots, a_n, b_1, \ldots, b_n$ are real. Show that the roots of

$$x^n + a_1x^{n-1} + \ldots + a_n = 0$$

are all real.
(C.S.)

126. From a population of N butterflies which are restricted to a certain habitat a sample of n is taken. Each butterfly is marked and released again in the habitat. A second sample of n is taken on the following day. Show that the probability P that exactly m butterflies in the second sample will be found to be marked is of the form Au_N, where A is independent of N and

$$u_N = (N-n)!/\{N!\,(N-2n+m)!\}.$$

Show that the value of N which makes P greatest is approximately equal to n^2/m. (C.S.)

127. Let $\mathfrak{M}$ denote the set of all 3×3 real matrices and define an operation $*$ between members of $\mathfrak{M}$ as follows:

$$\mathbf{X} * \mathbf{Y} = \mathbf{X}+\mathbf{Y}-\mathbf{XY}.$$

We define a matrix $\mathbf{X}$ to be *quasi-regular* if we can find a matrix $\mathbf{M} \in \mathfrak{M}$ such that

$$\mathbf{X} * \mathbf{M} = \mathbf{M} * \mathbf{X} = \mathbf{0}.$$

Prove the following results:

(i) $\mathbf{0}$ is quasi-regular.
(ii) If $\mathbf{X}$ is non-singular and quasi-regular, so also is $\mathbf{X}^{-1}$.
(iii) If $\mathbf{X}$, $\mathbf{Y}$ are both quasi-regular, then $\mathbf{X} * \mathbf{Y}$ is quasi-regular.
(iv) If $\mathbf{X}$, $\mathbf{Y}$, $\mathbf{Z}$, are quasi-regular, then $(\mathbf{X} * \mathbf{Y}) * \mathbf{Z} = \mathbf{X} * (\mathbf{Y} * \mathbf{Z})$.
(v) If $\mathbf{X}^n = \mathbf{0}$ for some positive integer n, then $\mathbf{X}$ is quasi-regular.

128. The polynomial $P(x)$ is defined by

$$P(x+1) \equiv P(x)+(1+x)^n,$$

where n is a positive integer. If $P(1) = 1$, find $P(0)$ and $P(-1)$.

Prove that:

(i) $P(-x) \equiv P(-x-1)+(-1)^n x^n$;
(ii) $P(x) \equiv (-1)^{n-1} P(-x-1)$.

Deduce that, if n is even, then the sum of the nth powers of the first k natural numbers is a polynomial in k, with integral coefficients, which is divisible by $k(k+1)(2k+1)$.

What can you say about the sum of the nth powers of the first k natural numbers in the case when n is odd?

129. In the tetrahedron $ABCD$, the perpendiculars from A, B to the opposite faces intersect; prove that the edges AB, CD are perpendicular.

Prove also that the perpendiculars from C, D to the opposite faces intersect.

130. The parametric vector equation of a line l through the origin in three-dimensional Euclidean space is

$$\mathbf{r} = t\mathbf{k},$$

where $\mathbf{k}$ is a constant unit vector and t denotes distance measured along l from the origin. A point P has position vector $\mathbf{s}$. Find the position vector of the reflection of P in l, i.e. of the point Q such that PQ is bisected at right angles by l.

If $\mathbf{r} = t\mathbf{k}_i$ $(i = 1, 2,)$ are two distinct lines through the origin, and S_i $(i = 1, 2)$ are the operations of reflection with respect to these lines, prove that $S_2S_1 = S_1S_2$ if and only if the two lines are perpendicular. (C.S.)

131. It is required to find the matrix **A** (relative to **i**, **j**, **k**) of the linear transformation representing a positive rotation θ about the axis

$$\frac{x}{l} = \frac{y}{m} = \frac{z}{n}.$$

Find the explicit form for **A** by proving the following results.

(i) If $\mathbf{u} = l\mathbf{i}+m\mathbf{j}+n\mathbf{k}$, $\mathbf{v} = l'\mathbf{i}+m'\mathbf{j}+n'\mathbf{k}$, $\mathbf{w} = l''\mathbf{i}+m''\mathbf{j}+n''\mathbf{k}$ are three mutually perpendicular unit vectors, then

$$\mathbf{Q} = \begin{pmatrix} l & l' & l'' \\ m & m' & m'' \\ n & n' & n'' \end{pmatrix}$$

is orthogonal.

(ii) If the point P has position vector **x** referred to **i**, **j**, **k** as base vectors, then it has position vector

$$\mathbf{X} = \mathbf{Q}'\mathbf{x}$$

referred to **u**, **v**, **w** as base vectors.

(iii) The orthogonal matrix

$$\mathbf{R} = \begin{pmatrix} 1 & 0 & 0 \\ 0 & \cos\theta & -\sin\theta \\ 0 & \sin\theta & \cos\theta \end{pmatrix}$$

represents a positive rotation of θ about the x axis.

(iv) $\mathbf{A} = \mathbf{QRQ}'$.

(v) $l'^2+l''^2 = 1-l^2$ and $m'n''-m''n' = l$.

(vi)

$$\mathbf{A} = \begin{pmatrix} l^2+(1-l^2)\cos\theta & ml(1-\cos\theta)-n\sin\theta & nl(1-\cos\theta)+m\sin\theta \\ lm(1-\cos\theta)+n\sin\theta & m^2+(1-m^2)\cos\theta & nm(1-\cos\theta)-l\sin\theta \\ ln(1-\cos\theta)-m\sin\theta & mn(1-\cos\theta)+l\sin\theta & n^2+(1-n^2)\cos\theta \end{pmatrix}.$$

132. If a, b are real and $a > 0$, and if the equation

$$z^3-az+b = 0$$

has a pair of complex roots $\lambda \pm j\mu$, prove that the point (λ, μ) is one of the points of intersection of the curves

$$3x^2-y^2 = a, \quad 2x(y^2+a) = 3b.$$

Draw rough sketches of these curves and deduce the condition for the given cubic to have complex roots.

133. The coefficients a, b in the quadratic polynomial

$$p(z) = z^2+az+b$$

are complex numbers and $a^2 \neq 4b$. Prove that the locus of the point z which varies in the Argand diagram in such a way that $p(z)$ takes real values is a rectangular hyperbola with asymptotes parallel to the real and imaginary axes. By a change of origin, or otherwise, determine the asymptotes.

Determine also the locus of z if it varies in such a way that $p(z)$ is purely imaginary. Sketch both loci on the Argand diagram and interpret algebraically their points of intersection.

Prove that there are two real values of z and two purely imaginary values of z for all four of which $p(z)$ is purely imaginary if

$$\{\mathrm{Re}(a)\}^2 > 4\mathrm{Re}(b) > -\{\mathrm{Im}(a)\}^2. \qquad \text{(O \& C)}$$

Bibliography

The following books may usefully be added to the more comprehensive bibliography of Volume 1.

Armitage, J. V. and Griffiths, H. B. *A Companion to Advanced Mathematics* (Cambridge). Surveys the underlying concepts of modern algebra and analysis: an exacting book for sixth-formers but could be read with profit by a student embarking upon a university course. Contains an excellent selection of questions.

Brand, T. and Sherlock, A. *Matrices: Pure and Applied* (Arnold). Illustrates many of the interesting applications of matrix theory.

Budden, F. J. *Complex Numbers and their Applications* (Longmans). An entertaining book illustrating applications of complex numbers in both pure and applied mathematics.

Cundy, H. M. and Rollett, A. P. *Mathematical Models* (Oxford). Gives instructions for making many models. Full of stimulating ideas: very strongly recommended.

Durran, J. H. *Statistics and Probability* (Cambridge). One of the best introductory texts available; contains an outstanding selection of examples: very strongly recommended.

Hardy, G. H. *A Mathematician's Apology* (Cambridge). A leading mathematician writes about what the subject means to him.

Hardy, G. H. and Wright, E. M. *An Introduction to the Theory of Numbers* (Oxford). A famous work on one of the most fascinating of all mathematical topics—the properties of the integers and their generalizations. Although the book goes well beyond school level it is charmingly written and contains much that will appeal to an intelligent sixth-former.

Henrici, P. *Elements of Numerical Analysis* (Wiley). A very good book on numerical methods, although rather advanced for sixth-formers.

Hunter, J. *Number Theory* (Oliver and Boyd). A more modern approach to the subject than in Hardy and Wright: the text is somewhat exacting as it is written for university students. Contains some good examples.

Kemeny, J. G. and Snell, J. L. *Finite Markov Chains* (van Nostrand). Applications of matrix theory to probability. An interesting but not always easy book.

Kline, M. and others. *Mathematics in The Modern World* (Freeman). Readings from 'Scientific American'. Contains a series of fascinating essays on a wide variety of subjects.

Maynard Smith, J. *Mathematical Ideas in Biology* (Cambridge). Among much other interesting material there is an account of the applications of probability to genetics. The same subject is dealt with in greater detail in D. S. Falconer's *Quantitative Genetics* (Oliver and Boyd).

Mendelson, B. *Introduction to Topology* (Blackie). Perhaps the most readable introduction available to topology, an important branch of modern mathematics.

Midonick, H. *The Treasury of Mathematics* (Pelican, 2 volumes). Selected writings of famous mathematicians—all within the understanding of the sixth-former.

Ogilvy, C. S. and Anderson, J. T. *Excursions in Number Theory* (Oxford). A very elementary introduction to number theory. Contains an interesting chapter on Fibonacci numbers.

Ore, O. *Number Theory and its History* (McGraw-Hill). Another very good introductory work: more advanced than Ogilvy and Anderson, less so than Hardy and Wright.

Rahman, N. A. *Exercises in Probability and Statistics* (Griffin). A formidable collection of problems, going far beyond school level but containing much challenging material suitable for a sixth-former. Also *A Course in Theoretical Statistics*, which takes the subject well beyond school level.

Ribbans, J. *Basic Numerical Analysis* (Intertext). A good elementary account of numerical methods.

Uspensky, J. V. *Introduction to Mathematical Probability* (McGraw-Hill). A classic: advanced but not too difficult.

Watson, W. A., Philipson, T. and Oates, P. J. *Numerical Analysis* (Arnold, 2 volumes). An elementary introduction to numerical methods: contains many worked examples.

Answers

CHAPTER 17

PAGE

381 **Ex. 3.** (i) $-1+2j$; (ii) $1-4j$; (iii) $7+4j$; (iv) $-8+6j$; (v) $-26-18j$.

382 **Ex. 5.** (i) $-5+18j$; (ii) $-99+20j$; (iii) $27+36j$.

Ex. 6. (i) $(2-j)/5$; (ii) $(5+j)/2$; (iii) $(4-3j)/5$.

383 **Ex. 8.** (i) 3, -2; (ii) 5, -12; (iii) $\frac{3}{13}$; $\frac{2}{13}$.

Ex. 9. b, 0.

Ex. 10. (iv) $8-j$; (v) $(2-j)/5$; (vi) $11-5j$; (vii) $8-8j$.

Ex. 11. R.

Ex. 12. (i) 5; (ii) 2; (iii) $1/\sqrt{2}$; (iv) 1; (v) $2\sin\frac{1}{2}\theta$.

Exercise 17(a)

1. (i) $7+6j$; (ii) $9+17j$; (iii) $4+3j$; (iv) $11+6j$; (v) $11+22j$; (vi) $1+32j$; (vii) 0; (viii) $(3-j)/5$; (ix) $-72+368j$; (x) $-6-12j$; (xi) $(7+j)/50$; (xii) $-(11+29j)/37$.

2. $\cos(\theta_1-\theta_2)+j\sin(\theta_1-\theta_2)$; $\cos 2\theta_1+j\sin 2\theta_1$; $\cos 2(\theta_1+\theta_2)+j\sin 2(\theta_1+\theta_2)$.

3. (i) $2\pm j$; (ii) $(1\pm j)/2$; (iii) $(5\pm j\sqrt{3})/2$; (iv) $\pm 1-j$.

384 **4.** $3-2j$; (i) $(1+j)/\sqrt{2}$; (ii) $2+j$. **5.** $3+2j$, $1-j$.

6. (i) $\frac{7}{5}$, $-\frac{4}{5}$; (ii) $2-j$, $2+j$.

7. $1-\cos\theta+j\sin\theta$, $(1+j\cot\frac{1}{2}\theta)/2$. **10.** z, pure imaginary.

12. (i) $x^2-4x+5=0$; (ii) $x^2-4x+13=0$; (iii) $x^2-4x+7=0$; (iv) $x^2-3(1+j)x+5j=0$; (v) $x^2-(5+j)x+4(1+j)=0$.

13. (i) $(x-1-j)(x-1+j)$; (ii) $(x+2jy)(x-2jy)$; (iii) $(x+j\sqrt{3}y)(x-j\sqrt{3}y)$; (iv) $(jx-1)(x+1-j)$; (v) $(jx-2y)(x+2jy)$; (vi) $(x+1)(2x-1-j\sqrt{3})(2x-1+j\sqrt{3})/4$; (vii) $\Pi(x\pm 1/\sqrt{2}\pm j/\sqrt{2})$.

14. 2, 3; (i) 2; (ii) $x+1$.

15. 1, j, -1, $-j$; 1, $1+j$, j, 0 according as $n=4m$, $4m+1$, $4m+2$ or $4m+3$.

385 **17.** -22.

388 **Ex. 21.** (i) 2, $\frac{1}{3}\pi$; (ii) $2\sqrt{2}$, $\frac{1}{4}\pi$; (iii) 1, π; (iv) 1, $\frac{1}{2}\pi$; (v) 1, $-\frac{1}{2}\pi$; (vi) 2, $-\frac{5}{6}\pi$; (vii) 2, $-\frac{1}{6}\pi$; (viii) $|\sec\alpha|$, α;

PAGE

388 (ix) $|\sec\alpha|$, $(\pi-2\alpha)/2$; (x) $2\sin\frac{1}{2}\alpha$, $(\alpha-\pi)/2$]
(in the last three parts the principal argument depends on α).

390 **Ex. 25.** $(\sqrt{3}-1)+j(\sqrt{3}+1)$, $\cos 75° = (\sqrt{3}-1)/2\sqrt{2}$, etc.

Exercise 17(b)

396 **1.** (i) 5, $-53°$; (ii) $\sqrt{13}$, 56°; (iii) $\sqrt{5}$, $-117°$; (iv) $\sqrt{5}/2$, 63°; (v) $\sqrt{29}$, 112°; (vi) $\sqrt{5}$, $-153°$; (vii) 13, $-23°$; (viii) 41, $-103°$.

2. (i) $\cos 3\theta+j\sin 3\theta$; (ii) $\cos(\theta+\phi)+j\sin(\theta+\phi)$; (iii) j; (iv) $(1+j)\sqrt{2}$; (v) j; (vi) $(1-j\sqrt{3})/2$; (vii) j;
(viii) $4\cos\frac{1}{2}\theta\sin\frac{1}{2}\phi(\sin(\theta+\phi)/2-j\cos(\theta+\phi)/2)$.

3. (i) $(\pi+2\phi-2\theta)/2$; (ii) $(-3\pi+2\phi-2\theta)/2$.

7. Circle, centre $2+j$, radius 1. **12.** $-j$.

397 **13.** (i) $(1+j)/2$; (ii) $(1\pm j\sqrt{3})/2$; (iii) j, $1-j$; (iv) $-1+\frac{1}{3}j$.

14. $9+6j$. **16.** (i) $-\frac{5}{3}$, $\frac{4}{3}$; (ii) semicircle, radius 1.

17. $(1-j\tan\frac{1}{2}\theta)/2$, $(1+j\cot\frac{1}{2}\theta)/2$. **18.** Square.

Miscellaneous Exercise 17

1. $(\sqrt{3}-j)(1+j)$, $2\sqrt{2}$, $\frac{1}{12}\pi$. **2.** 8, 2.

3. $1\pm\sqrt{3}+j(5\pm\sqrt{3})$.

398 **5.** $-1+2j$, $(x-1)^2+(y-\frac{1}{2})^2 = \frac{5}{4}$.

399 **14.** (i) $|z| = \sqrt{7}$; (ii) $2+j$, $3+2j$.

400 **15.** $r = 4$, $\theta = 147{\cdot}5°$.

CHAPTER 18

402 **Ex. 1.** x^3-9x^2-3x+4, 2.

Ex. 2. 25086278.

403 **Ex. 3.** 2600.

Ex. 4. $y = 1957x-7373$.

Ex. 5. 2412, 1957, 1190.

405 **Ex. 7.** -26, -8.

Ex. 9. $-0{\cdot}099$, $0{\cdot}357$.

407 **Ex. 13.** $2x^2-5x+3$.

Exercise 18(a)

1. (i) x^2-3x-5; 6; (ii) x^3-9x+4; 2;
(iii) $3x^3-5x^2-2x-7$; -10; (iv) $x^3+4x^2+16x+59$; 231;
(v) $7x^4-35x^3+175x^2-875x+4372$; -21865.

PAGE

408 **2.** (i) x^2-3x-7; 7; (ii) x^3-x+3; 0; (iii) $x^4-4x^3-x^2+3x+5$; 2;
(iv) $(16x^3-44x^2-14x+11)/8$; $\frac{5}{8}$;
(v) $(9x^4-63x^2+6x-40)/9$; $\frac{68}{9}$.

3. (i) 159·1; (ii) 31·4; (iii) 283·3.

4. (i) $(x-1)^3+4(x-1)^2+7(x-1)+5$;
(ii) $2(x-2)^3-5(x-2)^2-14(x-2)+8$;
(iii) $(x+3)^3-2(x+3)^2-21(x+3)+50$;
(iv) $(x+2)^4-8(x+2)^3+24(x+2)^2-32(x+2)+15$.

5. (i) $x^3-jx^2+4jx+2$; $1-j$;
(ii) $x^3+(1+j)x^2-(2-j)x+(1+2j)$; $-j$.

6. (i) 2·6; (ii) 3·0. **7.** $(x+4)^4+4(x+4)^2$; minimum; -4, -4, $-4\pm 2j$.

8. x^5-3x^3+2x+1; 1. **9.** $A=0$, $B=1$. **10.** $A=\lambda$, $B=4\lambda-17$.

11. -6, 139. **12.** 86, 68.

409 **13.** x^2-8x-8. **14.** $2x^3-4x^2-x+6$.

15. $n+15n(n-1)/2+25n(n-1)(n-2)/3+5n(n-1)(n-2)(n-3)/2$
$+n(n-1)(n-2)(n-3)(n-4)/5$.

Ex. 14. e.g. $a=5$, $b=-2$. **Ex. 15.** $a=3$, $b=3$, $c=-1$.

411 **Ex. 17.** (i) $4/(x-1)-3/(x+2)$; (ii) $3/2(x-3)-1/2(x+3)$;
(iii) $1/(x-a)-1/(x+a)$; (iv) $1/(x-a)+1/(x+a)$.

412 **Ex. 18.** (i) $\ln\sqrt{1{\cdot}5}$; (ii) $\ln(\frac{4}{3})$.

Ex. 19. (i) $4(x-2)^{-2}-6(x-3)^{-2}$;
(ii) $\{(x+1)^{-2}-36(x-6)^{-2}\}/7$.

415 **Ex. 21.** (i) $2/(x-2)-3/(x-2)^2-2/(x+1)$;
(ii) $1/(x+1)-4/(x+1)^2-1/(2x+1)$.

Ex. 22. (i) $1/(x-1)-x/(x^2+2)$;
(ii) $2/3(x-1)+(x-1)/3(x^2+x+1)$.

Ex. 24. (i) $1/(x-1)-1/(x^2-2)$;
(ii) $1/(x-1)-\{1/(x-\sqrt{2})-1/(x+\sqrt{2})\}/2\sqrt{2}$.

Exercise 18(b)

1. (i) $2/(3-x)+3/(1-2x)$; (ii) $2/x-3/(x+2)$;
(iii) $7/(x+3)-5/(2x-1)$; (iv) $1/2(1-x)-3/2(1+x)$;
(v) $2+3/x-5/(x+1)$; (vi) $1/(x-1)-3/(x+2)+2/(x-3)$;
(vii) $1/x-3/(1-x)-5/(1+x)$; (viii) $-x+3/(x-2)-7/(x+2)$;
(ix) $x^2+5x+19+81/(x-3)-16/(x-2)$;
(x) $1/x^2+1/x-2/(1+2x)$;
(xi) $1/(1-2x)^2-1/(1-2x)+1/(2+x)$; (xii) $x/(2+x^2)-1/(1+x)$;
(xiii) $(2x+1)/3(1-2x+4x^2)-1/3(1+2x)$; (xiv) $4/(x-3)^2+1/(x-2)$;
(xv) $1-1/3(x+1)+(x-4)/3(x^2+2)$;
(xvi) $1/(x-2)-1/x^4-2/x^3-2/x^2-1/x$.

2. $A=-13$, $B=13$, $C=-11$, $D=27$.

3. $2/(x-1)^4+6/(x-1)^3+4(x-1)^2+1/(x-1)-1/(x+4)$.

PAGE

416 **4.** $\frac{1}{2}-1/(n+2)$; $\frac{13}{36}-(3n^2+18n+26)/3(n+2)\,(n+3)\,(n+4)$.

5. (i) $1/3(1+x)-(x-2)/3(1-x+x^2)$;
(ii) $1/3(1+x)-(1+\surd 3j)/3(2x-1-\surd 3j)-(1-\surd 3j)/3(2x-1+\surd 3j)$.

6. $1/(x-2)-1/(x-1)$; $\frac{1}{2}-1/2(2y-1)+1/(y-1)$;
$\frac{1}{8}-1/8(4x-1)+1/(2x-1)$.

Ex. 25. 0·9802.

Exercise 18(c)

420 **1.** (i) $1+3x+9x^2+27x^3$; (ii) $\frac{1}{3}+\frac{1}{9}x+\frac{1}{27}x^2+\frac{1}{81}x^3$;
(iii) $1+4x+12x^2+32x^3$; (iv) $1+2x^2$;
(v) $\frac{1}{8}+\frac{3}{16}x+\frac{3}{16}x^2+\frac{5}{32}x^3$.

2. (i) $1+x+\frac{3}{2}x^2+\frac{5}{2}x^3$; (ii) $1-2x-2x^2-4x^3$; (iii) $2-\frac{1}{4}x-\frac{1}{64}x^2-\frac{1}{512}x^3$;
(iv) $\frac{1}{2}+\frac{1}{16}x+\frac{3}{256}x^2+\frac{5}{2048}x^3$; (v) $(\frac{1}{2}+\frac{1}{8}x+\frac{3}{64}x^2+\frac{5}{256}x^3)\surd 2$.

3. (i) $1+\frac{1}{3}x-\frac{1}{9}x^2+\frac{5}{81}x^3$; (ii) $1-8x+40x^2-160x^3$;
(iii) $1+\frac{2}{3}x+\frac{5}{9}x^2+\frac{40}{81}x^3$; (iv) $\frac{1}{32}+\frac{5}{64}x+\frac{15}{128}x^2+\frac{35}{256}x^3$;
(v) $2-\frac{1}{12}x-\frac{1}{288}x^2-\frac{5}{20736}x^3$.

4. (i) $1+3x+7x^2+15x^3$; (ii) $\frac{1}{6}+\frac{5}{36}x+\frac{19}{216}x^2+\frac{65}{1296}x^3$;
(iii) $\frac{1}{2}-\frac{5}{4}x+\frac{17}{8}x^2-\frac{49}{16}x^3$; (iv) $3x-5x^2+7x^3$;
(v) $1+8x+26x^2+64x^3$.

5. (i) $1-\frac{3}{2}x+\frac{15}{8}x^2$;
(ii) $1+x+\frac{1}{2}x^2$; (iii) $1+\frac{5}{2}x+\frac{51}{8}x^2$; (iv) $1-\frac{3}{2}x+\frac{11}{8}x^2$;
(v) $1-3x+\frac{13}{2}x^2$.

6. (i) $\frac{1}{2}-\frac{1}{4}(x-1)+\frac{1}{8}(x-1)^2$; (ii) $\frac{1}{3}+\frac{1}{9}(x+1)+\frac{1}{27}(x+1)^2$;
(iii) $1+(x+1)-\frac{1}{2}(x+1)^2$; (iv) $2+\frac{1}{4}(x-2)-\frac{1}{32}(x-2)^2$;
(v) $-2-\frac{7}{4}(x+1)-\frac{111}{64}(x+1)^2$.

8. (i) 0·99933; (ii) 1·0100; (iii) 2·0025; (iv) 0·92793.

421 **9.** 9·997. **12.** $1+\frac{1}{8}x+\frac{63}{128}x^2$.

13. $2/(2x-1)+(1-x)/(1+x^2)$. **14.** 2·009926.

Miscellaneous Exercise 18

1. $A=\lambda$, $B=$ arbitrary, $C=-\lambda-1$.

3. 1, $-(x+1)$, $1/(x-2)-(x+1)/(x^2+x+1)$.

422 **4.** $1/7(x-2)-(x+3)/7(x^2+x+1)$; $1/9(x-2)-1/9(x+1)-1/3(x+1)^2$.

5. $\frac{1}{2}, \frac{1}{8}$; $\frac{5}{24}-(4n+5)/8(2n+1)\,(2n+3)$.

7. (i) $\surd 3$; (ii) $(4)^{\frac{1}{3}}$; (iii) $2/\surd 7$; (iv) $(3)^{\frac{1}{4}}$.

8. $2x^3-13x^2+26x-13$, $4(\pi x-x^2)/\pi^2$.

9. $(7-x)/4$; tangent at $(-1, 2)$.

10. 249, $r^4-8r^3+24r^2-34r+21$.

CHAPTER 19

PAGE

423 **Ex. 2.** -2^9j.

424 **Ex. 4.** $j/2^9$.

Ex. 6. $2t(3-10t^2+3t^4)(1-15t^2+15t^4-t^6)$.

425 **Ex. 8.** $(10-15\cos 2\theta+6\cos 4\theta-\cos 6\theta)/32$.

Ex. 9. $5\pi/32$.

426 **Ex. 10.** $(1-x\cos\theta-x^n\cos n\theta+x^{n+1}\cos(n-1)\theta)(1-2x\cos\theta+x^2)^{-1}$.

Ex. 11. $x\sin\theta(1-2x\cos\theta+x^2)^{-1}$.

Exercise 19(a)

1. (i) -1; (ii) $-j$; (iii) j; (iv) $(\sqrt{3}+j)/2$; (v) 1; (vi) $16(-\sqrt{3}+j)/9\sqrt{3}$; (vii) $-(1+j\sqrt{3})/32$; (viii) $\sec^4\theta(\cos 4\theta-j\sin 4\theta)/16$.

2. (i) -1; (ii) 16; (iii) $-2^9(1+j\sqrt{3})$; (iv) $\cos(n+m)\theta+j\sin(n+m)\theta$.

3. (i) $1-8\sin^2\theta+8\sin^4\theta$; (ii) $5\sin\theta-20\sin^3\theta+16\sin^5\theta$.

4. (i) $1-8\cos^2\theta+8\cos^4\theta$; (ii) $16\cos^5\theta-20\cos^3\theta+5\cos\theta$; (iii) $6\cos\theta-32\cos^3\theta+32\cos^5\theta$.

5. $4t(1-t^2)(1-6t^2+t^4)^{-1}$.

6. $\cos\{(2k+1)/14\}$, $k = 0$ to 6; $\cos(2k\pi/7)$, $k = 0$ to 6.

7. (i) $(\cos 5\theta+5\cos 3\theta+10\cos\theta)/16$; (ii) $(\cos 4\theta-4\cos 2\theta+3)/8$; (iii) $(\cos 7\theta+7\cos 5\theta+21\cos 3\theta+35\cos\theta)/64$.

8. (i) $(\sin 5\theta-5\sin 3\theta+10\sin\theta)/16$; (ii) $(35\sin\theta-21\sin 3\theta+7\sin 5\theta-\sin 7\theta)/64$; (iii) $(2\sin 2\theta-\sin 4\theta)/8$.

9. $(\sin 9\theta-\sin 7\theta-4\sin 5\theta+4\sin 3\theta+6\sin\theta)/256$.

10. (i) $3\pi/16$; (ii) $3\pi/512$.

11. (i) $(\cos\theta+\cos n\theta-\cos(n+1)\theta-1)(2-2\cos\theta)^{-1}$; (ii) $(\sin\theta+\sin n\theta-\sin(n+1)\theta)(2-2\cos\theta)^{-1}$.

12. $(\cos\theta+2^{n+1}\cos n\theta-2-2^n\cos(n+1)\theta)(5-4\cos\theta)^{-1}$.

427 **13.** $2^n\cos^n(\theta/2)\cos(n\theta/2)$.

14. $\frac{1}{4}\operatorname{cosec}^2(\theta/2)\{\sin\theta+(n+1)\sin(n-1)\theta-n\sin n\theta\}$.

15. $(\sin^2\theta-\sin^{n+1}\theta\sin(n+1)\theta+\sin^{n+2}\theta\sin n\theta)(1+\sin^2\theta-\sin 2\theta)^{-1}$.

16. $(\Sigma t_1-\Sigma t_2t_3t_4)(1+t_1t_2t_3t_4-\Sigma t_1t_2)^{-1}$.

17. $(\sin 3\theta\sin^3\theta-\sin(2n+3)\theta\sin^{2n+3}\theta-\sin^6\theta+\sin(2n+1)\theta\sin^{2n+5}\theta)$
$\times(1-2\sin^2\theta\cos 2\theta+\sin^4\theta)^{-1}$.

PAGE

427 **18.** $a = \frac{1}{16}$, $b = -\frac{5}{16}$, $c = \frac{5}{8}$; $k\pi \pm \frac{1}{6}\pi$, $n\pi$.

428 **Ex. 14.** (i) ± 1, $\pm j$; (ii) ± 2, $\pm 2j$; (iii) $\pm(1 \pm j)/\sqrt{2}$.

429 **Ex. 18.** $-j$, $(\pm\sqrt{3}+j)/2$.

Exercise 19(b)

430 **1.** (i) 2, $(-1 \pm \sqrt{3}\,j)$; (ii) -1, $(1 \pm \sqrt{3}\,j)/2$; (iii) j, $(\pm\sqrt{3}-j)/2$;
(iv) $1+j$, $\{-(\sqrt{3}+1)+(\sqrt{3}-1)\,j\}/2$, $\{(\sqrt{3}-1)-(\sqrt{3}+1)\,j\}/2$.

2. (i) $2-j$, (ii) $1{\cdot}3+0{\cdot}79j$; (iii) $-0{\cdot}28-1{\cdot}8j$; (iv) $1{\cdot}4-1{\cdot}0j$.

3. $1+j$, $\{-(\sqrt{3}+1)+(\sqrt{3}-1)\,j\}/2$, $\{(\sqrt{3}-1)-(\sqrt{3}+1)\,j\}/2$;
$\cos \pi/12 = (1+\sqrt{3})/2\sqrt{2}$, $\sin \pi/12 = (\sqrt{3}-1)/2\sqrt{2}$.

4. $1{\cdot}1-0{\cdot}088j$.

5. (i) $\cos 3 - j \sin 3$;
(ii) $\cos (2\theta-\pi)/6 + j \sin (2\theta-\pi)/6$;
(iii) $\cos 2\theta - j \sin 2\theta$; (iv) $\operatorname{cosec}^{\frac{1}{4}} \theta \operatorname{cjs} (\pi - 2\theta)/8$;
(v) $\operatorname{cjs} (4\theta-\pi)/8$.

7. $(\pm\sqrt{3}\pm j)/2$. **8.** -1, $(\pm 1 \pm \sqrt{3}j)/2$.

9. $\frac{1}{2}j$, $-\omega j$, $-\omega^2 j$ where ω is a complex cube root of one.

10. $\tan k\pi/5$, $k = 0$ to 4.

11. $\cos \theta(1-\sin\theta)^{-1}$, $\theta = 2k\pi/n$.

431 **15.** $(a+b+c)\,(a+\omega b+\omega^2 c)\,(a+\omega^2 b+\omega c)$,
$(a+b+c)\,(a^2+b^2+c^2-bc-ca-ab)$;
$(a+b+c+d)\,(a-b+c-d)\,(a+jb-c-jd)\,(a-jb-c+jd)$

433 **Ex. 25.** (i) $j\pi$; (ii) $-\frac{1}{2}j\pi$; (iii) $e^{-\pi}$; (iv) $e^{\frac{1}{2}\pi}$.

Exercise 19(c)

1. $|ar+bs|$, θ. **2.** (i) $e^{x}(1+j)/\sqrt{2}$; (ii) e^{-nr}.

3. (i) $\ln 2 - j\pi/6$; (ii) $e^{\frac{1}{6}\pi}(\cos \ln 2 + j \sin \ln 2)$.

4. Moves along real axis from e^{π} to $e^{-\pi}$. **7.** $2c^2 \cosh v\,(\cos u + \cosh v)^{-1}$.

Miscellaneous Exercise 19

434 **1.** $\cos (2k+1)\pi/12 + j \sin (2k+1)\pi/12$, $k = 0, 2, 3, 5, 6, 8, 9, 11$.

3. $2^{-n} \cos 2n\theta(1+\cos\theta)^{-n}$.

4. (i) $2 \operatorname{cjs} (\pi/4)$, $8 \operatorname{cjs} (3\pi/4)$; (ii) $2 \operatorname{cjs} (11\pi/12)$, $2 \operatorname{cjs} (-5\pi/12)$.

5. $\pm(-0{\cdot}61+1{\cdot}2j)$, $\pm(-0{\cdot}61-1{\cdot}2j)$.

6. $\frac{1}{2}\{ac-bd+(a^2+b^2)^{\frac{1}{2}}\,(c^2+d^2)^{\frac{1}{2}}\}$.

9. (i) $\cos \{\alpha+\frac{1}{2}(n-1)\beta\} \sin \frac{1}{2}n\beta \operatorname{cosec} \frac{1}{2}\beta$;
(ii) $\sin \{\alpha+\frac{1}{2}(n-1)\beta\} \sin \frac{1}{2}n\beta \operatorname{cosec} \frac{1}{2}\beta$.

10. $35\pi/256$.

PAGE

435 **11.** (i) $2\cos\frac{1}{2}\theta \operatorname{cjs}\frac{1}{2}\theta$; (ii) $2\sin\frac{1}{2}\theta \operatorname{cjs}(\theta-\pi)/2$; (iii) $\sqrt{(\cot\frac{1}{2}\theta)} \operatorname{cjs}\frac{1}{4}\pi$.

13. $\phi = (2r+1)\pi/5$ **14.** $\tan\{4k-3)\pi\}/4n$.

15. $(\sqrt{3}+1)/2\sqrt{2}$, $1/(\sqrt{6}+\sqrt{2})$.

16. $-(3^{\frac{1}{3}}+3^{\frac{2}{3}})$, $-(3^{\frac{1}{3}}\omega+3^{\frac{2}{3}}\omega^2)$, $-(3^{\frac{1}{3}}\omega^2+3^{\frac{2}{3}}\omega)$.

437 **23.** $\frac{1}{2}n(n+1)$; $A = -C = \alpha(1-\alpha)^{-2}$, $B = (1-\alpha)^{-1}$.

24. $(x-1)\prod_{k=1}^{n}(x^2-2x\cos 2k\pi/(2n+1)+1)$.

CHAPTER 20

439 **Ex. 2.** $(2+3\text{j})z+(2-3\text{j})z^* = 13$.

Ex. 4. (i) 6; (ii) -3j.

Ex. 6. $zz^*-(1+2\text{j})z-(1-2\text{j})z^*-4 = 0$.

Ex. 7. (i) $1+3\text{j}$, 2; (ii) $\frac{1}{2}-\text{j}$, 1.

Exercise 20

447 **1.** (i) $|w-3| = 1$; (ii) $|w-\text{j}| = 1$; (iii) $|w| = 2$; (iv) $|w| = 1$; (v) $|w| = \sqrt{2}$.

3. -1, $\pm\text{j}$. **4.** $1+\text{j}$; $2+\text{j}$, $-1+\text{j}$.

448 **11.** $z(1-\text{j})+z^*(1+\text{j}) = 2$. Interior of 'shark's fin' bounded by real axis, $|z-\frac{1}{2}| = \frac{1}{2}$, $|z-\frac{3}{4}-\frac{1}{4}\text{j}| = \sqrt{6}/4$.

12. Circle, $2\text{j}/3$, $\frac{4}{3}$. **13.** Line. **14.** Line. **15.** Point j.

16. Circle, 2, $\sqrt{5}$. **17.** Line. **19.** $\arg\left(\frac{z-a}{z-b}\right) = \pi-\theta$.

449 **21.** $w = 3+\text{j}-1/z$. **25.** Arcs of three circles through origin, centres $\text{j}/2$, $1/2b$, $\text{j}/2(a-1)$ and imaginary axis.

26. (i) $3ww^*-2(w+w^*)+1 = 0$; (ii) $w+w^* = \frac{1}{2}$.

27. $w = (z-\text{j})(\text{j}z-1)^{-1}$.

CHAPTER 21

450 **Ex. 1.** $1+\text{j}$, $(-1+7\text{j})/5$.

454 **Ex. 2.** (i) $-\frac{19}{4}$; (ii) $-\frac{99}{8}$; (iii) $\sqrt{47}/2$; (iv) $\frac{6}{49}$.

Ex. 3. (i) $3z^2-10z+12 = 0$; (ii) $3z^2+z+1 = 0$; (iii) $9z^2-7z+9 = 0$.

Exercise 21(a)

455 **1.** 3, -7; (ii) 2, $5\frac{1}{2}$; (iii) $(1-\text{j})/2$, $-\text{j}$; (iv) $4-3\text{j}$, -1; (v) $(2-6\text{j})/5$, $-(1+2\text{j})/5$.

PAGE

455 **2.** (i) $z^2-z-6=0$; (ii) $6z^2-z-2=0$; (iii) $z^2-6z+4=0$; (iv) $z^2-(5+\surd 2)z+(2+\surd 2)=0$; (v) $z^2-6z+25=0$; (vi) $z^2-z+1=0$; (vii) $z^2-(2+j)z+(3+j)=0$; (viii) $z^2-(\surd 2+1)(1-j)z-3j=0$.

3. (i) 2·8, −1·3; (ii) 0·75 ± 1·7j; (iii) 0·42 + 0·86j, −1·42 − 1·86j.

4. (i) 3, −9, 27, $-\frac{1}{3}$, −972; (ii) $\frac{1}{3}$, $-\frac{5}{3}$, $\frac{31}{9}$, $-\frac{1}{5}$, $-\frac{230}{81}$; (iii) j, −1+j, 1−2j, (1−j)/2, −5+j.

5. (i) $-\frac{9}{4}$; (ii) $\frac{945}{32}$; (iii) $\surd 113/2$; (iv) $9\surd 113/4$.

6. (i) $(3k^2+2k-7)/3$; (ii) $\frac{134}{27}$; (iii) −4.

7. (i) $2(a^2+8ab+8b^2)$; (ii) $2(a+2b)/a^2$; (iii) $4|\{b(a+b)\}^{\frac{1}{2}}|$.

456 **8.** $-a^2mc/(b^2+a^2m^2)$, $b^2c/(b^2+a^2m^2)$.

9. (i) $2z^2+3z-6=0$; (ii) $z^2-5z-350=0$; (iii) $7z^2+z-2=0$; (iv) $4z^2-29z+49=0$.

10. (i) $2z^2-2(1-3j)z-5-9j=0$; (ii) $z^2-2jz+5-j=0$; (iii) $z^2-(2+14j)z-12+5j=0$.

11. (i) $4z^2+7=0$; (ii) $4z^2-(10+10j)z+9j=0$; (iii) $4z^2-19z+44=0$.

12. ±24j.

13. ±3(1+j).

14. (i) $(4+5j)z^2-(3+j)z-(2-j)=0$; (ii) $(2+j)z^2+(3-j)z-4+5j=0$.

19. $(ca^2)^{\frac{1}{3}}+(c^2a)^{\frac{1}{3}}+b=0$.

20. $b^2=2ca$.

Miscellaneous Exercise 21

459 **1.** (i) $11z^2-8z-400=0$; (ii) $z^2-5(1+j)z+47j=0$.

2. α^2, β^2. **7.** α/β, β/α.

CHAPTER 22

461 **Ex. 2.** $2x-3y+13=0$, $(5t-1, 3t-2)$.

462 **Ex. 5.** $b^2x^2-a^2y^2=a^2b^2$.

463 **Ex. 6.** (−1, 3), (−4, 4), (−2, 5), (−3, −4).

Exercise 22(a)

464 **1.** $4x^2+y^2=4$. **2.** $x^3=y^2$. **3.** $y=x^2-3x+3$.

4. $x^2+y^2=2$. **5.** $x^3+y^3=axy$. **6.** $x^2+y^2=x^2y$.

8. (i) (−1, 4); (ii) (2, 0); (iii) (2, 8); (iv) (4, −1).

9. $\pm\frac{1}{2}$.

465 **12.** (−2, 5).

PAGE

466 Ex. **8.** $P\hat{S}Z = 90°$, $S\hat{P}M$ bisected by tangent.

467 Ex. **11.** $4a$, a, $3a$.

Ex. **14.** $(7a/2, a)$, $x = 5a/2$; (i) $(-11a/4, 0)$, $x+5a/4 = 0$;
(ii) $(-a, a)$, $y+3a = 0$.

468 Ex. **20.** $q = -p-2/p$.

Exercise 22(b)

472 **1.** (i) 4, (3, 2), (4, 2), $x = 2$; (ii) 2, (3, −1), $(3, -\frac{1}{2})$, $2y = -3$;
(iii) 1, (1, −1), $(\frac{3}{4}, -1)$, $4x = 5$;
(iv) 3, (−2, −3), $(-2, -\frac{9}{4})$, $4y = 11$;
(v) 4, (5, −3), (6, −3), $x = 4$, (vi) 8, (2, 2), (2, 4), $y = 0$;
(vii) 2, (0, 4), $(-\frac{1}{2}, 4)$, $2x = 1$.

2. (i) $y^2-4x-2y+5 = 0$; (ii) $x^2+6x+2y+4 = 0$.

3. (i) $y^2-12x-6y+21 = 0$; (ii) $x^2+2x+3y+4 = 0$.

4. (i) $x+4y+9 = 0$; (ii) $x-6y-3 = 0$; (iii) $5x-7y+25 = 0$;
(iv) $8x+3y+30 = 0$.

5. (4, −2). **6.** (−4, 6). **7.** $-\mathbf{i}$, $-\mathbf{i}+\frac{1}{4}\mathbf{j}$.

473 **8.** $\frac{25}{8}$.

474 **18.** Same axis, focus; directrix is tangent at vertex of original parabola.

19. Parabola. **21.** $y^2 = 2ax$. **22.** Parabola, latus rectum a.

475 **28.** $y = 2a$, $y^2-2ax = a^2$. **30.** $y = 3x^2-4x$.

31. $y = 2x^2-x+2$.

Exercise 22(c)

478 **6.** $x = -3$, $y = 1$. **7.** (1, 1).

479 **10.** $2x-8y\pm 15c = 0$.

Miscellaneous Exercise 22

2. Y/a, $(Y^2-2aX)/2a^2$; $(b, 0)$, $(b+\frac{1}{2}a, 0)$.

481 **15.** $(3\sqrt{7})c^2$. **18.** $(2a, 2a\sqrt{2})$.

CHAPTER 23

484 Ex. **1.** 2 cjs $(2k+1)\pi/10$, $k = 0$ to 4, i.e. $\frac{1}{2}\sqrt{(10+2\sqrt{5})}+\frac{1}{2}(\sqrt{5}-1)\mathrm{j}$, etc.

Ex. **2.** 1, −2, −2.

Ex. **6.** $-1\pm\mathrm{j}$, ω, ω^2.

488 Ex. **10.** (i) 0, 0, $\frac{1}{3}(2\pm 2\sqrt{10})$; (ii) −1, −1, $\frac{1}{3}(5\pm\sqrt{10})$;
(iii) 2, 2, $\frac{1}{3}(-4\pm \mathrm{j}2\sqrt{2})$.

Exercise 23(a)

PAGE

488 **1.** $x^4-5x^3-4x^2+16x-8=0$. **2.** $x^4-3x^3+x^2+7x-30=0$.

3. $x^5+3x^4+4x^3+4x^2+3x+1=0$. **4.** 1, 1, $\pm j\sqrt{2}$.

5. -3, $3\pm j$. **6.** $\frac{1}{3}$, $-\frac{3}{2}$, $-\frac{3}{2}$, $-\frac{3}{2}$.

7. (i) $4z^3+z^2-1=0$; (ii) $4z^3-11z^2+10z-4=0$.

489 **8.** (i) $z^3-6z^2-12z-32=0$; (ii) $w^3-15w^2-15w-16=0$.

9. (i) $27z^3+3jz+1+j=0$; (ii) $z^3-2z-4=0$.

10. (i) $2z^3-13z^2+32z-21=0$; (ii) $2z^3+17z^2+52z+64=0$.

11. (i) $z^4+z-5=0$; (ii) $z^4-8z^3+24z^2-33z+13=0$;
(iii) $z^4-10z^2-z+25=0$.

12. 3, 3, $\frac{2}{3}$. **13.** $-1\pm j$, $-1\pm\sqrt{2}$.

14. $-j$, $\pm\sqrt{2}-j$.

17. 3 real roots if and only if $-11<k<16$.

18. 4 real if $-7<k<9$, etc.

490 **19.** 2 real if $k<-23$ or $-16<k<112$;
4 real if $-23<k<-16$; none real if $k>112$.

20. $G^2+4H^3<0$.

491 **Ex. 11.** $\frac{3}{2}$, $-\frac{5}{2}$, $\frac{13}{2}$.

Ex. 12. 0, 0, 4j, 1.

Exercise 23(b)

494 **1.** (i) 14; (ii) 46; (iii) 162; (iv) 10; (v) 17; (vi) 24; (vii) $-\frac{1}{2}$;
(viii) 22; (ix) -2; (x) 30.

2. (i) 16; (ii) -3; (iii) $\frac{1}{9}$; (iv) 67; (v) 32; (vi) 48; (vii) 18.

3. (i) $2p^2$; (ii) $(p^4+4pq^2)/q^4$; (iii) 0.

4. -3, $\frac{1}{2}$, 4.

5. $-\frac{5}{4}$, $-\frac{1}{2}$, $\frac{1}{4}$. **6.** $\frac{3}{4}$, $\frac{1}{2}$, $\frac{1}{3}$. **7.** -2, $2\frac{1}{2}$, 5.

8. 1, -2, -3.

495 **9.** 2, 3, -4. **10.** -1, 2, 3. **11.** $3abc=a^2d+2b^3$.

12. -3, $-\frac{1}{3}$, $\frac{3}{2}$.

13. (i) $z^3+2pz^2+p^2z-q^2=0$; (ii) $z^3-pz^2-q^2=0$;
(iii) $z^3+pz-q=0$; (iv) $z^3+4pz-8q=0$.

14. (i) $z^3+3qz^2+(3q^2+p^3)z+q^3=0$;
(ii) $z^3-2pz^2+p^2z-q^2=0$.

17. (i) $x^3+(3q-p^2)x^2+(3q^2-p^2q)x+q^3-p^3r=0$;
(ii) $x^3+(9q-3p^2)x-2p^3+9pq-27r=0$;
(i) G.P.; (ii) A.P.

496 **18.** $r=pq$.

497 **Ex. 15.** (i) $x^3-5x+9=0$; (ii) $2x^3+4x-7=0$.

Exercise 23(c)

PAGE

502 **1.** $-2,\ 1 \pm \mathrm{j}$. **2.** $-\frac{3}{2},\ (3 \pm \mathrm{j}\sqrt{7})/4$.

3. $2\cos \pi/9,\ -2\cos 4\pi/9,\ -2\cos 2\pi/9$.

4. $\cos\alpha,\ \cos(\frac{2}{3}\pi+\alpha),\ \cos(\frac{4}{3}\pi+\alpha)$ where $\cos 3\alpha = \frac{2}{3}$.

5. $3^{\frac{1}{3}}+3^{\frac{2}{3}},\ \omega 3^{\frac{1}{3}}+\omega^2 3^{\frac{2}{3}},\ \omega^2 3^{\frac{1}{3}}+\omega 3^{\frac{2}{3}}$.

6. $2-2^{\frac{1}{3}}+2^{\frac{2}{3}},\ 2-\omega 2^{\frac{1}{3}}+\omega 2^{\frac{2}{3}},\ 2-\omega^2 2^{\frac{1}{3}}+\omega 2^{\frac{2}{3}}$.

7. $5+6^{\frac{1}{3}}+6^{\frac{2}{3}},\ 5+\omega 6^{\frac{1}{3}}+\omega^2 6^{\frac{2}{3}},\ 5+\omega 6^{\frac{2}{3}}+\omega^2 6^{\frac{1}{3}}$.

8. $-4,\ -\omega,\ -\omega^2$.

503 **9.** $-2\mathrm{j},\ \mathrm{j}(1 \pm \sqrt{2})$.

11. $\frac{1}{2}\{\sqrt{2} \pm \mathrm{j}\sqrt{(2+4\sqrt{2})}\},\ \frac{1}{2}\{-\sqrt{2} \pm \mathrm{j}\sqrt{(2-4\sqrt{2})}\}$.

Miscellaneous Exercise 23

3. $\pm 8,\ \mp 288$.

4. 1. **5.** $-2 \pm \mathrm{j},\ \pm \mathrm{j}/\sqrt{2}$. **6.** 84, 708.

7. $-\frac{1}{2},\ \frac{1}{8}$. **8.** $1,\ -\frac{3}{4}$; $(1, 1, 1),\ (4, -\frac{1}{2}, -\frac{1}{2})$.

9. 3 roots if $-81 \leqslant k \leqslant 44$.

505 **10.** $-\frac{3}{2},\ -\frac{3}{2},\ 1 \pm \sqrt{5}$.

11. $q^2(y+1)^3 + p^3(y+2) = 0$.

CHAPTER 24

506 **Ex. 7.** $\pm 3\hat{\mathbf{c}}$.

508 **Ex. 9.** $\mathbf{i}+4\mathbf{j}-3\mathbf{k}$.

Exercise 24(a)

512 **4.** (i) $-2\mathbf{i}-\mathbf{j}+5\mathbf{k}$; (ii) $8\mathbf{i}-16\mathbf{j}+8\mathbf{k}$; (iii) $\mathbf{i}-\mathbf{j}+\mathbf{k}$; (iv) $-4\mathbf{j}+4\mathbf{k}$; (v) $-bc\mathbf{i}+ab\mathbf{k}$.

5. 0; coplanar with O. **6.** $-\mathbf{k},\ -\mathbf{j}$.

7. $(6\mathbf{i}+3\mathbf{j}+2\mathbf{k})/7,\ \frac{6}{7}$. **8.** $3\frac{1}{2}$, 1.

9. $\frac{1}{3}(\mathbf{i}+2\mathbf{j}+2\mathbf{k})$, 9, 9.

513 **11.** $|\mathbf{a}|^{-2}$. **14.** $\alpha = \mathbf{r}.\mathbf{a}/\{\mathbf{a}.(\mathbf{b} \wedge \mathbf{c})\}$.

Exercise 24(b)

518 **1.** (i) $3x-2y+z = 6$; (ii) $7x+2y-5z = 4$; (iii) $4x-4y+z = 5$; (iv) $x-5y+3z = 7$; (v) $bcx-3cay+2abz = 0$.

2. (i) $3x+4y-2z = -7$; (ii) $2x-3y+2z = 0$; (iii) $x+3y-4z = -5$; (iv) $2x-7y+4z = 1$; (v) $2x-5y-z = -2$.

3. (i) 6; (ii) $\sqrt{14}$; (iii) $2\sqrt{29}$.

PAGE

518 **4.** 0; lines intersect. **5.** $2x+3y-5z-21 = 0$.

6. $7x-3y+z-16 = 0$. **7.** $x+y-2z = 4$.

519 **8.** (i) $(4, -2, 3)$; (ii) $\mathbf{i}+2\mathbf{k}$; (iii) $6\mathbf{i}-5\mathbf{j}-3\mathbf{k}$;
(iv) $\frac{1}{6}(x-4) = -\frac{1}{5}(y+2) = -\frac{1}{3}(z-3)$.

10. $x = \frac{1}{3}y = \frac{1}{2}z$.

11. $\mathbf{r} = \mathbf{c}+\nu\{[\mathbf{b}_1 \wedge (\mathbf{a}_1-\mathbf{c})] \wedge [\mathbf{b}_2 \wedge (\mathbf{a}_2-\mathbf{c})]\}$.

12. $2\delta^3/(3\alpha\beta\gamma)$. **13.** $a\sqrt{6}/3$.

CHAPTER 25

523 **Ex. 2.** $\frac{3}{10}$, $f(x) = \frac{1}{100}$, $0 \leqslant x \leqslant 100$.

Ex. 3. 2×10^{-4}, $\frac{27}{100}$.

528 **Ex. 8.** $\frac{1}{2}$.

Exercise 25(a)

1. 1, $\frac{1}{2}$. **2.** 2, 0·117. **3.** $\frac{1}{2}\pi$, $\frac{3}{4}$.

4. (i) $(x-1)/x$ for $x \geqslant 1$; (ii) $1-e^{-2x}$ for $x \geqslant 0$;
(iii) $\frac{1}{2}(1-\cos \pi x)$ for $0 \leqslant x \leqslant 1$.

529 **5.** 1, 1, 1, $\frac{1}{5}$. **6.** $\frac{5}{8}$, $\frac{2}{3}$, 1, $\frac{13}{192}$.

7. 0, 0, 0, $(\pi^2-8)/4\pi^2$. **8.** $1/\lambda$, $(\ln 2)/\lambda$, 0, $1/\lambda^2$.

9. $F(x) = 0$, $x^2/4$, $(2x-1)/4$, $(6x-2x^2-5)/4$, 1;
$G(y) = 0$, $\frac{1}{4}y$, $\frac{1}{4}(2\sqrt{y}-1)$, etc.; $g = dG/dy$.

10. $g(y) = 2y^3$, $0 \leqslant y \leqslant 1$, $2y(2-y^2)$, $1 < y \leqslant \sqrt{2}$;
1, $\frac{1}{6}$; $8(2\sqrt{2}-1)/15$.

11. $\frac{97}{60}$, $\frac{77}{160}$.

530 **12.** 1, -4, 6.

13. e^{-kt_0}, e^{-3kt_0}, $6e^{-3kt_0}(1-e^{-kt_0})^2$.

14. 1·5, $4\frac{1}{3}$, 0·15.

15. $4/\pi$, $(2/\pi)\ln 2$, 0·0785, $\sqrt{2}-1$, 0·854.

535 **Ex. 12.** $\frac{1}{2}z^2$, $-1+2z-\frac{1}{2}z^2$.

Ex. 13. Triangular.

Exercise 25(b)

545 **1.** $\frac{7}{10}$. **2.** $1-1/\sqrt{(10\pi)}$. **3.** $\frac{1}{4}$. **4.** $\frac{9}{169}$.

5. 1, $\frac{1}{3}$. **6.** $\frac{1}{4}$. **7.** $4r/\pi$.

546 **8.** $2r^2-16r^2/\pi^2$. **9.** $l^2/4$, $l^4/48$.

10. $a^4\{4\pi-\pi^2-(\ln 4)^2\}/4\pi^2$. **11.** 1, 53, 160.

12. 0·149. **13.** 2·421, 2·641. **14.** 0·004.

15. 0·154. **16.** 2·19, 78%, $\sigma = 0{\cdot}156$.

17. 10·825, 98, 0·271.

547 **18.** 502 g.

Miscellaneous Exercise 25

PAGE

547 **1.** $8a^2/3$, $1/\sqrt{2}$. **2.** 18·6. **3.** $(1+\ln 4)/4$.

4. $1/(2\sqrt{y})$, $\sqrt{y}$, $\frac{1}{3}$, $\frac{4}{45}$.

548 **5.** 0·301, 0·092. **6.** $1+\exp[-\pi x/(\sigma\sqrt{3})]$.

7. 0·683. **8.** 0·1336, 1·630, 4·98.

9. λ, λ, 2λ. **10.** 183p.

549 **12.** $\frac{1}{3}$. **13.** $\frac{1}{2}$. **15.** $1-e^{-\lambda^2}$, $2\lambda^{-2}$.

17. $\frac{1}{2}(1+y)^{-\frac{1}{2}}$, $2/\pi\sqrt{(1-z^2)}$; $-\frac{2}{3}$, $2/\sqrt{45}$; $2/\pi$, $\sqrt{\{(\pi^2-8)/2\pi^2\}}$.

18. $\lambda/(\mu-\lambda)$.

550 **19.** (i) $p < 1/22{\cdot}4$; (ii) $p = 0$. **20.** $\frac{1}{4}$.

21. 4·471, 5·000, 5·029; 0·2280.

23. $\frac{7}{3}$, $\sqrt{65}/3$, 0·387.

CHAPTER 26

554 **Ex. 6.** Absolute: $2|y_1\epsilon_1|$, $|\epsilon_1/2\sqrt{y_1}|$, $(|\epsilon_1|+|\epsilon_2|)/2\sqrt{(y_1+y_2)}$.

Exercise 26(a)

559 **1.** $0{\cdot}7513 \pm 0{\cdot}0006$, $1{\cdot}1383 \pm 0{\cdot}0020$.

2. (i) $22{\cdot}36 \pm 0{\cdot}055$; (ii) $88{\cdot}75 \pm 0{\cdot}35$; (iii) $3{\cdot}333 \pm 0{\cdot}014$.

3. 0·31784. **4.** 0·043685(5). **5.** $5{\cdot}95 \pm 0{\cdot}10$.

6. $|\epsilon| \sec^2 y$. **7.** $|y\epsilon|/(1+y^2)^{\frac{3}{2}}$. **8.** (i) 0·011; (ii) 0·034.

9. 0·39. **10.** 0·23(1). **11.** 2·15(4). **12.** 1·15.

13. 1·94, −0·56, −1·38. **14.** 1·56, −2·56.

17. n even: no real roots; n odd: one real (negative) root.

Exercise 26(b)

569 **1.** 1·44. **2.** 4·64. **3.** 1·38. **4.** 5·14, −2·14.

5. 4·10. **6.** 4·13 (5). **7.** 2·30, −1·30.

8. 2·65, −1·42, −4·23. **9.** 4·193.

10. 1·092, 1·931. **11.** 0·8447. **12.** 0·7920.

13. 0·60. **14.** $1-(5k+3)^{-1}$. **15.** 0·877.

570 **16.** 0·838. **17.** 3·591.

Miscellaneous Exercise 26

1. 1·532. **4.** 60·3.

5. 0·88, 0·78, 0·69, 0·60, 0·50.

PAGE

571 **8.** 3·316625. **9.** 1·81713. **10.** 2·8.

572 **12.** 1·162(28). **13.** $K+1/2K^3-1/4K^4$. **15.** 0·567144.

CHAPTER 27

576 **Ex. 7.** $(\pm 3, 0)$. **Ex. 8.** $(2, -2), (0, -2)$.

577 **Ex. 10.** 0·205.

Exercise 27(a)

585 **1.** (i) $\frac{1}{2}$; (ii) $\sqrt{3}/2$; (iii) $\sqrt{2}/\sqrt{5}$. **2.** $\frac{2}{3}$, $(0, \pm 2)$, $y = \pm\frac{9}{2}$.

3. $\frac{1}{2}$, $(2\pm\sqrt{3}, -1)$. **4.** $1/\sqrt{2}$, $(-2, 1)$, $(-4, 1)$.

586 **5.** $\frac{1}{2}$, $(2, -1)$, $(2, -3)$. **6.** $\frac{24}{25}$. **7.** $x-2y = 7$.

8. $4x+y = 6$. **9.** $(-1, 2)$, $(-2, 1)$.

13. Mid-point of OS.

587 **22.** $a^2y^2+b^2x^2 = 4x^2y^2$. **23.** S, b.

27. $x\pm 3y = \pm 5\sqrt{2}$.

596 **Ex. 37.** Arc of a circle.

Exercise 27(b)

597 **1.** (i) $\frac{3}{2}$; (ii) $\sqrt{3}$; (iii) $\sqrt{(3/2)}$. **2.** $2\sqrt{3}/3$, $(\pm 2\sqrt{2}, 0)$, $\sqrt{2}x\mp 3 = 0$.

3. $\frac{5}{4}$, $(6, -1)$, $(-4, -1)$, $5x-21 = 0$, $5x+11 = 0$.

4. $\sqrt{\frac{5}{2}}$, $(-3\pm\sqrt{5}, 1)$, $x = -3\pm 4/\sqrt{5}$.

5. (i) $3x\pm 2y = 0$; (ii) $\sqrt{3}x\pm y = 0$.

6. (i) $x-2y = -3$, $x+2y = 1$; (ii) $x\pm\sqrt{2}y = \mp\sqrt{2}$.

7. $4x^2-y^2 = 7$. **8.** $9x^2-4y^2-18x-24y-32 = 0$.

9. $x+y-1 = 0$. **10.** $x+3y = 4$, $-\frac{17}{13}$.

11. $2x+y+3 = 0$. **13.** $x^2-y^2 = 8$.

598 **18.** $(1, 0)$, $y-2\sqrt{2}x+2(\sqrt{2}+1) = 0$.

19. $5x-2y-9 = 0$, $2x+5y-50 = 0$, $(20, -\frac{25}{2})$.

600 **Ex. 38.** (i) Line parallel to $\theta = 0$. (ii) Circle. (iii) Line at α to $\theta = 0$. (iv) Quadrant of annulus. (v) Interior of circle, centre $(a, 0)$.

Ex. 39. (i) $r(\sin\theta+\cos\theta) = 1$; (ii) $r = 2\sin\theta$; (iii) $r^2-2r(\sin\theta+\cos\theta)+1 = 0$.

Miscellaneous Exercise 27

605 **1.** $1/\sqrt{2}$, $(1, -3)$, $(-2, -2)$.

606 **3.** b^2/a^2, $x^2+y^2-ax\sec\alpha-by\operatorname{cosec}\alpha = 0$.

608 **15.** $2x(x_1-x_2)+2y(y_1-y_2) = x_1^2+y_1^2-x_2^2-y_2^2$.

16. $\frac{1}{4}\sqrt{(1+4e+3e^2)}$. **17.** Ellipse, parabola.

CHAPTER 28

PAGE

613 **Ex. 4.** 3, −2.

617 **Ex. 12.** $\frac{1}{\sqrt{2}}\begin{pmatrix}1 & -1\\ 1 & 1\end{pmatrix}$.

Exercise 28(a)

618 **1.** 2, 11, $\begin{pmatrix}2/\sqrt{13}\\ -3/\sqrt{13}\end{pmatrix}$, $\begin{pmatrix}1/\sqrt{10}\\ 3/\sqrt{10}\end{pmatrix}$, $3x+2y = 0$, $3x-y = 0$.

2. −4, 6, $\begin{pmatrix}1/\sqrt{5}\\ -2/\sqrt{5}\end{pmatrix}$, $\begin{pmatrix}\frac{3}{5}\\ \frac{4}{5}\end{pmatrix}$, $2x+y = 0$, $4x-3y = 0$.

3. 50, −25, $\frac{1}{5}\begin{pmatrix}4 & -3\\ 3 & 4\end{pmatrix}$. **4.** $\frac{1}{\sqrt{2}}\begin{pmatrix}1 & -1\\ 1 & 1\end{pmatrix}$.

5. $x+2y = 0$ and $x-y = 0$ invariant.

6. $x-4y = 0$ and $2x+y = 0$ invariant. **8.** Maps into $x = y$.

619 **9.** (i) $\begin{pmatrix}3.2^{n+1}-5.3^n & -5.2^{n+1}+10.3^n\\ 3.2^n-3^{n+1} & -5.2^n+2.3^{n+1}\end{pmatrix}$;

(ii) $\begin{pmatrix}-7.2^{n+1}+15 & 3.2^{n+1}-6\\ -35.2^n+35 & 15.2^n-14\end{pmatrix}$;

(iii) $\begin{pmatrix}4(-1)^n-3.2^n & 6(-1)^{n+1}+3.2^{n+1}\\ 2(-1)^n-2^{n+1} & 3(-1)^{n+1}+2^{n+2}\end{pmatrix}$;

(iv) $\begin{pmatrix}5.3^n-2^{2n+2} & -2.3^n+2^{2n+1}\\ 10.3^n-5.2^{2n+1} & -4.3^n+5.2^{2n}\end{pmatrix}$.

10. No. **11.** $-\frac{1}{12}$; maps everything close to $x = y$.

12. $\begin{pmatrix}11.2^{n+2}-21n-43 & -33.2^{n+1}+33n+66\\ 7.2^{n+2}-14n-28 & -21.2^{n+1}+22n+43\end{pmatrix}$,

$(n+1)$ and $(2^{n+1}-1)$.

626 **Ex. 30.** $\begin{pmatrix}1 & 0 & -28\\ -7 & 8 & 7\\ 0 & 0 & -27\end{pmatrix}$.

Exercise 28(b)

628 **1.** 1, 2, 8; $\begin{pmatrix}-5/\sqrt{33}\\ 2/\sqrt{33}\\ 2/\sqrt{33}\end{pmatrix}$, $\begin{pmatrix}1/3\sqrt{3}\\ 1/3\sqrt{3}\\ -5/3\sqrt{3}\end{pmatrix}$, $\begin{pmatrix}1/\sqrt{3}\\ 1/\sqrt{3}\\ 1/\sqrt{3}\end{pmatrix}$.

2. −2, 3, 0; A singular, $\begin{pmatrix}1 & 1 & 1\\ 3 & 1 & 1\\ 3 & 4 & 1\end{pmatrix}$.

PAGE

628 **3.** 1, −1, 2; det **A** = −2; $2\mathbf{A}^{-1} = \begin{pmatrix} 2 & -3 & 2 \\ 4 & -5 & 2 \\ 4 & -7 & 4 \end{pmatrix}$;

−1, 1, $\frac{1}{2}$; eigenvalues λ_i^{-1}.

4. 1, −1, −2; $\begin{pmatrix} 1/\sqrt{2} \\ 0 \\ 1/\sqrt{2} \end{pmatrix}, \begin{pmatrix} \frac{1}{3} \\ \frac{2}{3} \\ \frac{2}{3} \end{pmatrix}, \begin{pmatrix} 0 \\ 1/\sqrt{2} \\ 1/\sqrt{2} \end{pmatrix}$,

$\begin{pmatrix} 1 & 1 & 0 \\ 0 & 2 & 1 \\ 1 & 2 & 1 \end{pmatrix}$. $x+y-z = 0$, $y-z = 0$.

5. 1, 2, −3; $\frac{1}{3}\begin{pmatrix} 1 & -2 & 2 \\ 2 & -1 & -2 \\ 2 & 2 & 1 \end{pmatrix}$.

629 **6.** $y-z = 0$, $x-y = 0$, $15x+13y-8y = 0$.

11. $\lambda+\mu$, $\lambda\mu$. **16.** $\begin{pmatrix} 16 & 20 \\ -4 & -4 \end{pmatrix}$.

630 **17.** **A** if n odd, **I** if n even. **18.** $\begin{pmatrix} 370 & -455 \\ -260 & 240 \end{pmatrix}$.

20. $\mathbf{A}^3 = \begin{pmatrix} 10 & 9 & 23 \\ 5 & 9 & 14 \\ 9 & 5 & 19 \end{pmatrix}$, $\mathbf{A}^{-1} = \begin{pmatrix} 4 & -2 & -3 \\ 1 & 0 & -1 \\ -2 & 1 & 2 \end{pmatrix}$.

21. $(17\mathbf{I}+\mathbf{A}-\mathbf{A}^2)/52$.

22. $\begin{pmatrix} 2^{n+4}-15 & -2^{n+3}-4.3^n+12 & 2^{n+4}-4.3^n-12 \\ 10.2^n-10 & -5.2^n-2.3^n+8 & 10.2^n-2.3^n-8 \\ -10.2^n+10 & 5.2^n+3^{n+1}-8 & -5.2^{n+1}+3^{n+1}+8 \end{pmatrix}$.

631 **24.** 0, $\pm$j tan $\frac{1}{2}\theta$. **25.** $x+2y-6z = 0$;

$\mathbf{P} = \begin{pmatrix} \alpha_1 & \alpha_2 & 5 \\ \beta_1 & \beta_2 & 3 \\ \gamma_1 & \gamma_2 & 2 \end{pmatrix}$, where $\alpha_i+2\beta_i-6\gamma_i = 0$, det $\mathbf{P} \neq 0$.

632 **28.** $\frac{1}{2}x = \frac{1}{2}y = \frac{1}{3}z$, $x = y = z$.

CHAPTER 29

633 **Ex. 1.** $(-a, 0)$.
636 **Ex. 5.** $x \pm y = 0$.

Exercise 29(a)

637 **3.** $(2-t)/(2t+1)$. **4.** $p+q+r = 0$.
638 **8.** $(4ak^2, 4ak)$. **11.** $(at/(1+t^3), at^2/(1+t^3))$.

PAGE

639 **16.** 8:1.

640 **Ex. 7.** Ends of diameter.

Ex. 8. (1, 2, −1), 4.

641 **Ex. 10.** (2, 2, −1), 3.

644 **Ex. 12.** Elliptic cylinder.

Exercise 29(b)

652 **1.** (i) $x^2+y^2+z^2+2y = 0$; (ii) $x^2+y^2+z^2-4x+6y+2z-11 = 0$; (iii) $x^2+y^2+z^2-2ax-2by-2cz = 0$.

2. (i) (3, 1, −1), 4; (ii) (−2, 4, 1), 9; (iii) $(a, b, -c)$, $|a-b+c|$.

3. (i) $x^2+y^2+z^2-4x-2y+2z-32 = 0$; (ii) $x^2+y^2+z^2-2x-4z-16 = 0$; (iii) $x^2+y^2+z^2-ax-by-cz = 0$.

4. (i) (4, 1, 3), $\surd 6$; (ii) (5, −1, 2), 9.

5. $\frac{1}{2}\surd\{2d^2-(a-b)^2\}$. AB is diameter.

6. Centroid of $\triangle ABC$. $\Sigma PA_i^2 = d^2$; locus a sphere, centre at centroid of polygon $A_1 \ldots A_n$.

653 **8.** $(p^2+q^2+r^2)(ax+by+cz)^2 = (ap+bq+cr)^2(x^2+y^2+z^2)$.

10. (i) Circle; (ii) pair of straight lines; (iii) hyperbola.

11. $(\frac{1}{2}, 1, \frac{1}{2})$, $3\surd 2/2$.

13. $x-y+5z = 7$, (2, −2, 3).

654 **14.** 12, 70. **16.** Circle; rectangular hyperbola.

17. $(1/\surd 2, -1, -1/\surd 2)$. **18.** True. **19.** $c_1^2 : c_2^2$.

656 **26.** Ellipse; $2\surd 2/3$; $x-2y = 0$, $2x+y = 0$.

27. Rectangular hyperbola. $(\pm 5, \pm 1)$, $5x+y = 0$, $x-5y = 0$.

28. $\surd(\frac{3}{2})$; $x+(3\pm 2\surd 2)y = 0$.

Revision exercise C

657 **3.** $n > 0$ and odd; $n > 0$ only. **4.** −1, 0, 3. **5.** 8·2°, 90°.

6. 0·067. **7.** −1, 1. **8.** 1:1.

658 **10.** Cancellation.

11. $(\surd 2, -\frac{1}{4}\pi)$, $(\surd 2, \frac{1}{3}\pi)$; $\frac{1}{24}\pi$.

12. (i) $1/(x-1)-4/(x+2)$; (ii) $(2+3x)/(1+x^2)-3/x$; (iii) $1/(1-x)+1/(1-x)^2+1/(1-2x)$.

13. 12·67. **14.** $(2at^2, 3at)$.

15. (i) $\{-1 \leqslant x \leqslant 1\}$; (ii) $\{x > 1\}$; (iii) $\{x = 1\}$; (iv) $\{-2 < x < -1\}$; (v) $\{-1 \leqslant x \leqslant 1\}$; $E = B' \cap C'$, $F = B \cup C'$.

659 **16.** Line $x+y = a+b$. **17.** (i) $\frac{55}{96}$; (ii) $\frac{55}{144}$.

PAGE

659 **18.** $\begin{pmatrix} 0 & 0 & 0 \\ 2 & 0 & 0 \\ 0 & 1 & 0 \end{pmatrix}.$

20. $7\pi/12,\ 5\pi/4,\ 23\pi/12$.

21. $-1+\frac{1}{2}k,\ -1/k$. **22.** $x \leqslant -5$ or $-3 < x < 7$.

23. $1/(1-3x)-2/(1-2x);\quad -1-x+x^2$.

24. $|a|$, 0.

660 **25.** Plane perpendicular to OA. **26.** 1, −2, −2, 6.

27. $-1,\ (-1 \pm \sqrt{105})/2$. **28.** $\frac{1}{2}$, 0·215.

30. $\frac{1}{2}(6n^3-3n^2-n),\ \frac{1}{2}(6n^3+3n^2-n)$. **31.** 2:1.

661 **33.** (i) $60° < \theta < 109{\cdot}5°,\ 250{\cdot}5° < \theta < 300°$;
(ii) $0° < \theta < 30°,\ 150° < \theta < 199{\cdot}5°,\ 240{\cdot}5° < \theta < 360°$.

35. $\lambda = -2,\ (1, \frac{8}{5}, \frac{2}{5})$.

36. (i) $(2t+1)^3\,(3t+1)^2/432$; (ii) $\frac{181}{216}$; (iii) 3·5.

38. (i) $1/\sqrt{33}$; (ii) $3/\sqrt{33}$.

662 **42.** 3. **43.** $-2 < x < 0,\ 2 < x < 7$. **44.** (2, −3, 7), 2.

663 **47.** $\sqrt{2},\ 2\sqrt{2}$. **50.** 5·02, 103·26, no (1 % exceed 106·52).

51. 2, 6.

664 **53.** $\cot\theta - 2^n \cot(2^n\theta)$. **54.** 3/128.

665 **60.** (i) R; (ii) $\{x \in R: -\sqrt{2} \leqslant x \leqslant \sqrt{2}\}$.

61. $2\cos\frac{1}{2}\theta,\ \frac{1}{2}\theta;\ (2\cos\frac{1}{2}\theta)^n \sin\frac{1}{2}n\theta$.

62. $-1,\ -\frac{1}{4};\ \frac{1}{2},\ -1,\ -2;\ -\frac{1}{2},\ -1 \pm \sqrt{\frac{3}{2}}$.

666 **64.** $\frac{3}{32}$; 2, 0·89, 2; 0·374. **65.** $gw+g^*w^*+a^2 = 0$.

66. $\begin{pmatrix} 1 & 0 & 0 \\ 1 & 1 & 0 \\ 5 & -2 & 1 \end{pmatrix},\ \begin{pmatrix} 1 & -4 & 14 \\ 0 & 1 & -3 \\ 0 & 0 & 1 \end{pmatrix},\ \begin{pmatrix} -73 & -32 & 14 \\ 16 & 7 & -3 \\ -5 & -2 & 1 \end{pmatrix}.$

67. $\frac{1}{9},\ \frac{49}{12},\ \frac{6}{5},\ \frac{21}{5}$. **68.** $\frac{1}{6};\ \frac{1}{6}(\mathbf{4a}+\mathbf{3b})$.

667 **69.** $x = -y/5 = z/2$.

72. (i) $5\mathbf{i}+(\mathbf{i}+\mathbf{j})-5(\mathbf{i}+\mathbf{j}+\mathbf{k})$; (ii) $(3\mathbf{i}-2\mathbf{k})-(2\mathbf{i}+4\mathbf{j}+3\mathbf{k})$.

73. $(\frac{5}{6})^5,\ 1-(\frac{5}{6})^6-(\frac{5}{6})^5,\ (6\,!)\,(\frac{1}{6})^6,\ (\frac{5}{6})^{r-1}/6$; 4; 6, 66.

74. $-2\mathbf{i}+\mathbf{j}+3\mathbf{k},\quad y = 0$. **75.** 0·966.

668 **77.** $16y^2-24xz+7z^2 = 0,\ 4y^2 = 36x-63$.

78. $1+5x/2+39x^2/8,\ 5x-2y+2 = 0$, curve above tangent.

669 **83.** $\frac{1}{5}$. **84.** (i) (1, −2, 1); (ii) no solution; (iii) $(\lambda,\ -3\lambda,\ \lambda+\frac{1}{2})$; 8; 16.

85. $(-1)^{\frac{1}{2}n}$.

670 **87.** 0. **88.** (i) Line; (ii) circle; (iii) ellipse; (iv) parabola.

89. $c(\alpha+\beta)/2,\ c(\alpha+\beta)/(2\alpha\beta);\quad cYt^2-2XYt+cX = 0$.

90. $\frac{1}{6}(3n-1)\pi$.

PAGE

670 **92.** $\mathbf{x} = \lambda\mathbf{a} + (\mathbf{a} \wedge \mathbf{b})/|\mathbf{a}|^2$, $\mathbf{y} = (1-\lambda)\mathbf{a} - (\mathbf{a} \wedge \mathbf{b})/|\mathbf{a}|^2$.

671 **94.** $p^n/(p^n+q^n)$. **97.** $2x+9y+5z-12 = 0$.

672 **98.** 2·06. **99.** $5x+3y+4z = 0$, $y-z = 0$, $x-y = 0$.

100. Half-turn about $(0, \sqrt{3})$.

673 **102.** $\alpha^4-\alpha$, $\alpha^2-\alpha^3$, $\alpha^3-\alpha^2$.

104. Line in direction $3\mathbf{i}+2\mathbf{j}+\mathbf{k}$; plane $x+y-4z = 0$.

674 **110.** $-0{\cdot}96$, 2·88, 5·08.

675 **112.** $\binom{n}{r} \mu^r\lambda^{n-r}/(\lambda+\mu)^n$; 0, 0·0001, 0·0014, 0·0090; 7.

113. 130·25, 164·77, 215·49, 285·65.

114. $(N-1)(N+4)/(2N^3)$.

676 **118.** $\rho e^{\frac{1}{2}\alpha}$, $\rho^2(e^{2\alpha}-e^{\alpha})$. **119.** 1·26.

677 **121.** $N \geqslant 21$.

122. $8N^2(N+1)/(2N+1)^3$.

123. $\frac{7}{9}, \frac{11}{14}, \frac{15}{19}$.

678 **128.** Integral polynomial in k divisible by $k(k+1)$.

679 **133.** $2z = -\text{Re}(a)$, $2z = -\text{j}\,\text{Im}(a)$. Rectangular hyperbola. $p(z) = 0$.

Index